WHEN THERE
WERE BIRDS

WHEN THERE WERE BIRDS

The Forgotten History of Our Connections

ROY AND LESLEY ADKINS

Little, Brown

LITTLE, BROWN

First published in Great Britain in 2021 by Little, Brown

1 3 5 7 9 10 8 6 4 2

A CIP catalogue record for this book
is available from the British Library.

ISBN 978-1-4087-1357-0

Typeset in Caslon by M Rules
Printed and bound in Great Britain by
Clays Ltd, Elcograf S.p.A.

Papers used by Little, Brown are from well-managed forests
and other responsible sources.

MIX
Paper from
responsible sources
FSC® C104740

Little, Brown
An imprint of
Little, Brown Book Group
Carmelite House
50 Victoria Embankment
London EC4Y 0DZ

An Hachette UK Company
www.hachette.co.uk

www.littlebrown.co.uk

To our good friends

Jim and Kitty Bond

The pick of the flock

CONTENTS

CONTENTS

———— ·◆· ————

THE DAWN CHORUS

Just before dawn, at ten minutes past three on 7 June 1917, huge underground mines were detonated beneath the German lines by the British army. It was the start of the Battle of Messines at the Ypres Salient and was followed by a sustained bombardment:

> Even while the riot and clamour was at its height, the first flush of dawn crept rosy-red up the sky above Ypres. The sun, as it rose, was invisible behind the bank of smoke, but it flushed the sky above with red. It was a truly terrible dawn, most beautiful in its terror; and, if ever dawn did indeed come up like thunder, it was this. Then came the greatest miracle of all, for with the rose flush in the sky the whole bird chorus of morning came to life. Never, surely, did birds sing so – blackbird and thrush, lark, and black-cap and willow warbler. Most of the time their voices, of course, were inaudible, but now and again, in the intervals of the shattering noise of the guns, their notes pealed up as if each bird were struck with frenzy and all together strove to shout down the guns.[1]

This report was by Harry Perry Robinson, who sent stories to *The Times* throughout World War One and was the oldest correspondent to cover the conflict.

In the terrible conditions of the Western Front, combatants appreciated the presence of birds as a reminder of their former way of life and as a symbol of normality and hope. Nowadays, in times of peace, birds are part of the background of our everyday lives, but they were once centre stage, forming important elements of the history, traditions and sports of Britain and beyond, from which a legacy of literature, language and mythology developed. *When There Were Birds* is a social history charting the complex relationships between people and birds, set against a background of evolving tastes, beliefs and behaviour, as well as changing landscapes and a decline in numbers. It is a forgotten history made up of many disparate strands that form an intriguing, unexpected and significant part of our heritage.

No other group of animals has had such a complex and lengthy association with humankind, having been loved, feared, eaten, kept in cages, taught to speak and exploited as a source of fuel, feathers, food and fertiliser. In addition, they have been used to tell the time, warn of poisonous gases, send messages, predict the weather and treat unlikely ailments, while also being persecuted as pests, with some species tragically driven to extinction.

To past societies, birds possessed more than practical functions, because they seemed able to sense so much, including events in the future. Their behaviour was mysterious, even sinister, and with their ability to fly, they bridged the gap between the earth and the sky. They were therefore believed to possess a link to the supernatural, and so superstitions connected to birds were commonplace.

A growing curiosity about nature, in particular birds, is one strand of the story, and this eventually became the science of ornithology. The early ornithologists were nearly all educated or self-taught men, from a range of backgrounds. Some had private incomes and devoted their lives to studying birds. Their writings can illuminate their times, giving us a glimpse into their world, such as when the Scottish naturalist William MacGillivray, two centuries ago, suggested a trip at top speed in a horse-drawn carriage, seated on the outside so as to watch the countryside and its birds:

Few things can be more delightful ... than a rapid survey of a large extent of country, such as may be made in a vehicle proceeding at the rate of ten miles an hour over a succession of plains, valleys, and mountainous tracts, and along rivers, estuaries, and arms of the sea. But I must correct myself: the true naturalist travels outside, and is not content with observing nature through a square foot or so of dim glass.[2]

A variety of records, including those of ornithologists, casts light on the involvement with birds of a whole range of society – rich and poor, professional and labouring classes, male and female, children and adults – and many made a living from birds, including decoy-men, taxidermists, bird scarers, poachers and gamekeepers. For centuries, far more people lived in the countryside than in towns and cities, making them acutely aware of the natural world and the changing seasons.

Most people today have few opportunities of routinely experiencing the sight, sounds and smells of masses of birds of any kind, because of the slump in their numbers and the massive shift in the human population from rural to urban living. Birdwatching has nevertheless grown in popularity, along with membership of related organisations, while a general appreciation of attracting birds to gardens has increased. Even so, the pleasure of witnessing birds on a large scale tends to be the preserve of those in a position to travel to nature reserves and other protected landscapes, with far too many people unaware of what they are missing. Coupled with the degradation of the countryside, this situation can make it difficult to have any meaningful connection with birds or nature.

When There Were Birds is not a textbook about birds, and so we have kept their names as straightforward as possible, though the classification of birds is complex, with families of birds and sub-species. The same species of bird can have different names across the world and even within Britain, and numerous unofficial folk names were adopted throughout history, so that a chaffinch might be known as a pink, spink, shilfa, twink or chink chaffey. Further information about birds and their names can be found in innumerable specialist books, as well as on several websites, such as the Royal Society for the Protection of Birds (RSPB) and the British Trust for Ornithology (BTO).[3]

Birds have been exploited by human societies from earliest times. The tomb paintings of the ancient Egyptians provide a visual catalogue of many species that lived cheek-by-jowl with people of all classes. In prehistoric Britain, thousands of years ago, there would have been a similarly close relationship, though without any means of leaving a permanent record. The landscape would have been very different, with vast tracts of marshes and scrubland, all teeming with birds.

Although the number of birds has fallen markedly over time, at the beginning of the twentieth century their massed ranks in a north Hampshire field in winter could still astonish an experienced naturalist like William Hudson:

> a ploughed field of about forty acres was the camping-ground of an army of peewits [lapwings]; they were travellers from the north perhaps, and were quietly resting, sprinkled over the whole area. More abundant were the small birds in mixed flocks or hordes – finches, buntings, and larks in thousands on thousands, with a sprinkling of pipits and pied and grey wagtails, all busily feeding on the stubble and fresh ploughed land. Thickly and evenly distributed, they appeared to the vision ranging over the brown level expanse as minute animated and variously coloured clods – black and brown and grey and yellow and olive-green. It was a rare pleasure to be in this company, to revel in their astonishing numbers.[4]

World War One began a few years later, in August 1914, and armies soon became bogged down in the trenches of the Western Front in France and Belgium. Soldiers were very aware of birds, which provided a distraction from all the horrors, and some of their letters were actually published during the war. A Lancashire newspaper revealed the thoughts of one local soldier: 'It is curious how unconcerned the birds are, though War goes on beneath them and above them. On the Continent they go about their business quite as usual. Even on the battlefields it is no uncommon sight to see large flocks flying over the trenches in their annual migratory flight northwards, apparently undisturbed by the noises of fighting.'[5]

John Bryden was a young miner from Kirkconnel in Dumfries and Galloway, who had enlisted in the Cameron Highlanders. He also wondered how the birds could sing among the carnage and summed up the feelings of many: 'If it weren't for the birds, what a hell it would be.'[6] He was killed five months before the war ended, but with the passing years his words have become ever more relevant.

CHAPTER ONE

———— ·◆· ————

ABUNDANCE

To a country man, birds are a constant presence,
and though fewer are found in towns, it is probable
there are more than people think. I have a small
garden in Hull, almost surrounded by houses, yet
therein, I have either seen or heard thirty-three
different kinds of birds, including the night hawk
(night jar) and the heron.

JOHN NICHOLSON,
Hull headmaster, writing in 1890[1]

Birds have been a perennial source of marvel and mystery, and
even those accustomed to seeing them could be taken aback at
their sheer numbers, as happened to the radical journalist William
Cobbett. In September 1826, not long after the crops had been
harvested, he was riding his horse through the Gloucestershire
countryside, south of Cirencester:

Between Somerford and Oaksey, I saw, on the side of the road,
more *goldfinches* than I had ever seen together; I think fifty times
as many as I had ever seen at one time in my life. The favourite
food of the goldfinch is the seed of the *thistle*. This seed is just
now dead ripe ... they grow alongside the roads; and, in this
place, in great quantities ... the goldfinches were got here in
flocks, and, as they continued to fly along before me, for nearly

half a mile, and still sticking to the road and banks, I do believe
I had, at last, a flock of ten thousand flying before me.[2]

Nowadays, most people would be thrilled to spot a dozen goldfinches, one of the most colourful of British birds, but thistles in
fields and along grass verges have been largely eradicated.

Although Cobbett belonged to an era when it was commonplace
to see hundreds of birds in a single flock, unusual sights were
nevertheless worth recording, and in 1810, after a November day's
shooting near his Longparish estate in Hampshire, Peter Hawker
also expressed his amazement as he returned home across the
marshes, because they 'were literally swarming with starlings, of
which I killed a large number. These birds cared so little for being
shot at, that ... they returned and pitched within twenty yards of
me. They literally darkened the air, and the noise they made was
not to be described.'[3] On the Isle of Wight, the barrister and naturalist Cornwall Simeon was likewise surprised by a particularly large
gathering of starlings: 'I saw on the 18th of November, 1852, in the
Island, a flock of Starlings far exceeding in numbers any that had
ever before come under my own observation, or that of any of the
party who were with me. It would be impossible to form an estimate
of their numbers, but they blackened a very large extent (several
acres I should think) of the field on which they had alighted.'[4]

Precise numbers of birds were usually guesswork, and perhaps the most honest assessment was given a few years earlier to
Richard Lubbock, the rector of Eccles in Norfolk, when he was
looking into coots on the Norfolk Broads: 'A broad entirely devoid
of coots would be London without sparrows, or Newcastle without
coals ... [the coot] is found in immense flocks on Hickling broad,
Horsey mere, &c. On Hickling broad, a fen-man, to whom I put
the question, What quantity of coots might there be? returned for
answer, "About an acre and a half." '[5]

Birds that are now dangerously in decline, like lapwings, were
once a familiar sight. One autumn evening in the 1890s, in the village of Ablington in the Cotswolds, the young squire Arthur Gibbs
watched numerous lapwings, with their black, white and iridiscent

green plumage and wavering flight, able to change direction as if being directed by an unseen presence:

> a large flock of lapwings ... gave a very fine display – a sort of serpentine dance to the tune of the setting sun, all for my edification. They could not quite make up their minds to settle on a brown ploughed field. No sooner had they touched the ground than they would rise again with shrill cries, flash here and flash there, faster and faster, but all in perfect time and all in perfect order – now flying in long drawn out lines, now in battalions; bowing here, bowing there; now they would 'right about turn' and curtsey to the sun. A thousand trained ballet dancers could not have been in better time. It was as if all joined hands, dressed in green and white; for at every turn a thousand white breasts gleamed in the purple sunset.[6]

It was not just the sight of so many birds in a single flock that caused wonder, but also the sounds that they made, their different ways of flying and their place in everyday life. While he was an undergraduate at Cambridge university, William Bingley toured north Wales in the summer of 1798 and again three years later. On both occasions, he went to Priestholm or Puffin Island, an uninhabited island barely ¾ mile long, just off the east coast of Anglesey, where today only a handful of puffins breed:

> When we had arrived under the rock, and had cast anchor, we fired a swivel gun [a small cannon], to try the effect of the report round the island, when such a scream of puffins, gulls, and other sea birds, was heard, as beyond all conception astonished me. The immense multitudes that in a moment rose into the air, were unparalleled by any thing I had before seen. Here they flew in a thousand different directions, uttering as many harsh and discordant screams ... the air, the sea, and the rocks seemed alive with their numbers ... Upwards of fifty acres of land were literally covered with Puffins ... the number here must have been upwards of 50,000.[7]

The silence in the countryside was once profound, with an absence of noise from cars, tractors, motorbikes, aeroplanes, trains, chainsaws and all the other paraphernalia of modern rural life. In the past, the most noticeable sounds came from the wind and rain, from thunderstorms, the blacksmith's hammering, guns firing, dogs barking, church bells, sheep bells, sheep bleating and – depending on the season – birdsong. In order to defend their territory and attract a mate, each species of bird sings in some way, though what are termed 'calls' tend to be shorter and have a variety of other functions, such as warning of danger and keeping in touch while foraging or flying.

In his book on English songbirds that was published in 1737, Eleazar Albin commented: 'Singing birds are so pleasant a part of the Creation; whether we consider their variety, beauty or harmony; that the animal world does not afford more agreeable objects to the eyes, nor none that so sweetly gratifies the sense of hearing. They were undoubtedly designed by the Great Author of Nature, on purpose to entertain and delight Mankind, who, for the generality, are well pleased with these pretty innocent creatures.'[8]

Writing in 1845, Lubbock said that in former times in the month of May, 'the noise of marsh birds was incessant as you walked through a fen, – redshanks, reeves, and black terns, were perpetually dashing round you; and the various notes, – some pleasant, some discordant, – if heard for hours, took such hold upon the sensorium, as to haunt you afterwards.'[9] Few people found every single bird call and song delightful, and some can be annoying to the human ear, such as the harsh sounds produced by magpies and crows or the tedious notes of wood pigeons. After serving in the Royal Navy, Henry Davenport Graham moved to the island of Iona for a few years from 1848, spending his days outdoors studying ancient sites and birds. According to him, although the return in May of the corncrake with its 'unremitting and monotonous cry' was initially welcomed,

it soon becomes wearisome, especially when it is continued all through the night, which in these northern latitudes is scarcely dark even at midnight. I have sometimes been urged to turn out

at that unusual hour with my gun and favourite dog … and by
the ruddy light from the northern sky, where the sun seemed
to be rolling along only just beneath the horizon, have beat up
the long grass in front of the house to dislodge this murderer of
my sleep.[10]

As an accomplished mariner, he often spent all night out on the
water in a small boat and was enthralled by the sound of birdsong
at dawn, especially the skylark:

The praises of the lark have been so much celebrated on shore
by poets, that I will only add that the effect of its song is equally
exhilarating when heard on the water … We have only to rest
and watch the first rays of the morning sun gilding the steep
shores of the narrow inlet, and as we enter we are received by
a burst of choral music, thrushes and blackbirds responding to
each other from the opposite banks, and above all, high up in the
air, the larks singing their morning orisons. Or perhaps the song
of the lark is still more remarkably exhilarating when starting
off by boat on a fine clear morning bound for the distant purple
islands which hang upon the dark-blue horizon. As the boat
runs along the low coast, skimming over the crisp blue waves,
the larks spring up one after the other, continuing a succession
of merry carols; and when the last point is passed, and the boat
stands out into deep water … we still hear the jubilant chorus
of many larks filling the air above, though growing fainter and
fainter as the sea breeze now fairly fills the sail and the boat
settles down to her work.[11]

At the other end of the country, in Wreyland on the edge of
Dartmoor, John Torr was a farmer until his death in 1870. His
son, also called John, was for decades a London solicitor, while
his grandson Cecil became a barrister until he left London in 1913
and returned to his Devon roots to concentrate on antiquarian
pursuits. Cecil Torr said that in letters his grandfather frequently
commented on the enjoyment of listening to birds in their village:

26 January 1847, 'the home-screech [mistle thrush] singing merrily this morning' – 1 May 1850, 'the nut-hatch a cheerful singer' – 22 April 1864, 'how delightful and cheering is that old grey-bird [song thrush]' – and so on . . . Their singing was always a pleasure to him; and he writes to my sister, 10 March 1852: – 'I have often fancied that the thrushes know that I am pleased, when I am listening to them, from the cast of their little sharp eye down on me.'[12]

As a farmer, Cecil said, his grandfather always preferred the birds in the spring when they were singing, rather than in the autumn when they ate his fruit: 'Even in the spring he writes to my father, 29 April 1849: – "I certainly do like to hear them sing, but it is vexing to lose all the fruit . . . I loaded my gun; but, when I came out, one of them struck up such a merry note that I could not do it – so I suppose the fruit must be sacrificed to my cowardice, humanity, or what you may call it.".'[13]

While those living in Britain might quietly appreciate birdsong, visitors from overseas like Elihu Burritt from Connecticut were enchanted. A blacksmith by trade, he became a writer, lecturer and evangelist and lived in Britain for a decade from 1846, setting up the League of Universal Brotherhood and organising peace conventions. On a subsequent visit to Britain, from July 1863, he embarked on a walking tour, during which he was keen to record rural life:

There is no country in the world of the same size that has so many birds in it as England; and there are none so musical and merry. They all sing here congregationalwise, just as the people do in the churches and chapels of all religious denominations . . . I believe that everything sings that has wings in England . . . Every field is a great bird's-nest. The thick, green hedge that surrounds it, and the hedge-trees . . . afford nesting and refuge for myriads of these meadow singers. The groves are full of them and their music; so full, indeed, that sometimes every leaf seems to pulsate with a little piping voice in the general concert.[14]

Some two decades later, in June 1884, a large party of tourists was similarly spellbound by birdsong. They had set out on a journey from Charing Cross in London to the West Country in an old-fashioned coach drawn by four horses, at the invitation of Andrew Carnegie, the Scottish-born American industrialist (who later became renowned for his philanthropy). One guest was an American editor and writer, John Champlin, who subsequently published an account of the trip, which he dedicated to Carnegie. In it, Champlin referred to himself as the Chronicler, Carnegie as Maecenas and Samuel Storey, the Member of Parliament for Sunderland, as The M.P.[15]

Everyone was appreciative of the beautiful landscape, but most memorable was the hamlet of Fairmile in Devon, a few miles from Honiton, where they stopped for lunch in a meadow by the trout stream:

> Save the babble of the brook, the hum of bees, and the twitter of birds, scarce a sound broke the stillness of the scene. We could even hear the voices of the men talking on the hill-side behind us.
>
> 'Hark!' exclaimed Maecenas, looking upward, 'There's a lark!'
>
> 'There are two,' said The M.P., quietly.
>
> 'Three! Four! Five,' cried several of the ladies in a breath.
>
> A flood of melody came down from the clouds. We could just see the five songsters soaring in the sunlight far above us, looking so small – such mere specks against the sky – that we could but wonder how their song could reach us at all. But every note was clear and distinct, reaching our ears even after the tiny specks had faded away in the azure vault.[16]

Before they had finished eating, Champlin said, more birds joined in:

> The cup of delight of the lunchers at Fairmile was not yet full. As the voices of the larks died away, a cuckoo began to call its mate from the copper beeches over the brook, and a pair of wood-pigeons to purr in the willows by the road-side.

Maecenas could contain himself no longer. He clapped his hands in applause, and said:

'My friend, God is good to you! You have come three thousand miles to hear a lark sing, and you have heard what I never heard before – five larks singing at once, and cuckoos and wood-pigeons thrown in!'

The Chronicler might have wept tears of joy at this announcement, and probably ought to have done so; but while he was searching industriously for a clean pocket-handkerchief to do justice to the occasion, The M.P. announced that the champagne, which had been cooling under his supervision in the brook, was of the proper temperature for testing, and so the opportunity passed away forever.[17]

Perhaps the Chronicler would have wept tears of sorrow if he could have foreseen that, little more than a century later, a major highway, the A30, would be constructed between Exeter and Honiton, with Fairmile being the centre of high-profile protests.[18]

In the 1930s the rural writer John Massingham was also journeying through southern England and paused for a moment at a prehistoric barrow on Normanton Down in Wiltshire, barely half a mile from Stonehenge: 'Over my head passed a cloud of starlings, singing, and between them and the clouds in slow white sail above them, lapwings flew wailing and flashing their wings. The airy chorus chimed with the bright gongs of distant sheep-bells, and at this very moment a double rainbow flung its heavenly arch over the old grey ruin. The scene was a sudden hymn of the downs.'[19] In a joyful, poetic turn of phrase, he sketched what is now a lost landscape – the peaceful chalk downland filled with the sights and sounds of innumerable birds.

On another occasion, in late autumn, Massingham stood on the ancient ramparts of the Iron Age hillfort known as Alfred's Castle, 30 miles north of Normanton Down, and looked across the neighbouring estate. His attention was drawn to countless rooks returning to their roost: 'When I was there, the rooks were coming home to bed in the woods that bosom Ashdown House, and their cawing in concert was like the voice of the sea above the

green ocean of the encircling downs.'[20] The start of autumn is the
signal for many migrant birds to head off to warmer climates, but
others such as rooks are resident all year round. They make nests
in colonies high up in trees, but after the breeding season, these
distinctive rookeries are abandoned in favour of roosting sites,
where many flocks congregate to rest and sleep during the long
winter nights, after feeding in the daytime.

Lines of rooks flying many miles each evening in a straight line
to some distant roost gave rise to the phrase 'as the crow flies'. It
should be 'as the rooks fly', but typifies the failure to distinguish
between gregarious rooks and solitary crows. The crow or corvid
family consists of large birds, primarily black in colour, with sturdy
bills or beaks, including carrion crows, hooded crows, jackdaws,
ravens and rooks. They can be difficult to identify with certainty
from a distance, but especially so before the use of telescopes and
binoculars. Throughout history, the names 'rook' and 'crow' have
been applied at random, interchangeably.[21]

The behaviour of rooks puzzled James Cornish when he was
growing up in the 1880s in Childrey, a village not far from the
Alfred's Castle hillfort. He would see immense numbers of rooks
flying northwards each evening, but nobody knew where they
roosted – and there was no way of finding out. By chance, the
mystery was solved in 1892 when he became curate of the town of
Faringdon, some 6 miles north-west of Childrey:

To the north [of Faringdon] one has a splendid view over the
Thames Valley, and the hillsides here are clothed with fine
oak woods. To one of them, the Grove [Faringdon Grove] by
name, rooks came in thousands during the winter months ...
and it was interesting to walk along the church path and watch
flock after flock streaming in from every point of the compass
to their chosen sleeping place: a black river of rooks, perhaps
half a mile long, travelling steadily on their way when the light
was fading on 'a brief November day'. They came rather early
in the evening and held lengthy conversations just as starlings
do, before settling down to sleep.[22]

Because of their spectacular pre-roost aerial displays, starlings stand out among the various species of birds that have communal roosts. Like rooks, they are resident birds, but are joined in the autumn by many more from the Continent, and they can also fly miles in the evening to their roosting site. Just before settling down, they gather from time to time in enormous gyrating flocks called murmurations, a word that refers to the noise that is made rather than the spectacle. In about 1840, when walking over the Dorset downs with friends, the Reverend Charles Bingham was amazed at one display:

> we saw, at the distance of about a mile and a half from us, what we at first took to be the smoke of a lime-kiln, or of some great mass of burning weeds; but ... we perceived that it was a flock of starlings ... For half an hour we watched their evolutions with the greatest interest; and indeed I have seldom seen anything more graceful than the variety of their motions, tumbling and rolling over in the air in, what one might call, the most harmonious confusion ... we were struck by their resemblance to a light gauze scarf floating on the wind, – sometimes bellying out into transparency, and sometimes gathered up into an opaque mass.[23]

Half a century later, the ornithologist and writer Charles Dixon used different comparisons: 'The aerial movements of these flocks are beautiful and curious in the extreme, resembling the collapse and inflation of a huge balloon, or the spreading of a gigantic net.'[24]

Far fewer murmurations are seen today because starling numbers, like so many other species, have plummeted. It was once also routine during winter to come across flocks of thousands of starlings roosting together in woods and in reedbeds, where they could cause significant damage. Dixon described the gatherings of starlings near his home at Torquay in Devon: 'at dusk the birds congregate in chattering hosts, keeping up their noisy converse more or less intermittently throughout the entire night. Such a vast concourse of Starlings usually occupies the low chestnut and oak trees in the grounds fronting Torre Abbey.'[25] Also in the 1890s, Arthur Gibbs at Ablington mistook their noise for farm machinery:

> One night towards the end of September, whilst walking on the road, I heard such a loud, rushing sound in front, beyond a turning of the lane, that I imagined a thrashing machine was coming round the corner among the big elm trees. But on approaching the spot, I found the noise was nothing more nor less than the chattering and clattering of an immense concourse of starlings. The roar of their wings when they were disturbed in the trees could be heard half a mile away.[26]

Around this time, starlings began to roost in urban and industrial areas as well, selecting a variety of locations, from trees to bridges, gasworks, churches, museums and even monuments such as Nelson's Column in London. In October 1912, after commemorating the anniversary of the Battle of Trafalgar, the *Pall Mall Gazette* adopted a patriotic tone:

> Who, of the many who have this week paid their tribute to the memory of the hero of Trafalgar in the square that bears his name, have noted the starlings which come to roost at the top of Nelson's Column? For over three years the roost has grown in popularity, and the birds now come in flocks ... The fact that the first comers came singly or few together seems to show that the idea pertained to some original or adventurous spirits in the starling world, and there is a singular appropriateness in the boldest of the birds choosing a spot so honoured by the memory of a man of kindred qualities. As a matter of fact, none of our species so fully typifies the qualities on which we English pride ourselves – adaptability, pluck, and perseverance, cheerfulness under hard conditions, and adventurous spirit ... If any bird should be selected as the bird of the British *par excellence* it should be the starling.[27]

The starling roost on Nelson's Column did not last many years, but the birds did stay in the city, as *The Times* reported in 1928:

Not many years back they were chiefly to be seen at dusk, and heard far into the night, among the plane-trees, still in leaf, along the Victoria Embankment, in the Temple courts and groves, and other neighbouring coverts. Some slept, as they do still, among the carved foliage of the capital of the Nelson Column ... But in recent years the jubilation of their innumerable voices chides so vehemently from such large carved buildings as St. Paul's Cathedral and the British Museum that it overrides even the roar of modern traffic, and excites the attention of indifferent travellers on open omnibus tops.[28]

Other species of birds made their presence known when they crossed the country or migrated from the Continent in flocks, during which they frequently called to each other. The whimbrel is a large wading bird that overwinters in Africa, and they were often heard flying across Britain when returning to their northerly breeding grounds. Charles Dixon in Torquay said that they were always aware of them in May:

Great numbers of these birds pass over the Tor Bay district by night, as may be proved by their well-known cries uttered whilst on passage. On the still May nights we often stand and listen to the passing flocks, sometimes flying just overhead, at other [times] at heights so remote that their cries can scarcely be heard. They fly low on overcast, misty nights; high when the moon and stars are shining brightly. The flocks succeed each other at intervals of a few minutes, but sometimes much longer periods divide them. First the well-known notes may be heard sounding faintly from afar, louder and louder becomes the rapid chorus of tittering cries, until the birds are directly overhead, then they grow fainter and fainter as the flock pursues its way almost directly north.[29]

A few years later, during World War One, the popular writer and naturalist William Hudson put in a letter that he was spending the autumn of 1915 at Lelant in Cornwall as a convalescent, then mentioned one nocturnal incident: 'Here I have the estuary of the Hayle before the window and can hear all day and night the piping and thrilling conversation of the curlews going on. A few nights ago we had a flock of wild geese to visit us too; their cries woke me up some time after midnight.'[30]

During the nineteenth and twentieth centuries, there was an increased awareness that birds were disappearing. The economist and philosopher John Stuart Mill had stated in 1848 that it was not sustainable across the world for the human population density to increase, because society needed solitude as the cradle of thoughts and aspirations. It would, he said, be deeply unpleasant to see all the land used for the cultivation of food for human beings, with 'every flowery waste or natural pasture ploughed up, all quadrupeds or birds which are not domesticated for man's use exterminated as his rivals for food, every hedgerow or superfluous tree rooted out, and scarcely a place left where a wild shrub or flower could grow without being eradicated as a weed in the name of improved agri-culture.'[31] Even Mill could probably not have imagined that in less than two centuries, the natural world would have changed so much.

The present-day scarcity of birds is hardly surprising considering the pressures that have resulted from the enormous population growth in Britain and beyond. Over a period of a thousand years, the number of people in England alone has risen from about 1½ million (deduced from the Domesday Book survey of 1086) to some 56 million today. By 1348, the figure may have reached five million, at which point half the population was wiped out by the Black Death. The numbers crept up to about four million by 1600 and five million by 1700. In 1801, when the first census yielded more reliable figures, England had in excess of 8,300,000 people.[32]

The next two centuries experienced a population explosion. In the 1901 census it was reckoned there were 30½ million people, more than three times the number a century earlier, while the 2011 census recorded more than 53 million people, and the numbers continue to soar. A change has also occurred in the balance between the countryside and urban areas. In the medieval period, only about one in ten people lived in towns and cities in England, rising to one in seven by 1801. Over the next two hundred years, urban areas expanded, leading to a dramatic reversal, so that only one in twenty people now live in the countryside. Incredibly, the population of London today is roughly the same as all of England in 1801.[33]

Britain's dwindling numbers of birds are not entirely due to local factors. Conditions across the globe affect migrating birds, and so a combination of complex reasons is responsible, driven by the vast rise in the human population worldwide. Increasingly, though, there is much at fault within Britain itself, and successive generations have voiced concerns about wildlife, pointing to the unsustainable numbers of birds shot for sport, eradicated as vermin, killed by cats, and taken for food and fashion, while also highlighting the effects of expanding towns, cities and roads, the loss of woodland and wetland habitats, soil compaction, pesticides and other chemicals, industrial agriculture, motor vehicles and aircraft, the effects of wind and solar power stations, climate change and the terrifying loss of insect life.

Birds cannot survive if there is no food. In about 1837, Archibald Hepburn, an inhabitant of the village of Linton, some 20 miles east of Edinburgh, described in one sentence what was once a typical rural scene in mid-May: 'In the pasture fields, where thousands of insects are disturbed by the cattle, how beautifully they [the swallows] skim along.'[34] Such a scene is now rare after decades of disastrous agricultural policies. The fact that vehicle windscreens no longer become covered in insect casualties during a long drive is proof that something is seriously amiss.[35]

Particularly from the late eighteenth century, millions of acres of fens and marshes were drained to create land suitable for producing food for the increasing population. This reclamation, as it was inappropriately called, included extensive areas of wetland like the Fens in East Anglia and the Somerset Levels, where a network of ditches and channels was dug. Glastonbury is in the eastern part of the Levels, and in 1810 William Holland, who lived several miles to the west, commented briefly in his diary: 'My wife & I dined together on two snipes, which are now become rarities.'[36]

In the 1820s, when the naturalist John Knapp was living with his family at Alveston in Gloucestershire, he expressed his worries: 'Some of our birds are annually diminishing in numbers, others have been entirely destroyed, or no longer visit the shores of Britain. The increase of our population, enclosure, and clearage of rude and open places, and the drainage of marshy lands, added to the noise of our fire-arms, have driven them away, or rendered their former breeding and feeding stations no longer eligible to many.'[37] The scarcity of snipe, a wading bird, became a constant lament, because it was very good to eat. Knapp further lamented that 'the extensive marshes of Glastonbury, which have afforded me the finest snipe shooting, are now luxuriant corn farms'.[38]

The Reverend Murray Mathew grew up in north Devon, and after becoming vicar of Buckland Dinham in neighbouring Somerset in 1888, he collaborated with William D'Urban, who had been the first curator of the Royal Albert Memorial Museum in Exeter. They co-authored *The Birds of Devon*, which was published in 1892, and one section described how snipe had been affected by drainage schemes:

the very great diminution in its numbers during the last half-century is one of the characteristic changes ... which the enclosure and drainage of moorlands, salt-marshes, and wet highland farms have helped to bring about ... Many a moor on which Snipes once nested numerously has disappeared to give place to corn- and turnip-fields. Many a meadow, formerly ornamented all over with rushes and full of soft splashlets, is now through subsoil drainage sound and dry ... In old days there was not a warm ditch by the side of a hedge which did not harbour in the winter-time some half-dozen Snipe.[39]

Not only did birds disappear, but entire landscapes were altered, along with the lives of those whose existence had been formerly intertwined with their watery environment. To them, the drainage of the wetlands was not reclamation, but the destruction of a way of life. At the start of the twentieth century, the journalist William Dutt spoke of the impact that drainage was having on the Norfolk Broads and included a sorrowful verse in the local dialect:

> several of the Broads marked on old maps of the district have wholly disappeared. Over the sites of some of them the mowing machines pass, cutting down hay crops where the reeds used to wave in the wind. As time goes on the reed and rush cutters will find their occupations gone, and they will have to seek other means of livelihood. This will be no easy matter, for the men of the rivers and marshes do not take kindly to change. 'A Norfolk Lament' in the Eastern Counties Magazine well expresses the views of one of them. In it an old marshman is made to say –

> *They're a sluggin' an' widenin' miles o' the deeks;*
> *I reckon the whool mash 'ull sewn ha'e to goo;*
> *An' us old mashmen 'ull fare like the freeaks*
> *I see last summer at Barnum's shoo.*
> *Paarson say, 'Bill, yew dew narthin but growl!'*
> *Well, bor! I reckon theer's plenty o' rayson.*
> *Whass become o' the flightin' fowl?*
> *I never see fifty the whool o' last sayson.*
> *Whass become o' the gre't deeak eeals?*
> *Shillun a stoon we wuz glad to get for 'em.*
> *Gone! with the mallards an' snipes and teeals!*
> *If ye want any now yew ha' got to sweat for 'em.*[40]

The drainage of wetlands was only one method of bringing more land into agricultural production. The medieval system of agriculture was based on open fields, commons and woodland, with few hedges, but following the Black Death of 1348–9, land began to be divided by formal agreements into small fields enclosed by hedges, enabling farmers to graze sheep independently. From about 1750, the Great Enclosures took over, with hundreds of private Acts of

Parliament allowing many more open fields, commons and so-called wasteland to be enclosed and turned into arable land and pasture, at times in the face of opposition and violence.[41]

This land grab was effectively privatisation, enriching the new landowners and depriving the poor of their communal rights to fuel, foraging and grazing, as well as putting small farmers out of business and disrupting the wildlife. Walls and fences were erected, new farms were constructed in previously isolated areas, and roads were laid. In addition, about two hundred thousand miles of hedges were planted, which actually proved beneficial to the bird population, helping to mitigate the loss of wasteland. With the enclosures, a chequered pattern of small fields developed, which has become symbolic of the traditional English landscape.

Only a few decades after this spate of enclosures, protective legislation was passed in 1791 that paid a bounty for exporting cereals when the price was low at home, which encouraged farmers to remove their old and new hedges in order to increase the amount of their land under cereal cultivation. Hedges did not just act as boundaries, but were important for timber for fuel and a whole range of other functions, from fences to shipbuilding. In February 1792, during a parliamentary investigation into the availability of suitable timber for the Royal Navy, several instances were cited of hedges being removed, including in the county of Hampshire: 'The grubbing up of hedge rows is become general, and the growth of timber in them is therefore totally destroyed, owing to the great price given for corn since the bounty took place for exporting of corn and beer, which gives every farmer encouragement to grub hedge rows up, and convert them into corn land.'[42]

The old hedges tended to have many plant species, whereas the new ones were far less attractive to wildlife. When, in 1808, a survey was compiled for the Board of Agriculture on the county of Cheshire, one of the correspondents noted:

The inclosures in Cheshire are generally fenced by hedge and ditch. The ditches are usually three feet wide, and of sufficient depth to act as a drain. The soil and other stuff taken from the ditch is generally formed into a regular cop [heap], upon which the sets are planted (most commonly white, or haw-thorn), which are guarded in their infancy with rails and posts on one side, and sometimes on both … This is the common mode of making fences, where new enclosures are wanted … The ancient fences are composed of hazel, alder, white or black-thorn, witch-elm, holly, dogwood, birch, &c. &c.[43]

Enough decent hedges existed for Elihu Burritt to be captivated by the scenery a few decades later, in the summer of 1863, and he thought that they were an asset and a valuable link to America: 'The three great, distinctive graces of an English landscape are the hawthorn hedges, the hedge-row trees, and the everlasting and unapproachable greeness of the grass-fields they surround and embellish. In these beautiful features, England surpasses all other countries in the world. These make the peculiar charm of her rural scenery to a traveller from abroad.' He reckoned that all those people who had left the Motherland for other countries still clung to these distinctive features of the landscape, as 'memories that are transmitted from generation to generation; which no political revolutions nor severances affect; which are handed down in the unwritten legends of family life in the New World, as well as in the warp and woof of American literature and history.'[44]

When being shown round a farm at Great Bardfield in Essex, though, Burritt was fearful for the future, because hedges were being replaced here by wire fences:

Locomotive steam engines, on broad-rimmed wheels, may be met on the turnpike road, travelling on their own legs from farm to farm … It is an iron age, and wire fencing is creeping into use, especially in the most scientifically cultivated districts of Scotland … Iron wire grows faster than hawthorn or buckthorn. It doubtless costs less. It needs no yearly trimming, like shrubs with sap and leaves. It does not occupy a furrow's width as a boundary between two fields … It is not a nesting place for destructive birds or vermin.[45]

It saddened him to see hedges being removed, and he worried that the mature trees in these boundaries would also be felled:

> Those little unique fields, defined by lines and shape unknown to geometry, are going out of the rural landscape ... *'Once destroyed, never can be restored.'* And destroyed they will be, as sure as science. As large farms are swallowing up the little ones between them, so are large fields swallowing those interesting patches ... There is much reason to fear that the hedge trees will, in the end, meet with a worse fate still. Practical farmers are beginning to look upon them with an evil eye – an eye sharp and severe with pecuniary speculation; that looks at an oak or elm with no artist's reverence.[46]

In 1860, exactly three years before Burritt, Walter White, who worked in London as a library assistant for the Royal Society, spent his summer holiday touring eastern England, during which he visited the model farm at Tiptree in Essex. The new methods of farming were impressive, but he too felt uneasy:

> I thought there was a look of nakedness, which, however, might be occasioned only by the fewness of hedgerows ... Already we see foretokens of the way in which the work will be carried out: by thousands of steam-engines and ingenious machines in the fields; by implements of ever-increasing efficiency in the barns and farm-yards, until the operations which have formed for ages the especial charm of rural life shall have become prosaic and mechanical as in a shovel factory. For this, trees and hedgerows are grubbed up, banks are levelled, ditches filled, and large naked fields contrived; and farms are becoming mere factories of roots and grain. But here also is a matter for debate. Will our climate be affected by the destruction of foliage? will birds and small wild animals be as well able to play their part in Nature's great scheme as heretofore?[47]

White's next stop was at Broomfield Hall farm in the picturesque village of Broomfield in Essex:

When we came to the fields I could not help expressing my
regret at the sight of fallen trees and up-grubbed hedges. The
argument that steam-ploughing would not pay except in large
level fields, afforded me but little comfort, for what will an
English landscape be when deprived of trees and hedges? Is so
much exposure to the sun good for the land; and will singing
and warbling fill the air as now when the covert for nests shall
have been cut down as a wasteful encumbrance?[48]

His comments were both perceptive and prophetic, as Broomfield
itself is now a northern suburb of Chelmsford, while across the
country industrial towns such as Birmingham and Manchester
rapidly devoured the rural landscape. The capital city of London
grew uncontrolled, especially with acres of shoddy dwellings put
up by speculative builders. Only a few years after White's tour, *The
Leisure Hour* described London's unwelcome expansion:

One of the unavoidable consequences of the monstrous and
ever-increasing spread of the metropolis is that girdling fringe
of waste land which encompasses it on all sides, and forms such
an unpicturesque introduction to the capital ... The first active
measures taken for the transformation of the waste into a populous suburb is the making it a worse waste by converting it into
a brick-field ... involving as it does the utter annihilation of one
pleasant and picturesque nook after another. To the north of
Islington, stretching away towards the east, there is (we ought
rather to say there *was*) a pleasant shady district, which has
been the occasional holiday bourn of hundreds of thousands of
Londoners for fifty years past, and known under the designation
of the 'Green Lanes' ... The route they once trod between the
tall overshadowing hedges, where the wild-flowers grew and
the linnet sang, now leads between the rows of bricks and the
smothering reek of kilns.[49]

Another major change to both urban areas and the countryside
was the construction of miles of canals and railways. Gervase

Mathew, who was brought up with his older brother Murray in north Devon, worked as a clerk for the Royal Navy at Plymouth. One frosty morning in early December 1869 he was back at Instow on the River Torridge estuary, which he knew well: 'The oozes, which formerly at this time of the year were frequented by vast flocks of ringed plovers, dunlins, curlews, turnstones, &c., were to-day, comparatively speaking, deserted.' He remarked that many other types of bird had also disappeared, such as knots, curlew sandpipers and terns. 'The cause of this is easily explained,' he said: 'railways, embankments and drainage are the foes that have driven our birds away. Hundreds of acres of salt-marsh, such excellent feeding-ground for all the Tringae [waders], have been reclaimed, and are under cultivation; a railroad from Barnstaple to Bideford runs close to the river, and the passing trains frighten the birds away as soon as they arrive.'[50] This railway line opened in 1855, closed in 1965 and is now part of the Tarka Trail for walking and cycling.

While the railways destroyed habitats, birds did become accustomed to them. The line that passed through Lustleigh in Devon was opened in 1866, and Cecil Torr at nearby Wreyland could hear the reaction of tawny owls: 'If there was no shunting, the train just waited at the station till the specified time was up. The driver of the evening train would often give displays of hooting with the engine whistle while he was stopping here, and would stay on over time if the owls were answering back.'[51]

The railway companies adopted electrical telegraphy so as to control signalling and introduce a standard time across the country, rather than use local time, which varied. Messages could be transmitted through wires that were placed underground, though from the 1840s they were carried above ground on wooden telegraph poles or even slung between buildings in urban areas. This communication system spread out across the country, not only for railway lines, but also post offices, and from the mid-1850s it was faster to send short messages as telegrams between two post offices rather than by post. Insulated cables that were placed underwater also transmitted messages worldwide.[52] This revolution in communication led to untold numbers of birds being 'telegraphed' by the wires – killed or injured – as they flew at speed into these unexpected obstacles.[53]

In 1866 twenty-two-year-old Thomas Gunn, who would develop the largest taxidermy business in Norwich, reported:

> I have observed several instances of the destruction of birds by telegraph-wires within the last two or three years. A specimen of the kingfisher was picked up a short time since by a man on the railway line at Lakenham. My friend Mr. Arthur Taylor picked up specimens of the cuckoo and wryneck, laying both together on the embankment of the railway near Somerleyton Station, in Suffolk, about a month since. I remember finding examples of the blackbird and redwing by the roadside, in the neighbourhood of Wymondham, two years ago. All ... were killed in precisely the same manner, *viz.*, across the throat, in one or two instances the windpipe being nearly severed.[54]

From the mid-1850s Robert Gray from Glasgow made many excursions in Ayrshire with his father-in-law Thomas Anderson, and they noticed skylarks being killed in this manner: 'Some winters ago immense flocks of larks appeared during hard weather in some fields close to the town of Girvan [Ayrshire]. On rising from the ground, the cloud of birds appeared so dense as to obscure objects in the line of their flight. Large numbers were killed on the telegraph wires, and after the flocks passed it was found that many birds had been mutilated, their wings being torn off by the wires.'[55] In America in 1876, Dr Elliott Coues, an army surgeon and naturalist, witnessed the destruction of horned larks:

> I recently had occasion to travel on horseback from Denver, Colorado, to Cheyenne, Wyoming, a distance of one hundred and ten miles, by the road which, for a considerable part of the way, coincided with the line of the telegraph ... The bodies lay in every instance nearly or directly beneath the wire. A crippled bird was occasionally seen fluttering along the road ... I began

to count, and desisted only after actually counting *a hundred* in the course of one hour's leisurely riding ... a distance of three miles. Nor was it long before I saw birds strike the wire, and fall stunned to the ground; three such cases were witnessed during the hour. One bird had its wing broken; another was picked up dying in convulsions from the force of the blow ... sometimes two or three lay together, showing where a flock had passed by, and been decimated.[56]

High-voltage lines are still lethal, but over time, telegraph wires proved beneficial for some birds. Henry Stevenson was a respected Norfolk ornithologist, based in Norwich, and to him the sight of thirty to forty house martins sitting in rows upon the telegraph wires was curious:

Whether or not our British *Hirundines* believe that the telegraph wires were erected for their special accommodation, undoubtedly one of the strangest points in their modern history is the manner in which they avail themselves of these novel resting places. Indeed, when beholding, as I have done in autumn, each wire lined with their little bodies, more especially in the vicinity of our rivers and broads, one wonders, almost, how they managed without them, since no other perch seems half so suitable for their tiny feet, or affords so great facility for launching themselves upon the wing.[57]

Each modern development inevitably brought new dangers to the bird world, including telegraph wires, power lines, artificial lighting, illuminated masts, oilrigs and high-rise buildings. Numerous lighthouses were constructed around the perilous coast of the British Isles, mostly in the nineteenth century, introducing powerful beams of artificial light into a completely dark landscape. From 1880, Murray Mathew spent eight years as curate at Wolf's Castle in Pembrokeshire, where he examined the reports of bird strikes maintained from 1879 by the local lighthouse keepers: 'In fine weather the birds fly wide of, or high above, the light-houses, but in stormy, or misty weather, they flutter about them during the night and the early hours of the morning in a bewildered manner, and hundreds perish from dashing themselves violently against the lanterns.'[58]

Apart from lighthouses, very few lights would have been visible anywhere at night two centuries ago, until gas lamps began to be installed along urban streets and in buildings, luring migrating birds from their course. Stevenson reported on the death of a woodcock that was migrating by night over Norwich: 'A fine woodcock in my own collection . . . was found on the 23rd of October, 1858, against the wall of Mr. Towler's residence, adjoining Park Lane . . . no doubt, the lamps on Unthank Road, which passes in front of the house, had so dazzled it that the next instant it dashed with full force against the front of the building.'[59] He was especially concerned about the glare of gas lamps in the city, and a few years after this fatality he talked about golden plovers moving over Norwich by night:

> Probably many of my readers have remarked, at such times, the melodious notes of these plover, which would seem to be uttered incessantly in order to keep the whole body together; and as this always occurs when the nights are extremely dark, I believe the birds, once drawn within the radius of the city lights, become perfectly bewildered, and fly round and round for hours, till, at day-break, the spell is broken, and they resume at once their direct course of flight.[60]

CHAPTER TWO

——— •◆• ———

VERMIN

Away, away, away birds,
Take a little bit, and come another day, birds;
Great birds, little birds, pigeons and crows,
I'll up with my clackers, and down she goes.

Traditional bird-scaring chant[1]

Renowned worldwide as a Roman historian, the Oxford university tutor William Warde Fowler was equally passionate about birds, and in the latter part of the nineteenth century he lived in a cottage at Kingham in Oxfordshire, from where he could see the village allotments:

> here all the common herd of blackbirds, thrushes, sparrows, chaffinches, and greenfinches help to clear the growing vegetables of crawling pests at the rate of hundreds and thousands a day, yet the owners of the allotments have been accustomed since their childhood to destroy every winged thing that comes within their cruel reach. Short-sighted, unobservant as they are, they decline to be instructed on matters of which they know very little, but stick to what they know like limpets.[2]

Such customary, systematic destruction of birds actually began centuries earlier under Henry VIII, at a time when he was pushing for the annulment of his marriage to Catherine of Aragon. This led to cataclysmic religious upheaval in England, with legislation being passed in 1532 that marked the first steps in transferring the authority of the Pope to the English Crown and severing ties with Rome. In January 1533 Henry VIII married Anne Boleyn, who was pregnant, and a few weeks later his marriage to Catherine was annulled.

During this turmoil, legislation about wild birds was passed, though it was not intended to protect wildlife. The population then barely exceeded three million, but the sheer number of birds meant that grain crops risked being plundered, causing poor harvests and famine, and the legislation tried to address that problem. The Act was the first state-sanctioned order to slaughter those birds blamed for the worst depredations – rooks, crows and choughs, though the word 'choughs' was probably used for what are now called jackdaws: 'Forasmuch as innumerable number of rooks, crows and choughs do daily breed and increase throughout this realm, which rooks, crows and choughs do yearly destroy, devour and consume a wonderful and marvellous great quantity of corn and grain of all kinds; that is, to wit, as well in the sowing of the same corn and grain, as also at the ripening and kernelling of the same.'[3]

This was the earliest reference to the destruction of vermin, and the three named birds were also blamed for wrecking the protective straw or reed thatch on the roofs of houses, barns, hay-ricks and stacks of corn and hay. Instructions were therefore given to take measures to destroy these birds or else face a penalty: 'in every parish township hamlet borough or village within this realm wherein is at the least ten households inhabited . . . during ten years next . . . make or cause to be made one net commonly called a net to take choughs, crows and rooks'.[4] A bounty was to be paid by every farmer or landowner for each bird taken. It was specifically forbidden to kill pigeons, even though they raided grain crops, because pigeons were bred in dovecotes as a source of food for the wealthier classes.

Anne Boleyn fell from Henry VIII's favour after giving birth to a daughter. In 1558 that daughter became Queen Elizabeth I, and during her long reign, further legislation added mammals such as foxes, badgers and hedgehogs to the vermin list, all requiring destruction, along with unbroken eggs and a range of other named birds that included ravens, magpies, jays, starlings, buzzards, bullfinches, kites, ospreys and kingfishers. Any bird that was believed to compete with humans for fish, fruit, vegetables and grain, either by eating or damaging it, was to be treated as vermin. The financial burden gradually shifted to parish churchwardens and overseers, who paid an agreed figure for the heads of specific vermin, funded through the church rates.[5] Parish officers simply did what was thought to be effective and lawful, and much wildlife was treated as the enemy that needed eradicating. Cruelty was routine, because only the heads of vermin had to be presented for payment, encouraging children to decapitate birds.

Surviving churchwardens' accounts show that payments were given for the massacre of hundreds of thousands of birds, as at Chipping Camden on 30 April 1689: 'Paid for 25 dozen of Sparrowes heads.' In the Yorkshire parish of Dent, the account books recorded that twopence was normally paid for each raven's head, with thirteen in 1713 and further annual payments noted up to 1750. When Thomas Norwood was vicar of Wrenbury in Cheshire at the end of the nineteenth century, he examined the old accounts: 'In 1755 and 1758, the owls were cleared out of the Church Tower. In 1786, the Parish paid for 708 dozen of birds' heads ... This havoc went on for ages. A motion to stop it was carried in the vestry in 1842, but rescinded the next year, when bullfinches were 3d. a head, and sparrows 3d. a dozen.'[6]

Payments were also made for gunpowder, guns and shot, as at Milton Bryan in Bedfordshire, with approval being given for crows and pigeons to be scared off: 'for Gunpouder as was delivered to Nicholas [Clarke] to fright the Crowes and piggons in the wheatefield in the year 1697, for three quarterns of Gunpouder the 13 day of July 1s. 0d., for 1 lb of Shott the 16 Day of July 2½s.'[7]

Bullfinches voraciously consume the blossom and buds of fruit trees and bushes, and the increasing number of orchards from the seventeenth century led to these birds becoming a key target. In his

1737 book on songbirds, the naturalist Eleazar Albin described how bullfinches fed on the buds of fruit trees 'and by that means do great damage to the gardeners; who, therefore hate and destroy them, as a great pest of their gardens. They say, in some part of the kingdom, a reward is given by the church-wardens for every *Bullfinch* that's killed.'[8] Numerous examples of such payments appear in church-wardens' accounts, including Ashburton in Devon, where £3 10s. 4½d. was paid for 1,661 bullfinches between 1761 and 1820.[9]

With the enclosure of much open land in the eighteenth and nineteenth centuries and miles of hedges planted as boundaries around the new fields, sparrows were able to thrive.[10] They became by far the most common birds killed for a reward, and in 1743–4 at Westoning in Bedfordshire, £1 3s. 4d. was paid for '138 dozen of sparrows', while at the village of Cam in Gloucestershire, a total of £37 9s. 6¾d. was paid for 22,620 sparrow heads from 1819 to 1837.[11]

In 1836 the writer William Howitt moved with his family from Nottingham to Esher in Surrey, where he published *The Boy's Country-Book* that included an account of him killing sparrows as a child in Derbyshire at the end of the eighteenth century:

> Many a summer evening we went round the village with a ladder, visiting the eaves of almost every cottage, and even of the church itself, for sparrows' nests. The eggs, young ones, and even the old birds, that we sometimes caught on their nests, we sold according to a parish custom, to the overseer of the poor. Three eggs for a penny, two young birds for a penny, and one old bird's head for the same price ... It was reckoned a good joke with many of them to watch the parish officer after they had sold him a quantity of sparrows [heads] and their eggs, for it was not unusual for him, when the lads appeared to be out of sight, to throw the sparrows into the street; having observed this, they would go presently and pick them up, and sell them two or three times over; and some lads have been known to put amongst them the heads of hedge-sparrows, buntings, and larks, for which the parish allowed nothing, but which the overseer did not always know from real sparrows.[12]

Farmers also tried to control birds in their fields, relying on the cheap employment of children as bird scarers, who started work as young as five or six years of age, with their meagre earnings often being critical for a family's survival. Bird scaring took place over several months of the year, seven days a week, and without these children, much of the sown seed, as well as the young plants and ripening crops, fell victim to rapacious birds. Because birds were so abundant, the idea of protecting them for their own sake was inconceivable, and to admire and preserve wildlife was rarely an option.

Young boys especially, but also women, girls and elderly men, trudged up and down fields to frighten rooks, crows and pigeons, a task known variously as rook-starving, crow-clapping and bird-keeping. Until his death in 1900, John Atkinson was for forty-five years vicar of Danby in the North Riding of Yorkshire, where he also enjoyed being an antiquarian, but while growing up in Essex in the 1820s, he had dreamed of becoming a bird scarer:

> What a jolly life I used to think the little village boys who were set to 'keep the crows' in that then wheat-growing county of Essex must lead. No tiresome school, dame or boarding, no multiplication table, or, worse still, pence-table, to learn, no work of any kind – for bird-nesting and cutting the bark off long switches in alternate rings clearly was not work – but just to halloa *ad libitum* from time to time when the crows might be coming, or the master or the foreman be within hearing – and what boy does not like kicking up a hullabaloo of that kind when it suits him?[13]

Atkinson simply yearned to create a racket, search for birds' nests in the hedgerows and, most of all, discharge a gun now and then, but the reality was a harsh, lonely and impoverished existence.

In the early 1830s, while still living in Nottingham, Howitt often came across young bird scarers during his travels in the surrounding countryside:

I have seen little boys set to drive birds from a corn-field just sown, in the early spring. Afar off in the solitary fields they watched and wandered to and fro, from early dawn to nightfall, till their task became insupportably weary. Not a soul had they to exchange a word with; they had their dinner in a bag, a clapper to drive away the birds, and they would be found making a miserable attempt with turfs, and sticks, and dry grass, to raise a sort of screen against the wind and rain. I have found such a one cowering under a shed of this sort, or under a high bank, in the midst of Sunday, while the village bells sounded at a distance merrily, and told of people assembling happily together.[14]

The wooden clappers comprised a rectangular piece of wood with a handle, like a butter pat, to which a loose piece of wood was attached either side with wires or cord. In Wales, another device for frightening crows was the rhugylgroen, a rattle made from a dried sheepskin containing pebbles.[15]

When riding through farmland on one occasion within the desolate expanse of Sherwood Forest, Howitt noticed a lone boy:

It was a cold, raw, foggy day in February; the wet hung in myriads of drops on the hedge, and the dampness of the air clung about you with a dispiriting chillness . . . As I passed a tall hedge, I heard a faint, shrill cry, as of a child's voice, that alternating with the sound of a wooden clapper, sung these words: –

> We've ploughed our land, we've sown our seed,
>   We've made all neat and gay;
> So take a bit, and leave a bit,
>   Away birds, away!

I looked over the hedge, and saw a little rustic lad apparently about seven years old, in his blue carter-frock [smock], with a little bag hanging by his side, and his clapper in his hand. From ridge to ridge of a heavy ploughed field, and up and down its long furrows, he went wading in the deep soil, with a slow pace, singing his song with a melancholy voice, and sounding his clapper.[16]

The sight of the boy distressed Howitt: 'There was something in the appearance of that little creature in that solitary place, connected with his unvaried occupation, and his soft and plaintive voice that touched powerfully my heart; and, as I went on, I still heard his song, fainter and fainter, in the deep stillness.' On his return journey that evening, in twilight, he again caught sight of the boy. He discovered that his name was Johnny and this was his first day of work: 'Never did I feel a livelier pity for any living thing! ... a neighbour who passed this afternoon and asked him how he liked his task, said he was crying; and that he said, the silence frightened him, and he wished himself at home again.'[17]

Because the boys were dispersed across different fields, the loneliness, tedium and severe treatment were difficult to bear. Joseph Arch was the son of a shepherd and later in life became an agricultural trade union leader and politician. He recalled the hardships of his first paid employment at Barford, near Warwick:

I was a youngster of nine when I began to earn money [in 1835]. My first job was crow-scaring, and for this I received fourpence a day. This day was a twelve hours one, so it sometimes happened that I got more than was in the bargain, and that was a smart taste of the farmer's stick when he ran across me outside the field I had been set to watch. I can remember how he would come into the field suddenly, and walk quietly up behind me; and, if he caught me idling, I used to catch it hot. There was no sparing the stick and spoiling the child then! This crow-scaring was very monotonous work, and many a time I proved the old adage about Satan finding mischief, for idle hands. My idle hands found a good deal to do, what with bird-nesting, trespassing, and other boyish tricks and diversions.[18]

Looking back, he admitted that his lot was far better than that of a coal miner's son, who from an early age would have been toiling down a pit. Even so, the long hours were exhausting, and children like him often fell asleep, as is evident in the similar rhymes and songs that were chanted across the country:

> *Fly away, blacky-cap,*
> *Don't you steal my master's crap*
> *While I lie down an' take a nap.*[19]

In this example, 'blacky-cap' refers to a crow and 'crap' is West Country dialect for 'crop'.

In evidence given to the Poor Law Commission in 1843, John Peirson, who was a landowner and farmer at Framlingham in Suffolk, reported: 'Children come at eight, bring their dinner, rest from twelve to one, leave work at five or six, according to season of year. They earn not less than 1s. 6d., nor more than 2s., by crow-keeping. Every person in the parish employs them in that way. I dare say at one time we had 50 or 60 children employed as crow-keepers.'[20] A few years later John Glyde, a self-educated social commentator, addressed the plight of these young bird scarers in Suffolk:

> There is a general complaint of the ignorance, and almost unreasoning existence of our agricultural population. How can it be otherwise, when a boy is taken from school before he can read with ease ... Boys of tender age often go into the fields at day-break to drive away the birds from growing crops, or to attend cattle and sheep. Thus engaged, he watches the same flock of sparrows, walks the boundaries of the same field, leans against the same gate, and sits under the same hedge, for months together. Intellectual stagnation eats into the very soul of the child in performing his monotonous duty.[21]

John Bailey Denton was a land surveyor and engineer, and he continued this argument in a paper presented to the Society of Arts in London on agricultural labourers. Most children left school at the age of ten, he explained, and if a boy could not write, he is consigned to the farm either as a bird-keeper, crow-clapper, ploughboy, pig-keeper, cowman's boy, or shepherd's boy ... he may exercise his voice in singing while bird-keeping, but as this involves an effort, he only does it when he sees his master coming. This literally and truly has been, and still is, the occupation of nearly every village boy during that time of his life when his mind is most susceptible of useful impressions. Though not altogether idle, his mind is an empty one.[22]

The 1880 Education Act subsequently brought about the virtual cessation of child labour, making school attendance compulsory for those between the ages of five and ten, rising to fourteen by 1914. Once cheap child labour was no longer available for scaring birds, a greater reliance was put on scarecrows, mawkins or dudmen – grotesque images of humans, dressed in cast-off clothing, that were positioned in fields to deter marauding birds. The adventure novelist Rider Haggard had two farms on the Norfolk–Suffolk border, near the town of Bungay, and writing in 1898 he said that scarecrows were supposed to unnerve the rooks:

Hood [his steward] promises to set up some mawkins to fright them, but the mawkin nowadays is a poor creature compared with what he used to be, and it is a wonder that any experienced rook consents to be scared by him. Thirty years or so ago he was really a work of art, with a hat, a coat, a stick, and sometimes a painted face, ferocious enough to frighten a little boy in the twilight, let alone a bird. Now a rag or two and a jumble-sale cloth cap are considered sufficient.[23]

Haggard also advised boosting the message of these lacklustre scarecrows by 'a dead rook tied up by the leg to a stick'.

William Sterland described a similar method of scaring rooks on farmland near Ollerton in Nottinghamshire, on the edge of Sherwood Forest where he grew up. In 1869, he published a book about birds of the region, in which he pointed out that although farmers seemed to understand that rooks were by and large beneficial, they did shoot the occasional one, 'who, with wings extended by two split sticks, is placed *in terrorem* in the centre of a corn or potato field; and a very effectual scarecrow he makes – his constrained attitude is understood at a glance by his wary brethren, and they need no other hint.'[24]

Over a century earlier, in 1748, Pehr Kalm, who was a naturalist from Sweden, spent a few months in England before heading to North America. That July, he travelled round Gravesend and Rochester in Kent, where he saw numerous cherry orchards. He found that birds were kept off the fruit by suspending numerous dead jackdaws, rooks, crows and magpies from tree branches, producing a stench of rotting corpses.[25]

Growing up in the early twentieth century at Rottingdean in Sussex, the countryman and folk singer Bob Copper saw how rooks pillaged the fields each spring, when 'five or six hundred sooty, ragged-winged robbers, from the rookery in the neighbouring village of Stanmer over the hill . . . used to fly down daily in droves . . . to raid the freshly sown fields'.[26] Haggard likewise described the damage on his farm:

> I noticed that the rooks are playing havoc on the three acres of mixed grain which we drilled a few days ago for sheep food . . . They are congregated there literally by scores, and if you shout at them to frighten them away, they satisfy themselves by retiring to some trees near at hand and awaiting your departure to renew their operations. The beans attract them most, and their method of reducing these into possession is to walk down the lines of the drill until (as I suppose) they smell a bean underneath. Then they bore down with their strong beaks and extract it, leaving a neat little hole to show that they have been there. Maize they love even better than beans; indeed, it is difficult to keep them off a field sown with that crop.[27]

In Bob Copper's opinion, rooks grew accustomed to the sound of clappers and so needed a gun fired at them to teach them a lesson: 'When-ever Grand-dad shot a rook I had to run and fetch it to him and he would tie it on to a thatching-rod and make me stick it into the ground some distance away. Here the unfortunate bird would swing from its gibbet as a grim reminder to those that got away.'[28]

Apart from birds being scared away by children, scarecrows and dead trophies, other ingenious devices prevented predatory incursions. In his *The Book of the Farm*, published in 1844, the agricultural writer Henry Stephens stated that rooks loathed the smell of gunpowder (probably because of the association with guns being fired at them). To protect potato fields, he recommended burning matches, which were slow fuses of cord coated with gunpowder paste and dried: 'the burning of gunpowder matches here and there and now and then along the windward side of the field, the fumes of which sweeping across the surface of the ground, being smelt by the rooks, put them in constant trepidation, and at length to flight'.[29] A more complex device was what he called a 'rook-battery', composed of a large circular piece of wood on a tripod, with slow matches arranged to fire a series of small brass cannons, aided by a woollen rag coated in gunpowder that produced smoke.[30]

A few tips on protecting vegetables, fruit bushes and trees were given in *The Garden* magazine of December 1872, one of which involved an imitation hawk: 'Take a good sized sound Mangold Wurzel, and with a pocket-knife roughly shape it into something like the body of a bird, stick some turkey's (or any large dark-brown) feathers in on each side for wings, not forgetting to add a good wide-spread tail.' Using string and a flexible rod, this 'hawk' was suspended from a branch of a fruit tree so that it resembled a hovering bird of prey. Another creative suggestion was made: 'A stuffed cat is also an excellent bird scarer, and, if placed in a conspicuous position, will protect a large batch of seed beds, or a number of trees or bushes. The skin may be stuffed in the rudest manner with hay, putting sticks into the legs. With a little care it will last for years. It should not be forgotten that before it is stuffed, it must be well salted or otherwise prepared against putrefaction.'[31]

Magazines and newspapers constantly advertised the latest contraptions, but Rider Haggard warned against sparrow-traps:

Of late, patent basket sparrow-traps have been largely advertised, and with them testimonials from gentlemen who say they have caught great numbers by their means. I purchased one of these wicker traps for five shillings, but the result showed that I might as well have kept my money in my pocket, as not one single sparrow have I been able to catch with it. I suppose therefore that the race must be more artful about here than in the neighbourhood of those gentlemen who give the testimonials.[32]

He complained bitterly about his adversary: 'the bold, assertive, conquering sparrow, that Avian Rat, as someone has aptly named him, pursues his career of evil almost unchecked, producing as many young sparrows as it pleases him to educate. Indeed, I do not understand how it comes about that we are not entirely eaten up with these mischievous birds.'[33] In July 1898, flocks of about a hundred were targeting fields close to his village: 'A tenant of mine, whom I met this morning, tells me that he is "fairly crazed" with them, and that in one plot they have nearly stripped the ringes [rows] which are nearest the hedge. It is not so much what these vermin eat that does the mischief, as what they destroy. I believe that for every grain they swallow, they throw down six.'[34]

Sparrow clubs were established years before the Education Act limited the availability of children for bird scaring, and in December 1862 a letter was written by 'A Real Friend to the Farmer' about a club at the small market town of Crawley in Sussex. It was published in *The Times* under the heading 'Sparrow Murder':

Sir, – I think the following exploit of the 'wise men' of Crawley ought to be shown to the world in your widespread journal; it speaks for itself and requires no comments on my part. It is taken from a country paper of this week: –

'CRAWLEY SPARROW CLUB.

'The annual dinner took place at the George Inn on Wednesday last. The first prize was awarded to Mr. J. Redford, Worth [south-east of Crawley], having destroyed within the year 1,467. Mr. Heavsman took the second, with 1,448 destroyed. Mr. Stone third, with 982 affixed. Total destroyed, 11,944. Old birds, 8,663; young ditto, 722; eggs, 2,559.'[35]

On reading this same letter, William Sterland commented that too many farmers were ruthlessly destroying birds: 'I do not know where Crawley is, but I feel ashamed of the profound ignorance and inhumanity of its inhabitants, and especially of the three individuals who carried off the prizes in their sparrow club.'[36]

At the village of Ablington in the Cotswolds in early November 1896, Arthur Gibbs received a postcard with a printed message: 'A meeting will be held at the Swan Hotel, Bibury, on Friday, November 13th, at 6.30 P.M., to arrange about starting a *Sparrow Club* for the district.' On asking a good friend of his, a labouring man, what a sparrow club was, he learned that it was for killing sparrows when they became too numerous,

> paying boys a farthing a head for every bird they catch, and giving prizes for the greatest number killed. Boys may often be seen out at night, with long poles and nets attached to them, catching sparrows in the trees … I was disappointed that the 'Sparrow Club' for which a great public meeting had to be convened, was not of a more exciting nature. One was led to believe by the importance of the printed postcard that some good old English custom was about to be revived.[37]

Richard Jefferies, who had died a few years earlier, in 1887, wrote for many years about rural England, including sparrow catchers: 'If you lived in the country you might be alarmed late in the evening by hearing the tramp of feet round your house. But it is not burglars; it is young fellows with a large net and a lantern after the sparrows in the ivy. They have a prescriptive right to enter every garden in the village. They cry "sparrow catchers" at the gate, and people sit still, knowing it is all right.' In his view, this practice made cruelty acceptable: 'The agriculturists in some southern counties give the boys in spring threepence a dozen for the heads of young birds killed in the nest. The heads are torn off, to be produced like the wolves' of old times, as evidence of extinction.'[38]

By the end of the century, with a depression in agriculture, Rider Haggard said that in Norfolk sparrow clubs had disappeared, and the only effective way of controlling these pests was to use poisoned wheat, even though it became illegal in 1863:

> but this, unless spread with great care in places frequented by sparrows alone, such as eaves and water-troughings, is highly dangerous to all life. Also the sale of it is illegal; indeed, we have convicted men for this offence before my own Bench. Still, farmers use it a good deal under the rose [secretly], and, I am sorry to say, not for sparrows only, but for pigeons and rooks also, with the result that a great deal of game and many harmless birds are poisoned.[39]

Gamekeepers and shepherds also hated and persecuted certain birds that were a menace to gamebirds and livestock, especially sheep and lambs. When touring the Highlands of Scotland in the summer of 1769, the naturalist Thomas Pennant passed Braemar Castle: 'Eagles, peregrine falcons, and goshawks breed here: the falcons in rocks, the goshawks in trees ... These birds are proscribed; half a crown is given for an eagle, a shilling for a hawk, or hooded crow.'[40] A few decades later, the journalist Alexander Shand said that the Highland shepherds still worried about eagles: 'if a sheep is crippled or ailing, the eagle is always on the look-out, guided by the ravens and hooded crows. For the eagle is the most voracious of gluttons, and the best chance of the shepherd taking his revenge, is when he weathers on him when gorged to the beak with drowned mutton. Then the prince of the air and the mountains may be knocked senseless with the staff.'[41]

An old shepherd on the South Downs in Sussex said that he once shot an eagle, the only one he ever saw. Usually, he was on the look-out for foxes and marauding birds: 'If a yoe [ewe] happened to get overturned on a lonesome part of the hill, the ravens and carrion crows would come and pick out her eyes before she was dead. This happened to two or three of my yoes, and at last I got an old gun and shot all the crows and ravens I could get nigh.'[42]

John Atkinson mourned the fact that ravens had been killed off during his forty years of being in the Yorkshire parish of Danby. For much of his life he was interested in birds, but had sadly noticed many changes: 'The barn owl – they used to breed in the church-tower – had gone a few years before ... The beautiful merlins, too, are comparatively little seen ... And I have not seen a harrier or a buzzard these thirty years ... but the hawk I miss the most is ... the kestrel or windhover.'[43] He complained bitterly that gardeners and farmers were overrun by mice, and yet they slaughtered kestrels and owls, the very birds that would have reduced their numbers.

A rare early voice in support of birds was John Byng (who later became Viscount Torrington), and during his summer travels through Cambridgeshire in 1790 he stayed at the Haycock Inn at Wansford Bridge. One morning he was appalled to see nests of house martins beneath the eaves being wantonly destroyed. Later that day, he rode to Chesterton, where he deplored the new fashion for felling groves of trees in villages to open up the views, and so he was driven to pen a diatribe against the destruction of birds:

The rook (in my opinion) does one shillingsworth of advantage, by destruction of the slug, for a pennyworth of damage sustain'd by the corn. The blackbird, and bullfinch destroy the worms and insects inimical to greens, and fruits; and richly deserve their small reward of peas, and cherries. The sparrow hardly earns his poor pickings from the stack, by his destruction of the caterpillar. I once ask'd a lady, if she loved to be stung by gnats? She answer'd with horror No; 'Why then, madam, do you order the swallows [house martin] nests to be pull'd down from the window, who wou'd keep them away from your chamber?'[44]

When Byng was writing, at the end of the eighteenth century, the war waged by cultivators against birds had barely begun.

With a tendency to be pessimistic, he was sure that all birds were under threat:

> The race of birds will be quickly extinct ... for as ignorance is the offspring of barbarity, the farmer and gardener are taught to believe that all birds rob and plunder them, and that they should be destroy'd by every engine, and scared away by every invention. By these means the country is stript of a chief beauty; and the contemplative man misses a prime satisfaction. On the few remaining tall trees the rook is forbid to build! The blackbird and bullfinch are shot in the kitchen garden! The sparrow is limed in the farm yard! And the swallow is driven from 'the coigne of vantage'![45]

Birds were frequently killed in ignorance as vermin, when in reality many caused little harm and were even beneficial, preying not just on mice, but also eating prodigious quantities of insects, caterpillars, slugs, wireworms and other damaging grubs, something that ornithologists proved time and again when they dissected birds they had shot. In the mid-nineteenth century, John Blunt, an ecclesiastical historian, was a curate in East Anglia, and when his church was being restored, he climbed up a ladder to an oak beam, from which he extracted some lumps of lead, labelling them as 'slugs from the guns of Cromwell's soldiers'. Shortly afterwards, he was told that they were lead shot fired by the old sexton, John Wilkins, when slaughtering sparrows inside the building. Blunt criticised the folly of paying parishioners to kill sparrows: 'while the farmers were thus expending the Church Rate the grubs must have laughed from the furrow into the faces of their ploughmen: and the wireworms must have sung merrily as they bored into the very hearts of their turnips'.[46]

Without modern pesticides, episodes occurred when insects or rodents caused considerable problems, and on occasions birds would flock to the scene as if from nowhere, something the landowner and ornithologist Hugh Gladstone recorded for cuckoos on his estate in Dumfriesshire: 'In some summers, as in 1885 and

1903, this species is exceptionally abundant. In the latter year the oak trees at Capenoch (Keir) were infested with caterpillars innumerable and never in my life have I seen so many Cuckoos gathered together. They destroy the *larvae* of the antler-moth ... or "hill-grub" which is so injurious to upland hill-pasture.[47]

Field voles also caused damage to pasture land in Scotland, as happened in 1890–2. Gladstone's friend Robert Service, a nurseryman from Dumfries, witnessed how short-eared owls thrived on such an abundance of food:

> The vole plague of 1875–76 saw a big movement of these Owls, and they were accompanied by Rough-legged Buzzards, while the plague of 1890 and subsequent years brought an altogether marvellous immigration. In Dumfriesshire alone not less than 500 pairs of breeding birds were estimated as present. Probably there were three or four times that number on the vole-infested farms ... It was a never-to-be-forgotten experience to see on mile after mile of the moorlands always a dozen of them in sight at once ... The finest sight of all was when at midsummer of 1892 I had a chance of going along the hills at midnight. The night was bright and clear, and very still. The Owls were on all sides, flying like no other birds that I ever saw. The voles were scurrying hither and thither, squeaking and rustling as one stepped over and amongst them. The unfeathered owlets had left their nests, and were sitting blinking their eyes and contorting their bodies in groups on almost every hillock. The parents never troubled to alight amongst their offspring – they simply flew past, and flung the dead voles at their young in the byegoing. It was a weird and impressive bit of bird life, one my poor powers of description cannot do justice to.[48]

Service's account is a reminder that birds can flourish in unlikely conditions and that the presence of these owls was a positive aspect of an otherwise devastating vole plague.

CHAPTER THREE

—·◆·—

HUNTING

It is most curious how some of the birds have been caught; some are held by a leg alone, others are wound round and round in a complete tangle, the fine twine of the net being buried in some places deep under the plumage. Some birds are caught by a wing alone, others by both; whilst yet again many are held by the neck and the wing. Many of the birds are drowned.

CHARLES DIXON on birds caught
in flight nets on the Wash mudflats[1]

From prehistoric times, birds have been killed for a great variety of reasons, not just to eradicate them as pests, but also for food, medicine, fuel, feathers, fertiliser for fields or for specimens that collectors were seeking. Live birds were also captured to be fattened for eating or to be put in aviaries and cages as pets. Until guns became predominant, the most practical weapons were slingshots, catapults, bows and arrows and crossbows, while devices such as simple snares, nets, traps and birdlime also enabled birds to be taken. Whatever the method, it was essential to lure birds within range, which meant employing decoy birds and call birds, lights and reflective mirrors.

For thousands of years, sticky birdlime will have helped catch all manner of birds, though in Britain its use was finally made illegal in 1925. The word is derived from the Old English 'lim', an adhesive substance, which came to be written as lym, lyme and lime. On several occasions Shakespeare used the term 'lime' or 'limed', meaning ensnared or setting a trap, as in *Henry VI, Part II*, written around 1591–2. The Duke of Suffolk was aiming to influence Queen Margaret, his lover, by speaking about Duke Humphrey's wife, who she saw as her rival:

> *Madam, myself have lim'd a bush for her,*
> *And plac'd a quire of such enticing birds.*
> *That she will light, listen to their lays,*
> *And never mount to trouble you again.*[2]

In 1621 the prolific writer Gervase Markham published *Hungers Prevention*, a book that described the art of fowling by water and land – the word 'fowl' was then used for all birds. Intricate instructions were given on making birdlime:

> To make then the best and most excellentest bird-lime, you shall take at Midsomer the barke of holly and pill [pull] it from the tree, so much as will fill a reasonable bigge vessell, then put it to running water, and set it one [on] the fire, and boyle it till the gray and white barke rise from the greene, which will take for the most part, a whole day or better in boyling, then take it from the fire and seperate the barke after the water is very well drained from it, which done, take all the greene barke and lay it one [on] the ground in a close place and a moist store, as in some low vault or celler, and then with all manner of greene weedes, as dockes, hemlock, thystels [thistles] and the like, cover it quite over a good thicknesse.[3]

The mixture was left for up to twelve days, turning filthy and slimy, after which it was pounded in a mortar and taken to a stream to be thoroughly washed: 'then put it up in a very close earthen pot, and let it stand and purge for divers days together, not omitting but to skum it and clense it as any foulness rises, for at least three or four daies together, and then perceiving no more skum will arise, you shall then take it out of that pot and put it in another cleane earthen vessell and cover it close, and so keepe it'.[4]

Birdlime proved effective because birds are extremely light relative to their size, so that only the largest birds could break free. It could be smeared on bushes or else on straw, string, twigs or rods that were then placed in suitable locations such as hedges, open fields and backyards, but just before use, Markham said, more preparation was needed: 'Now when you have occasion to use your lyme, you shall take of it such a quantitie as you shall thinke fit, and putting it into an earthen pipkin, with a third part of hogges-grease, capons-grease or goose-grease, finely clarified (but capons-grease or goose-grease is best) and set it on a very gentle fire, and there let them melt together, and stirre them continually, till they be both incorporated together.'[5] He also gave advice for freezing conditions: 'Now if it to fall out that the weather doe prove so extreme sharpe and frostie, that your lyme-roddes doe freeze ... then when you mix your grease and lyme together, you shall take a quater [quarter] so much of the oyle of Peter (which the pothecaries call Petrolium) as you do of capons-grease, and mixing them together well ... that no frost how great or violent ... shall by any means anoy or offend you.'[6]

In the late seventeenth century, while serving as the manciple for St Edmund College in Oxford, Nicholas Cox, who was also a bookseller and author, published *The Gentleman's Recreation*, a book on field pursuits aimed at the young gentlemen who attended Oxford's various colleges. Part of it was lifted wholesale from *Hungers Prevention*, including birdlime, and Cox's own book was in turn plagiarised for many decades, reflecting how fowling traditions changed very little.[7] In the late nineteenth century, though, in his book on taxidermy, Montagu Browne wisely said that birdlime was no longer worth making, as by then it could be purchased from any professional birdcatcher, but for those who wished to do so, he recommended boiling linseed oil, which gave a similar product.[8]

After setting up a sticky trap, it was advisable to stay fairly close and lure in the birds by mimicking their call or using an artificial calling device. Another method was to attract birds to the spot by placing a decoy bird or stale close by, the word 'stale' being derived from the Anglo-Saxon for a pigeon that was used to lure a hawk into a net. A living bird was tied to a pole, pinned down or put in a cage, or else a stuffed bird was displayed, giving the illusion of the safe presence of other wild birds. Sometimes owls or bats were employed as stales, because in the daytime they were perceived as predators, and so birds tended to mob them.[9] Markham suggested placing two live bats near a lime-bush, saying that the birds would come down to gaze and wonder at them, but he also found owls effective: 'Now as these night battes, so the owle is of like nature, and may be employed after the same manner; and by reason that he is lesse stirring and more melancholly than the batte, as also of greater quantity [size], and sooner to bee perceived, she is a better stale than the bat.'[10] He then explained how to make a stale:

> if you have not a live owle or a live batte, if you can get but the skins of either, and stoppe them with wool or flockes, they will serve as well as if they were alive, and continue (with carefull keeping) twenty yeares and better. I have seene some that for want of either of these hath had an owle so lively cut out in wood, and so artificially painted, that it hath served him for this purpose, as well as any live one could doe, and he hath taken byrdes in wonderfull great abundance therewith.[11]

Many more birds could be captured with nets than by using individual weapons such as slingshots. The three main types were clap-nets, which were usually laid on the ground and had two panels of nets that could envelope the birds; drag-nets, which were dragged over the ground at night to capture birds like skylarks; and flight nets, which were long lines of vertical nets stretched between poles across flight paths.[12] On the mudflats of tidal rivers, along coastlines and at some inland locations, flight nets were probably popular from prehistoric times.[13] In December 1862, John Gurney, then a young, enthusiastic ornithologist, described a day spent wildfowling with Frank Cresswell, an alderman and banker from King's Lynn who was skilled in using flight nets:

> Well protected with wraps, for the cold was intense, I went on board the 'Wild Duck,' which is the name of his yacht, and all night we rode at anchor in a sea so tempestuous that she lurched like a drunken man, in order to be early at the nets in the morning; but it was well worth the trouble to see the singular spectacle which so many varieties of birds dangling in the meshes presented. There is nothing like a pitch-dark, blustering night, and the catch was good.[14]

Eighty birds were caught, and Gurney described their fate:

> The nets, which are about five feet high, are generally placed at high-water mark. All of them together reach at least a third of a mile. They are fatal to everything between a Lark and a Shelduck. If a Dunlin so much as touches with the tip of his wing it is wound round in an instant, and there he hangs until he is taken out and killed. The majority of the birds are taken out alive, and many small waders so caught, especially Knots, have been presented by Mr. Cresswell to the Zoological Society.[15]

Some two decades later, Charles Dixon was on the Lincolnshire coast, thankful that the Wash still had miles of mudflats for wildfowl. Although the autumn months saw an abundance of bird life, the landscape was too exposed for the gunner to stalk his quarry, and the punter had no chance of approaching on the water. Instead, birds were caught by flight nets. The most productive time was during rough and stormy weather, with a new moon, and on one occasion he accompanied the local coastguard, John Stiff, formerly of the Royal Navy, to check the nets near the village of Friskney, before the crows got to them first:

> Biting cold is the breeze on this November morning as we hasten along over the sea banks to the wastes of mud and water

beyond ... The tide has not yet fully ebbed, and we have to wade knee-deep through the pools in many places, and take long jumps across the narrow dykes and trenches. Far away before us are the nets, but we are too distant to make out whether they contain any birds ... We are soon at the first long reach of netting, and the scene before us is a most novel and curious one. Several gulls are hanging by the wings in one portion of the net, alive, but apparently philosophically resigned to their fate. Farther on in the same net we find a pair of fork-tailed petrels and a fulmar.[16]

The second net contained only one dunlin and a few wigeon, but the third net had many more birds: "Twenty knots are entangled in the meshes, some of them so intricately that it is a tedious task to extricate them. A pair of short-eared owls and a sky lark hanging side by side tell us that migration has been going on during the night ... In the last net we find three golden plovers, a curlew, two godwits, a redshank and several dunlins. In some places the nets are torn showing where a flock of geese or ducks have passed clean through."[17]

Friskney is also a place where duck decoys or decoy ponds were once commonplace. First known in the Netherlands, they possibly developed out of the practice of driving moulting ducks – when they were unable to fly – into extensive tunnels of nets. Under Henry VIII, in 1534, legislation was passed making it illegal to use nets and other devices to catch wildfowl between 31 May and 31 August, while they were moulting, to ensure a continuing plentiful supply.[18] In order to avoid the moulting season, duck decoys were constructed in wetland areas of England and Wales, operating mainly from October to February.

Over two hundred duck decoys once existed, where shallow pools of water were artificially created or existing pools modified. Leading off these pools were a few curved stretches of water-filled ditch, known as pipes, so that different pipes could be used whatever the wind direction. Each ditch was covered over by netting (and later by wire mesh) supported on arched wooden or metal hoops, effectively forming a tunnel that became increasingly narrow and low, terminating in a long detachable net. Screens of reeds down the length of the pipes hid the decoyman and his dog (the 'piper').

Initially, tame call-ducks (decoy ducks) lured the wildfowl (mainly mallards, wigeon and teal) to swim from the pool towards the pipes. Wildfowl cannot resist following dogs, so the dog then took over, appearing to the wildfowl at successive gaps in the screens, until the decoyman was able to force them into the long net, from which, one by one, they were removed and killed. This process was summarised in 1886 by Sir Ralph Payne-Gallwey, the author of the first book about duck decoys: 'This is a knack, requiring a certain amount of skill … which consists in breaking the necks of the captured fowl artistically.'[19] Thousands of birds were sent to London and other markets to be sold, and they were particularly popular because, unlike game killed with guns, they did not contain the hazard of lead shot (small lead balls or pellets).

The decoymen were a strange mix of countrymen and watermen, utterly in tune with their surroundings. Andrew Williams, who died on 18 April 1776 at the age of eighty-four, worked for sixty years as the decoyman at Aston Hall in Shropshire. He retired to a plot of land at nearby Whittington, and the church register recorded his epitaph, which started:

> *Here lies the Decoyman who liv'd like the otter*
> *Dividing his time betwixt Land and Water*

Williams was so amphibious that Death was said to have lured him to retire to dry land in order to stand a chance of turning him to dust:

> *So Death turn'd Decoyman, & decoy'd him to Land,*
> *Where he fix'd his abode 'till quite dried to his hand,*
> *He then found him fitting for crumbling to dust,*
> *And here he lies mould'ring as you & I must.*[20]

Numerous decoys were damaged by drainage schemes and by the increased enclosure of land from the mid-eighteenth century. Lincolnshire was especially important for decoys, and in 1829 Edmund Oldfield described how Friskney was affected after enclosure legislation was passed in May 1809: 'Prior to which period, from time immemorial, the low grounds of this parish were in general flooded near six months in the year, the waters seldom subsiding entirely until the month of May or even later. So effectually, however, has the parish been drained in consequence of the measures adopted ... that scarcely an acre of land is now flooded, in the wettest seasons.'[21] Over 1,500 acres were turned into cultivated land at Friskney, and as happened all across Britain, both wildlife and the old traditions disappeared, as Oldfield described:

Great as are the advantages arising from the inclosure and drainage, they have in some measure been counterbalanced, as it respects this parish, by the loss sustained by the decoys, and the almost total failure of the cranberry harvest. Friskney was at one time noted for the number and magnitude of its decoys, and for the immense quantity of wild fowl caught in them. London was at that period principally supplied with ducks, wigeon and teal, from the decoys in this neighbourhood. In one season, a few winters prior to the inclosure of the Fens, ten decoys, five of which were in this parish, furnished the astonishing number of 31200 for the markets of the metropolis. Since the inclosure the number caught has been comparatively small. Only three decoys remain, two in Friskney and one in Wainfleet Saint Mary's, and the decoymen consider 5000 birds as a good season.[22]

Peace and quiet were essential for enabling duck decoys to work properly, as otherwise the wildfowl would fly off. Payne-Gallwey said that the once flourishing decoy on Middle Moor at Aller in the Somerset Levels ceased operation in the 1860s because the railway came too close and nearby 'wasteland' was cleared:

The widow of a former lessee, now aged 86, states that she well remembers her husband getting sometimes 20 dozen fowl in a day, and that 'hagglers' [itinerant dealers] used to come from all parts and sometimes paid £20, for what they carried off. Another old resident in the parish recollects seeing the Decoy covered with wildfowl 'like the dressing spread over a field.' He had seen cartloads carried away, and passing to or from the Decoy could often see 1,000 ducks over his head. The present rector of Aller, the Rev. J.Y. Nicholson, who kindly favoured me with this information, adds that he remembers seeing the Decoy in 1858, when it was still worked, but its yield had then greatly declined. It is now numbered with things of the past.[23]

Another decoy, at Great Grovehurst Farm in Kent, was described by Payne-Gallwey as a 'pool about 2 acres in extent, exclusive of the surrounding cover of willows and alders, and had four pipes'.[24] Built in the early eighteenth century, it was situated just north of Sittingbourne, a mile from Swale Creek and close to marshland. William Gascoyne purchased this farm with the decoy in 1847, and Payne-Gallwey was permitted to examine a book of expenses and catches that George Chapman had kept from his time as decoyman there: 'I find that the largest take in one season during his time was 2,500 fowl. The best day's catching resulted in 140 fowl, and the greatest number at a drive was 80 duck and mallard ... all the district surrounding this decoy consisted of extensive marsh lands, intersected by numerous creeks and arms of the sea, and so very favourable to attract wildfowl.'[25]

The opening of the railway line from Sittingbourne to Sheerness in 1860 came within 250 yards of the duck decoy, and an all-too-familiar decline began – cement works were built nearby, the population increased, and the area was drained by some 4 feet, so that acres of shallow pools and marshland dried out and were changed to arable farming and orchards, over which a huge distribution centre on an industrial site has since been built.[26]

The Norfolk journalist William Dutt commented on decoys in his 1903 book on the Norfolk Broads: 'The difficulty in preserving that absolute quietude essential to the successful working of a decoy is one of the chief reasons why so many of the old pipes have fallen into disuse.' In his view, the land-locked Fritton Lake, just 5 miles south-west of Great Yarmouth, was the loveliest part of Broadland. Surrounded by woodland, the east end was still being used as a duck decoy, though it was barely known by anyone cruising for pleasure on the nearby waterways:

> he can only imagine what the lake is like when its bordering woods are white and gleaming with hoar-frost, and every breath of breeze sets ice-crystals tinkling among the reeds. This is an aspect of Fritton with which the favoured few are familiar; but a decoyman will tell you that there are times when, in the excitement of watching the wild-fowl paddling up the pipe of his decoy, he can feel neither the nip of the frost at his finger-tips nor the chill which is creeping into his bones. And if you ask him, he will say that it is in winter, and in frosty weather, that Fritton is seen at its best; for the sight of the wild-fowl swarming on the lake is far more pleasant to him than the singing of nightingales and reed birds and the shimmering of green leaves in the sunlight.[27]

Another method of catching birds was by falconry, also called hawking. This ancient sport of hunting wild animals with trained birds of prey (raptors) was originally introduced into Britain by the Saxons. The falconer and the rest of the party, along with spectators, usually on horseback, would follow a bird of prey while it pursued the quarry over open countryside. It was largely a male sport, though in wealthy circles women also flew falcons.[28] Falconers with birds of prey would accompany armies on campaign, and nobles routinely took falcons on journeys and even into church. Falconry proved so popular for many centuries that numerous words and phrases entered the English language.

For the nobility and royalty, falconry was a costly and obsessive pastime – a way of life that required falconers, horses, dogs and grooms, while those lower down in society trained their own birds. In the mid-eighteenth century, Thomas Pennant was not impressed by the tales of medieval and Tudor England, because 'during the whole day our gentry were given to the fowls of the air, and the beasts of the field: in the evening they celebrated their exploits with the most abandoned and brutish sottishness: at the same time the inferior rank of people, by the most unjust and arbitrary laws, were liable to capital punishments, to fines, and loss of liberty, for destroying the most noxious of the feathered tribe.'[29]

A hierarchy of species reflected society. Falconry was all about status, as outlined by James Edmund Harting, a librarian, ornithologist and keen hawker: 'In Shakespeare's time ... every one who could afford it kept a hawk, and the rank of the owner was indicated by the species of bird which he carried.'[30] These ranged from a gyrfalcon for the king to a merlin for a lady and a kestrel for a knave or servant. Many hawks were taken from the nest (and called eyesses), and others were caught as young birds, known as passage hawks or passagers before their first moult, or haggards if they were taken after their first migration. Some falcons were imported, either for purchase or as high-status royal gifts, most notably gyrfalcons.[31]

Valuable falcons and hawks had lightweight bells attached to their legs so that they could be tracked by the falconer, as the hawking enthusiast Francis Salvin and his co-author William Brodrick wrote in 1855: 'Bells are of the greatest use in finding a Hawk that has killed its game, at some distance from the Falconer, particularly where the ground is rough, as amongst brackens, turnips, &c.; also for discovering the haunts of a strayed Hawk; and, when good, they can be heard at a considerable distance, particularly at night.'[32] A strip of leather, known as a jesse, was also secured to each leg just above the foot, to which was attached a vervel, a ring through which a detachable leash was passed for tethering falcons to a perch or block. Usually of silver, vervels were inscribed with the owner's name, and many have been found by metal-detector users.[33] Salvin and Brodrick pointed out that they were rarely used by the mid-nineteenth century, since they proved too heavy and the jesses could become entangled in trees.[34]

Because the birds and the vervels often went missing, Edward III, who was obsessed by falconry, made it an offence for anyone to keep or conceal stray falcons: 'That every person that findeth any faulcon, tercelet, laner or laneret, or any other hawk that is lost . . . he shall bring him to the sheriff of that county.'[35] The Act in 1360 inflicted a punishment of two years' imprisonment, but in 1363 it became a felony, liable to both imprisonment and confiscation of property. Almost three centuries later, the lawyer Edward Coke advised that this Act referred to falcons, which are long-winged hawks that can fly substantial distances and therefore be lost. It was not, he said, intended for short-winged hawks, like the goshawk and sparrowhawk.[36]

What Edward III's Act highlighted was the confusion of names, because in falconry the terms 'falcon' and 'hawk' were applied haphazardly and interchangeably to any trained bird of prey, with little rigorous ornithological classification. Further confusion arises, because not only do the names of birds in the past not always match those of today, but the females were usually referred to as falcons and the males as tiercels. In 1826, Sir John Saunders Sebright, a politician and agriculturalist, expressed his views in a small book on hawking: 'The females of almost every kind of hawk are considerably larger than the males. In the language of falconry, the former are called *falcons*, and the latter *tiercels*. These terms are applied to almost every species of hawk. It is to be regretted that this language should prevail, as it has led to many mistakes.'[37]

Female peregrine falcons were known by many other names including falcon, slight falcon, falcon gentil and falcon gentle – gentil or gentle denoted a noble or excellent falcon (not that they were favoured by gentlemen). Their smaller male counterpart was known as a tiercel gentle or tiercel, with the word tiercel (or tercel, teircel and even tassel) supposed to mean one-third smaller than the female.[38]

The wealthy employed falconers to train and look after the

birds, as it was a very time-consuming process, and in 1368 Edward III was employing forty of them. For training, hoods of stuffed leather were initially placed over the heads of falcons, which made them docile (or hoodwinked). They were also hooded when carried from place to place on a 'cadge', a wooden frame that was suspended from the shoulders of the cadger, a term that came to mean an itinerant hawker who was laden with goods in a similar way. When falcons were moulting, they were housed in a structure or building known as a mew or mews (from the French 'muer', to moult), usually of timber-framed construction, with a thatched or shingle roof. The word mews was later applied to stables for horses, and some falconry mews were actually converted to stables when sporting tastes changed.[39]

The birds were trained to pursue and attack their 'quarry' and then to return to the falconer when called, attracted by his 'lure', which was an imitation bird with a piece of fresh meat attached. A falcon dives or 'stoops' on a victim with immense force from a considerable height, known as the 'pitch'. Shakespeare's plays are full of hawking references, so that the king in *Richard II* says: 'How high a pitch his resolution soars!', meaning how very determined he is.[40] When diving, peregrine falcons, with a wingspan of over 3 feet, are the world's fastest creatures, capable of more than 200 miles an hour.[41]

It was the spectacle of falcons and hawks taking birds of all kinds that was admired. Kites, herons, cranes and ducks were highly esteemed as prey, though cranes became increasingly scarce, as did red kites, even though they had once been protected as scavengers in towns and cities. Sebright was especially fond of watching magpies being pursued, which is a curious choice considering the amount of superstition attached to them. Two hawks were recommended, 'for the magpie shifts with great cunning and dexterity to avoid the stoop; and when hard pressed, owing to the bushes being rather far apart, will pass under the bellies of the horses, flutter along a cart rut, and avail himself of every little inequality of the ground in order to escape'.[42] As well as a falconer, Sebright said, five assistants were needed:

They should be well mounted, and provided with whips; for the magpie cannot be driven from a bush by a stick; but the crack of a whip will force him to leave it, even when he is so tired as hardly to be able to fly. Nothing can be more animating than this sport: it is, in my opinion, far superior to every other kind of hawking. The object of the chace is fully a match for its pursuers ... It is not easy to take a magpie in a hedge. Some of the horsemen must be on each side of it; some must ride behind, and some before him; for, unless compelled to rise, by being surrounded on all sides, he will flutter along the hedge, so as to shelter himself from the stoop of the falcon.[43]

The hunting of grey herons was thrilling, according to Salvin and Brodrick: 'The heron has, at all times, been considered the most noble quarry at which the falconer could test the qualities of his favourite birds; the height to which it will rise in the air when pursued, together with the powerful weapon of defence it carries, being such as to try to the utmost the courage and endurance of the boldest falcons.'[44] The ideal time was in the spring, when the herons were flying some distance for food for their young, and Sebright thought that a perfect location was Didlington in Norfolk, the seat of Colonel Robert Wilson, which had a large heronry on the river:

> The heron disgorges his food when he finds that he is pursued, and endeavours to keep above the hawks by rising in the air; the hawks fly in a spiral direction to get above the heron ... The first hawk makes his stoop as soon as he gets above the heron, who evades it by a shift, and thus gives the second hawk time to get up and to stoop in his turn. In what is deemed a good flight, this is frequently repeated, and the three birds often mount to a great height in the air.[45]

Eventually the heron would be forced to the ground, and if unharmed, the falconer attached to one leg a thin copper plate with the date and name, before releasing it. If captured repeatedly, some herons acquired more than one marker, and Salvin and Brodrick mentioned a particular ring:

> one of the authors saw [it] upon a heron that was taken during the spring of 1844 near Hockwold in Norfolk, by Mr. Newcome's Hawks. This heron, at the time it was taken, was on its 'passage' to the Didlington Heronry, and from the engraved date, had evidently been taken fifteen years before [about 1829], near the same locality, by Colonel Wilson ... This bird having been also taken the previous year by Mr. Newcome, was again released with three rings upon its legs, an honour which probably few herons would covet.[46]

Falconry declined from the mid-seventeenth century, on which Harting commented: 'To those who have ever taken part in a hawking excursion, it must be a matter of some surprise that so delightful a pastime has ceased to be popular. Yet, at the present day [1871], perhaps not one person in five hundred has ever seen a trained hawk flown.'[47] The reasons for the decline of falconry were complex after so many centuries, but changes in society and the rising population numbers made the emphasis on symbolism and status less relevant. The continued enclosure and drainage of the countryside, with more cultivation, meant that much land was no longer suitable. Heronries were also diminishing in size, and the red kite was almost wiped out. Instead, fox-hunting became more fashionable and was accessible to the new wealthy, while technological advances in firearms meant that shooting proved more exciting for the participants than watching a bird of prey.[48] The hereditary title of Grand Falconer of England is nevertheless retained by the Duke of St Albans, a reminder of an almost forgotten sport.

Cormorants are found around the coast and on inland waters, and these aquatic birds were at times trained to fish. Johann Faber, a German physician and botanist who lived in Rome from 1598, wrote about their training in England during the reign of James I:

When they carry them out of the rooms where they are kept to the fish-pools, they hood-wink them, that they not be frightened by the way. When they are come to the Rivers, they take off their hoods, and having tied a leather thong round the lower part of their Necks that they may not swallow down the fish they catch, they throw them into the River ... When they have done fishing, setting the Birds on some high place they loose the string from their necks, leaving the passage to the stomach free and open, and for their reward they throw them part of the prey they have caught ... This kind of fishing with *Cormorants* is it seems also used in the Kingdom of China.[49]

James I kept trained cormorants so that he could take them when travelling round the country in order to fish. He set up a new office of 'Master of the Royal Cormorants', held initially by John Wood, and one writ of 5 April 1611 recorded: 'To John Wood, the sum of £30, in respect he hath been at extraordinary charge in bringing up and training of certain fowls called cormorants, and making of them fit for the use of fishing, to be taken to him of His Majesty's free gift and reward.'[50] By the eighteenth century, this official position no longer existed. When Francis Salvin received an untrained cormorant in 1847, he decided to apply his falconry skills to these birds to try to revive the sport. For three years he raised and trained cormorants, and when fully grown, he disabled (pinioned) their left wings to prevent them flying away. They took all kinds of fish from rivers, as he described: 'In the summer of 1849, I made a most delightful tour in the northern counties with four cormorants, upon which occasion I took in twenty-eight days 1,200 good sized fish.'[51]

Before guns, other weapons used against birds included the bow and arrow and the crossbow. Written towards the end of the fourteenth century, *Le Ménagier de Paris* advised ladies to hunt birds for food in late September by using a combination of falconry and archery: 'And if you can go on foot and have a bow and blunt arrow, when the blackbird is driven into a bush and dares not leave because of the hawk that is overhead and watching him, the lady or damsel who knows how to shoot can kill him with a blunt arrow.'[52] Hunting bows were smaller and less powerful than military ones, and arrowheads known as blunts could kill small birds and stun larger ones. An arrow with a single point might not prove lethal when dealing with waterfowl and large gamebirds, but a swallowtail (or forked head) arrowhead with two points could cut off their wings or even heads and was also less likely to stick into tree trunks and branches.[53]

Crossbows were much more powerful and fired bolts or quarrels from a bow attached crosswise to a wooden stock or tiller. In woodland, blunted bolts were often used, which were especially suitable for killing young rooks in rookeries. Such crossbow shoots were held as sociable gatherings, and in about 1800, as a child at his grandfather's house in Heanor, Derbyshire, William Howitt took part in the annual ritual:

> the house was screened on the north by a tall wood, where the rooks built by thousands ... and many a day's entire occupation did we find there in seeing the young rooks shot in spring with a cross-bow; in running to catch them as they fell; in climbing to get at them when they got entangled in the boughs of some lower trees ... it was grand fun to mount, and shake their long boughs where the dead rooks had lodged, and send them tumbling to the ground.[54]

In 1824, an author using the pseudonym of Fletcher Beaumont satirised the well-connected people who gathered for the spring rook shooting at an imaginary estate in Essex:

> It was, indeed, a sort of holiday-time with the younger branches of the Landed Interest; and such of their refined London cousins as might condescend a visit at the particular period. The youths and maidens were as gregarious as the rooks themselves ... The gentlemen armed with cross-bows, bird-bolts, and bullets; and the Dulcineas of the vicinity with far more fatal weapons ... If the reader cannot handle a cross-bow, he can, at least 'follow to the field'.[55]

Everyone congregated at the avenue of trees, not far from the house, and initially one young rook was singled out:

The archer has already marked him for a most beautiful shot – drawn up the bow-string to its revolving catch – laid the fatal bolt in the groove of the stock, and taken his station for the work of death ... The bow is now raised perpendicularly upon his shoulder – his aim taken, and there is a cry to the bye-standers to look out for the direction of the arrow, lest it should be lost. Then in a moment as the trigger is touched, the bolt leaps with a dull *whirr* from its resting place, and, chucking the young rook under the chin, tumbles him headlong from his 'high estate,' to fall from branch to branch, with a dying flutter and a dead weight, at the very feet of the sportsman![56]

Beaumont tracked the bolt: 'Immediately there is a shout of applause, mingled with a cry of 'HEADS!' for the arrow is coming down almost as straight as it went up, drops in the meadow just beyond the widest bough of the oak, and is picked up and returned to the owner.' Although this particular bolt was retrieved, he explained that other archers were not as skilled, 'for many a bolt has erred from its object, or glancing from a bough above the mark it had struck, has passed into the air obliquely, to fall at last in the almost unsearchable underwood of the neighbouring grove; a few also are left sticking on the cross branches of the oak tree; and half-a-dozen others are irrecoverably buried in the mud of the garden moat.'[57]

One or two hundred rooks were likely to be shot on such an occasion, most of which were given to servants for rook pies. By the start of the nineteenth century, most birds were shot with guns, but William Daniel, writing about countryside sports, said that shooting rooks with crossbows was an old custom that was kept up by gentlemen, because modern crossbows could fire bolts that were extremely accurate.[58]

At Old Hunstanton Hall in Old Hunstanton, Norfolk, payments were listed in the early accounts of the Lestrange family for various birds purchased from fowlers using nets or killed with hawks or a crossbow, as in 1527: 'a bustard & ij mallards kylled with ye crosbowe ... viij malards, a bustard & j hernsewe [heron] kylled wt ye crosbowe'.[59] But in September 1533 a significant entry was made: 'A goos & a crane kylled with the gon', which was followed in November by 'a watter hen kylled wt the gun', and then in December by 'a cranne kyllyd wt ye gunne'.[60] Guns and gunpowder had arrived, and the shooting landscape was about to change.

CHAPTER FOUR

———— ·◆· ————

SHOOTING

I fell in with the Red-throated Pipit one fine sun-
shiny day at the beginning of the month of April,
1880, flying up and down, singing and feeding
along the ploughed furrows behind my plough . . .
I was attracted by the bird being alone, and
returned with my gun and shot it.

WALTER PRENTIS on his farm
at Rainham in Kent[1]

For thousands of years, the methods of taking birds were relatively
primitive, but gunpowder and guns completely transformed the
way animals, including birds, were hunted and killed. This new
technology led to a change in countryside sports, the rise of gamekeepers and big shooting estates and an increase in the depletion
and even extinction of different bird species. Early guns were
unwieldy, especially as it was thought that the longer the barrel,
the better the gun, and being expensive, it was a long time before
they replaced other methods of killing birds.

In Europe the first use of gunpowder was for cannons, from
which hand-held firearms developed. The earliest ones had a long
barrel or tube, a matchlock and a wooden stock, and they were
muzzle loading, which meant that gunpowder, wadding, shot and
more wadding were loaded into the barrel through the mouth
or muzzle. The matchlock was the mechanism that made the

gunpowder explode, which was achieved by igniting some priming powder in a pan with a slow match (a cord treated with saltpetre). A flame then spread through a touch-hole to the main gunpowder charge, and when that exploded, the shot was fired from the barrel.

Most advances in firearms occurred on the Continent. From about 1500, the wheel-lock appeared, which had a steel wheel that rotated against a piece of pyrites to create sparks, but it was complicated and unreliable. The snaphance flintlock mechanism was subsequently invented, but the most significant advance was in the early seventeenth century with a more developed flintlock gun. This design proved so successful that it was used for more than two centuries.

Flintlock guns were fired by a mechanism that comprised a cock and a hammer. The cock had a clamp or jaws holding a gunflint (a wedge-shaped piece of flint), and the so-called hammer was an L-shaped cover-plate for the small pan or flash-pan that contained the priming gunpowder. By pulling on a trigger, the flint struck the hammer, forcing it up to expose the priming powder, while simultaneously creating sparks that ignited the powder and caused the main gunpowder charge to explode.[2] One incident in about 1835 demonstrated what could be done if no flints were available. John Haining, who was then about twelve years old, said that his father was away from home when a mute swan flew over their farm to the south-east of Castle Loch, near Lochmaben: 'The ploughman induced me to get an old flint lock gun out of the house and we charged it. The gun, however, was minus the flint, so I went to the house and brought a piece of live peat from the kitchen fire. The ploughman then levelled the gun at the bird over the bank of the river, and I touched the powder in the flash-pan of the old gun with the live peat, and the swan was shot dead.'[3]

In the half-cock position, which was a type of safety mechanism, a gun would not fire, and it was usual for sportsmen to carry loaded guns in that way to prevent accidents. According to Robert Southey, when Captain Horatio Nelson was ashore, on half-pay and waiting for an appointment to another ship, he returned to Norfolk with his new wife Fanny and stayed at his father's parsonage in

Burnham Thorpe. From the summer of 1788, he amused himself there with various pursuits: 'Shooting, as he practised it, was far too dangerous for his companions; for he carried his gun upon the full cock, as if he were going to board an enemy; and the moment a bird rose, he let fly, without ever putting the fowling-piece to his shoulder. It is not, therefore, extraordinary, that his having once shot a partridge should be remembered by his family among the remarkable events of his life.'[4]

Improvements to guns and gunpowder enabled sporting guns to have shorter barrels, as well as double barrels. Gunpowder was carried in horns or flasks, and it was manufactured in different sizes of grain for different guns. Loading a hot gun with gunpowder was risky, and guns were prone to bursting, especially if a barrel in a double-barrelled gun was inadvertently loaded twice. Injuries such as the loss of fingers could be caused, though on 14 September 1796, one sportsman had a lucky escape: 'as Francis Lefevre, Esq. of Great Ormond-street, Queen-square, was shooting with his double-barrelled gun, about four inches of the right-hand barrel, close to the hand, with part of the lock, was completely blown off, but he fortunately received no injury'.[5]

Others were not so lucky, and newspapers relished reporting such incidents from far and wide. In February 1803 an East Anglian newspaper published details of an accident in Oxfordshire: 'Wednesday morning as Mr. McLory, of Abingdon, was shooting at larks, the gun burst, and shattered his left hand in so dreadful a manner as to oblige him to have it immediately amputated.'[6] Over a year before, the *Kentish Gazette* featured a calamity in Scotland: 'As Lord Douglas, and his son, the Hon. Archibald Douglas, now 27 years of age ... were last week on a shooting party near Aberdeen, the Hon. A. Douglas's double barrelled gun burst on the first shot, and completely tore off the fore finger of his left hand, shattered the others, and also his arm in so dreadful a manner ... The explosion was so great, that the two locks, barrels, and stock, were also shattered to pieces.'[7]

From the sixteenth century, guns were produced in London, though some decades later the town of Birmingham became an important manufacturer. English gunmakers began to lead the way from the late eighteenth century, where previously they had followed continental designs. William Greener's father had his own business at Birmingham, and he himself was a gunsmith and writer, publishing in 1881 a book on the history of guns, which went through several later editions. The bore of a gun was worked out by taking a lead ball of the same diameter as the barrel and then calculating how many balls of that size weighed a total of one pound (1 lb.). The smaller the number, the bigger the diameter of the barrel, as Greener explained: 'There is no actual limit, perhaps, to be set upon the capabilities of any weapon until trial has been made – a customer of the author's once shot a couple of snipe with an 8-bore elephant rifle – but ordinarily a gun is made for some special purpose ... The collector who requires humming-birds, and the wild-fowler who thinks of getting wild geese, will arm themselves very differently.'[8] He said that the 12-bore (or 12-gauge) was the standard calibre for shooting game and pigeons, with twelve balls of the diameter of the barrel weighing a total of one pound.

The sportsman would not use balls of this size, but instead fired small lead balls, also referred to as shot or pellets, that were suitable for the occasion. Military muskets fired a single musket ball big enough to incapacitate or kill a person, but in order to slaughter birds, guns were loaded with hundreds of small lead pellets, which is why numerous birds could be struck by a single shot. In October 1825, the sportsman Peter Hawker was staying at Alresford in Hampshire: 'I turned out at five in the morning ... in order to storm an enormous number of starlings, into which I blew off the great double gun with 30 ounces of small shot, just before sunrise. What I killed it is impossible to say ... my spectators guessed at least 500 ... What I bagged at the time, however, was 243 starlings at one shot.'[9] In 1816, he had published in an instruction manual for young sportsmen: 'Many select their shot, *in proportion to the size of the bird*, when it ought to depend *more* on that of the *caliber*, for it is not so much the *magnitude of the pellet*, as the *force with which it is driven*, that *does the execution*. For instance, a common sized gun ... will shoot N° 7 better than any other shot.'[10]

A few years earlier, William Daniel said in his *Rural Sports* that he often asked an old gamekeeper of his 'why he was so partial to *small Shot?* (for he generally used Nos. 6 and 7 mixed) and his reply was, "Sir, they go between the feathers like pins and needles; whilst the large Shot you use, as often glance off as penetrate them." '[11] In a table of pellets, Hawker advised: 'The shot of different manufacturers varies much in size: for example, an ounce of Nº 7, from Messrs. Walker and Maltby, amounts to 341 pellets; and the same weight from Mr. Beaumont (late Preston), 398.'[12] Pellets were manufactured commercially by pouring molten lead through a sieve at the top of a shot tower. While falling down the shaft and into water, the lead cooled and formed globules. Misshapen ones were supposed to be melted down again. There were ways of hardening the shot, but, as Greener said, 'Hard shot is disagreeable to the teeth; but so is soft shot.'[13]

Flintlocks were unreliable in windy weather, which Richard Hayes found in March 1764 when he was at the marshes alongside the River Thames, not far from his home at Cobham in Kent:

I went this afternoon down to Chalk marshes to see if there was any wild fowl up, and on going across ye marshes to ye Thames Wall I heard the cackling of wild fowl, and behold up comes about 50 wild geese, and I being near a gate squatted down, and when they were within 8 rods of me I rose up to shoot at them, and behold (by the winds blowing so strong) my gunpowder missed the fire, upon which I immediately cocked and presented again, and my gun then went off.[14]

The flash in the pan forewarned birds, but wet weather was another problem, as the priming powder could not be kept dry. Consequently, the Reverend Alexander John Forsyth in Scotland invented a 'scent bottle' lock, so-called because of its shape, which contained an explosive powdered fulminate and a striker or firing pin. When the gun's trigger was pulled, a hammer (rather than a cock) hit the striker, causing the fulminate to ignite. He patented his system in 1807, and Hawker heard of its success in July 1811. At the end of May, owing to injuries sustained in the Peninsular War, he had returned to England and was back in his Longparish estate in Hampshire, but by July felt well enough to sail to the Isle of Wight to shoot cormorants at the Needles. 'They dive so quick that if you fire at one on the water', he said, 'he will generally be down at the flash ... I was told by the boatmen that a man completely outmanoeuvred them (a few days since) by one of Forsyth's patent locks, which never failed to kill them on the water.'[15]

From Forsyth's invention came percussion caps or primers, which were small steel tubes containing an explosive fulminate and with a thin copper cap. They were used from the 1820s and proved much better in all weather conditions, so that the flint-lock gradually disappeared.[16] The explosive fulminate was more dangerous to handle in large quantities, and in July 1841 Hawker heard that William Eley, a maker of gun cartridges, had died: 'He was blown to atoms by fulminating powder at his factory in Bond Street, where I was in the habit of repeatedly going to give him hints and advice; and often have I advised him not to extend his practice to the dangerous trade of making percussion caps. His loss will be irreparable to the sporting world.'[17]

Shooting was controlled not just by technology but by legis-lation. Over the centuries, the function of laws regarding wild animals and birds was to protect the rights of the landed classes to indulge in field sports, to prevent birds being taken that were regarded as exclusive sources of food and sport for the wealthy and to stop birds destroying crops. Laws and privileges grew out of the Norman forest laws until an Act was passed in 1390 under Richard II that limited the hunting and killing of game to landowners or the clergy.[18] Subsequent acts dealt with issues such as closed seasons (when birds were not permitted to be killed), the sale of game and also poaching, which tended to be carried out by the poor. While animals were believed to have been created for the benefit of man-kind, the lower classes were generally treated like a species apart. Their lot was to serve their superiors, and they had few rights of taking birds for food, let alone sport.

In 1671, under Charles II, a Game Act was passed that took

account of guns. It formed the basis of game laws for over a century and a half, until 1831, and made shooting a sign of landed status. Anyone taking game, which was defined as the wild animals of hare, pheasant, partridge, red grouse and black grouse, had to own land and buildings, either a freehold with a minimum yearly value of £100 or a leasehold of £150, or else be the son and heir of a person of repute. Unless warned off, anyone so qualified had the right to shoot on other people's land. A person's income was immaterial, because what mattered was land ownership.[19]

Each lord of the manor (and there were thousands of them) was now authorised to employ a gamekeeper, whose function was to preserve game for shooting, and as such he was allowed to control vermin and had wide powers to confiscate guns, dogs, snares, nets and other apparatus from unqualified people, especially poachers.[20] Gamekeepers were also permitted to kill game, and yet small landowners, including farmers, were excluded from shooting, even on their own property, nor was a tenant entitled to the shooting rights.

A letter to *The Sporting Magazine* in 1794 was from an outraged farmer at Diss in Norfolk, who said that he paid between £300 and £400 each year to rent farmland and that he also owned a freehold worth about £60 to £80, which was below the qualification:

> [I] dare not keep a greyhound to follow at my heels, about my land, nor a gun to shoot a partridge or a pheasant for my longing wife, but shall be severely trounced by my next great neighbour; whilst his game-keeper, who is one generally picked out for one of the best shots in the county, shall load his table with game, and some to spare for your town poulterers; but, my poor son Tom, if found with a fowling-piece in his hand, though it is only to kill a crow that is pecking my lambs eyes out, his gun shall be taken from him by this saucy game-keeper, and severely rebuked in the bargain by the squire.[21]

Hawker also despaired of the absurd laws and in 1816 set out the qualifications: '100*l.* per annum, clear of all deductions, in own or wife's right, charged upon lands or tenements, or other estate of inheritance. 150*l.* per annum for life, or on lease for life, or ninety-nine years. Eldest sons of esquires, or persons of higher degree.' He added, with derision: 'Notwithstanding the eldest *son of an esquire is qualified*, yet the *esquire himself may not* be qualified! Such is the *consistency* of the game laws!'[22]

The tenant farmer especially could suffer at the hands of those who were qualified to shoot on his land, and the worst damage occurred each year on 1 September, the start of the partridge shooting season, because in wet weather the crops may not have been harvested, resulting in them being trampled by the sportsmen and their dogs.[23] Hawker recorded the situation at Longparish for 1 September 1812: 'So much corn was standing, and so execrably bad was the prospect of sport for this year, that many first-rate sportsmen declined going out, and several of those who did came back home with empty bags.'[24] In his diary two seasons earlier, William Holland wrote: 'This is a terrible day for the partridges, yet very little barley cut, so that they will have shelter against their enemies.'[25]

As vicar of Over Stowey in Somerset, he was qualified to shoot game, but in 1791, when a new law increased the tax for an annual stamped certificate or licence from 2 to 3 guineas, he refused to pay up, grumbling in a diary entry for 20 August 1800: 'This is pout [young game birds] shooting day, and I can see several persons on the hills [Quantocks], but I do not think it worthwhile to give three guineas for my certificate, and not be able perhaps to kill one bird afterwards.'[26] Instead, he relied on gifts of game from his wealthy neighbours, and on Christmas Eve 1808 his family had hare for dinner:

> though I am qualified to kill game, and we have game around us, and poachers are continually destroying game, yet not having taken out a licence I do not go out, and no one sends me game, at least not often ... I think the act about licences a very severe one, all the game by this means falls into the hands of poachers and night hunters, for unqualified persons take out licences,

and others go out without licence or qualification, so that these persons snap up all the game in one way or another. I cannot think government take such a very large sum by this act, and it is hard that qualified persons should not be allowed to have their amusement without restraint.[27]

Aware of his status in the community, he commented: 'As to the unqualified persons, they in fact cannot afford to idle away their time in this manner.'

One of Holland's main concerns was trying to ensure that his parishioners attended church, especially on a Sunday, which until recent decades was widely observed by most sectors of society. It was not a day of leisure but a holy day, with church attendance, a ban on recreation, minimal work and no hunting or shooting. Many felt that birds, especially rooks, had an uncanny knowledge of it being Sunday, though they may simply have been responding to the lack of guns trained on them. Henry Stevenson asked: 'Do rooks know Sundays from week-days? and if not actually capable of smelling gunpowder, do they, or do they not, know a gun from a walking stick? Often I have been led to ask myself these two questions, and . . . the very actions of these birds, in both the instances I have cited, leads irresistibly to the conclusion that by some means or other they can and do discriminate in either case.'[28] Towards the end of World War Two, when the travel writer William Palmer was exploring Surrey, he too noted that in some towns, 'the rooks are reputed to be keepers of the Sabbath and even at their busiest and noisiest, find time for Sunday afternoon rest'.[29]

Inevitably, such religious observance became entwined with superstition, with bad luck resulting for those who broke the rules. Early in 1795, *The Sporting Magazine* was not surprised by one story from the Lake District:

an inhabitant of Nether Wasdale went to Wass Water [the deepest lake in England] in search of wild fowl; but after staying the whole day, returned to his home without the smallest success. The next, being Sunday, and a holiday of course, he resolved to try again. Fortune, who had cruelly jilted him the preceding day, seemed now to be in a better humour. There *was* game. He fired – his gun burst – and he returned from his second excursion with the loss of a finger and thumb. The sufferer, we hear, has piously determined to relinquish *Sunday shooting* in future.[30]

From time to time, it was frustrating for sportsmen like Peter Hawker not to be able to use a gun. In January 1837, he was at his shooting cottage in Keyhaven, Hampshire, when he spotted 'a splendid eagle' on the beach near the remote Hurst Castle. He was desperate to shoot this rarity, but it was scared away, and six days elapsed before it reappeared: '*15th.* – The eagle came again on the beach, as if he knew it was Sunday.'[31] Hawker may have been obsessed by shooting, but to him the Sabbath was sacrosanct, and he had to leave it in peace.

At Lochgilphead in Argyllshire, Henry Graham caught sight of a splendid male velvet scoter. He wanted to shoot this large black seaduck for his collection and described how it 'paraded himself under my garden wall, his jetty plumage flashed in the sunshine ... but it was Sunday. I offered to use a loose leaf out of an old bible for wadding, but I could not get a dispensation from the authorities (Mrs G.) to shoot, though this would have been a new method of diffusing the Scriptures ... so I could do nothing but watch him sailing about within thirty yards.'[32]

Although all birds were once called fowls, the term now tends to be restricted to wildfowl – ducks, geese and swans. These are found in both freshwater and coastal locations, especially in winter, feeding on grasses, seeds, berries, fish, molluscs and amphibians, using their broad, relatively flat bills.[33] In order to shoot them, hunters needed to get as near as possible, and one way was by being out on the water. William Daniel knew East Anglia well, and he described how, in the late eighteenth century, wildfowlers went out on the water in punts with a gun like a small cannon: 'at the bottom of the *punt* he lays upon his belly, and gets as near the *rout* of fowl that are upon the water as possible'. After deliberately making a noise, the birds would rise up, and 'at that moment he pulls the trigger, and cuts a *lane* through their ranks; he instantly follows the direction of his shot, and gathers up those that are killed ... he then charges his gun, and drifts further down the river, in hopes of a second, third, and successive shots'.[34] James Padley was once the county surveyor of roads and bridges in Lincolnshire, and in 1881, at the age of eighty-nine, he recalled some of the fenmen hiding in their boats, 'of which numbers might be seen from the Fen-sides, drifting like logs of wood and only showing signs of being occupied by their reports and smoke from the guns'.[35]

On one occasion in December 1813, Hawker stayed at Poole Harbour for several weeks. He was highly accomplished at shooting, but found the night-shooting frustrating as so many other shooters were present, 'buried in casks or floating in canoes'.[36] Sinking a wooden cask into the mud to hide from the wildfowl was a popular technique, which he practised himself on New Year's Day: 'Buried myself in an old sugar cask in the mud, where I remained from ten till two, reading, and waiting for the geese, which were coming in immense force precisely where I wished them, till some scoundrel in a canoe rowed after them to no purpose, and spoiled me a shot.'[37]

He found wildfowl shooting along the coastal creeks thrilling, especially when out on the water in a punt or canoe, and he explained how best to use a Poole canoe:

> Sit down, on some straw or rushes, with your gun by your side, and a Newfoundland dog *abaft*. Cruise about, till you can see or hear a flock of wildfowl on the mud, and when you have rowed within three or four gun shots of them, take in your oars, and reconnoitre the creeks. Having ascertained which is likely to be the best, lie down and *push along* with a stick (called a *set*), and, from the mudbanks standing so high above the channel, you are so completely hid, that you will seldom fail to get a shot.[38]

The next stage was to make everything secure, including the oars, which might be lost at the recoil of the gun:

Having *made all fast*, rise up and fire, either *sitting* (*at the heads of the first birds*) or flying; and then put on your *mud pattens* (without which you would be *lost* to *eternity*), and assist your dog in *securing* the *killed and wounded*. The *gunner* here, generally, calculates on bringing home the half only of what he shoots, from the difficulty of catching the whole of his winged [wounded] birds, which he calls *cripples*, and those, that (to use the pigeon phrase) *fall out of bounds*, which he calls *droppers*.[39]

Walking about on the mudflats collecting dead birds was perilous without pattens, which were square wooden boards tied to each foot, but his came adrift on one occasion a few years later: 'Gale and rain continued ... Fought our way out in the punt ... got 1 brent goose and 8 wigeon. My pattens came untied on the mud, and I came home in a miserable mess, that took us four hours to clean up and rectify.'[40]

Some of the punts on the Fens were fitted with a reed screen that concealed the fowler and came to be called stalking horses, taking their name from the real horses on land that were used for hiding behind in order to approach a bird for a shot.[41] Those horses were trained to tolerate the firing of guns, and Gervase Markham in the early seventeenth century suggested using an old horse that was happy to walk slowly alongside rivers and brooks or on heathland, eating grass and other vegetation, and 'without taking any affright at the report of the Peice [gun], you shall shelter your selfe and your Peice behind his fore shoulder, bending your body downe low by his side, and keeping his body still full between you and the fowle'.[42] Models of horses to hide behind were also effective, and he said that a fowler could make a canvas model: 'he may take any pieces of oulde canvasse, and having made it in the shape or proportion of a horse with the head bending downeward, as if hee grased, and stoping [stuffing] it with dry strawe, mosse, flocks, or any other light matter, let it be painted as neere the colour of a horse as you can devise; of which the browne is the best, and in the midst let it be fixt to a staffe with a picke of iron in it to sticke downe in the ground at your pleasure.'[43]

In about 1808, Robert Agars, a gamekeeper, shot several great bustards on the Yorkshire Wolds, hastening them to extinction in Britain, as related to his grandson: 'the weight of the gun ... was more than the strongest man could hold without a rest ... The horse was a big bay coaching mare properly trained as a stalking-horse; his (own) coat was generally made of a bay horse-skin tanned with the hair on.'[44]

As guns improved, there was less need to hide behind stalking horses, though as late as the mid-nineteenth century Robert Gray saw one when he was rambling along the sands at the mouth of the River Tyne, on Scotland's east coast: 'I noticed a man and a boy, with a horse and cart, stalking dunlins and other small shore birds. The pony had been trained to walk slowly towards a group of birds, while the owner laid concealed in the cart ready to fire his long-barrelled gun over the side; and after waiting until I had seen him literally mow down a flock, I had an opportunity of turning over the contents of his bag, in which I found a fair proportion of ... sandpiper.'[45]

For many decades after the adoption of guns, birds were shot on the water or on the ground, using various techniques of concealment. A new technique was adopted in the late seventeenth century of 'shooting-flying' – firing at the birds as they flew away after being disturbed. It was done by 'walking up', usually with one or two men roaming over the land, their dogs in front, and when birds such as partridges took off, they fired at them over the dogs.[46] In 1727 George Markland wrote *Pteryplegia: Or, the Art of Shooting-Flying*, a poem giving tips to British sportsmen on how to shoot flying birds as skilfully as the French. For partridges, he wrote:

> *There sprung a single Partridge—ha! she's gone!*
> *Oh! Sir, you'd Time enough, You shot too soon;*
> *Scarce twenty Yards in open Sight!—for Shame!*
> *Y'had shatter'd Her to Pieces with right Aim!*
> *Full forty Yards permit the Bird to go,*
> *The spreading Gun will surer Mischief sow;*
> *But, when too near the flying Object is,*
> *You certainly will mangle it, or miss.*[47]

Pheasants were probably brought to Britain by the Romans. They are barely known in the Saxon period, but have been found in several archaeological excavations of medieval date.[48] By the early nineteenth century, they were still not particularly common, but useful to shoot, and a poem of 1733 described one trying to fly away before being successfully shot by a fowler:

> *His Birding-Piece the wily Fowler takes,*
> *And War upon the feather'd Nation makes.*
> *Whirring, the Pheasant mounts and works his Way,*
> *'Till Fate flies faster, and commands his Stay:*
> *He falls, and flutt'ring, pants away his Breath,*
> *What boots his Beauty in th' Embrace of Death!*[49]

Another way that sportsmen shot flying birds was to walk side-by-side in a line, with their dogs in front, over which they fired as birds were flushed out, but such organised events were not to William MacGillivray's taste. He was a Scottish ornithologist and naturalist who found great satisfaction in hunting for brown ptarmigan, usually known as 'red grouse', and what he enjoyed was the pleasure of roaming over tracts of moorland and mountains with a friend and a dog, shooting two or three braces (pairs). He deplored 'grouse-shooting' when undertaken by gentlemen shooters accompanied by their servants who looked after the guns: 'it is a pitiful and barbarous sport, as pursued by a regularly equipped and legally qualified slaughterer, who, even without the labour of charging his gun, still less of carrying home the produce of his idle industry, destroys as much game in one day as might serve for a dozen'.[50]

The Edinburgh portrait painter and ornithologist William Smellie Watson wrote to MacGillivray about his excursions in the 1830s to Dumfriesshire and Galloway to shoot black grouse, including one incident:

> I killed a very fine Blackcock in a rather singular way. I happened to be near to an old plantation, and my friend at least six

hundred yards into the moor, when he called *Mark!* I turned round, and observed a cock flying directly towards me. Raising my gun, I waited until I judged him sufficiently near, when I fired, and brought him dead to the ground. It is very difficult to shoot a bird flying directly towards you, and I have no doubt that in this instance his eyes were directed towards the dogs, and not to me.[51]

Unexpectedly, he had just achieved a method of shooting that would soon become the height of fashion.

From the 1830s, cartridges filled with the main charge of gun-powder were being used in muzzle-loading guns, but in the second half of the nineteenth century improved cartridges and advances in guns made breech-loading guns viable, though they had been developed decades earlier. The lead shot and gunpowder could now be loaded far more easily and quickly at the rear (breech) end of the gun, rather than through the muzzle, which led to significant changes in how birds were shot.[52] Sportsmen could aim at a steep angle without the danger of loose priming powder, but the biggest change was that instead of individual sportsmen spending all day in search of birds, it was now the beaters who did the walking, driving pheasants and grouse towards the stationary shooters, who fired at them as they flew overhead.

The sport not only became organised, but also faster and more exciting, and the number of birds shot increased dramatically, so that country gentlemen turned to hatching and hand-raising game by artificial means, a costly business that involved more game-keepers. The battuc – a mass slaughter – was also adopted from the Continent, which appalled many in Britain, who considered it highly unsporting. The pheasant was ideal for this purpose, because immense numbers could be produced, while upland moors were intensively managed by burning heather and killing predators to increase the number of grouse.[53] Henry Stevenson was of the opinion that pheasant shooting should be tolerated because it was a display of shooting skills:

> There is every reason ... to believe that the majority of those writers who are loudest in their denunciations against the 'battue,' and can find no milder epithet than 'blood-thirsty and unsportsmanlike,' to mark their abhorrence of it, are either practically unacquainted with the working of the system, or are deficient themselves in that necessary coolness and skill, without which even pheasants, big as they are, will escape from a perfect volley of double barrels.[54]

He explained that at the top shooting estates, only the best shots were invited, and that gamekeepers spent months preparing for one or two days of sport, 'and if any one is inclined to despise the amusement on the ground that pheasants are so easy to kill, let him try his hand, late in the season, at a few old cocks, flushed some two hundred yards from the post of the shooter, so that the bird is in full flight when he passes over; the pace is then tremendous!'[55]

Stevenson did admit that excessive numbers were at times killed, not just in his home county of Norfolk, but across the country: 'it would seem as though, of late years, the enjoyment of sport had become subsidiary, on the part of our larger game preservers, to the desire to outvie one another in the amount killed on their respective estates. The rivalry of the masters extends to the keepers, till, in many cases, the impression on the minds of the latter appears to be that game preserving is the end and aim of existence.'[56] The heyday of the battue, for pheasants and grouse, was from 1880 to the outbreak of World War One, and it was aided by the ease of travelling by railway. Writing at the end of the nineteenth century, Arthur Gibbs in the Cotswolds was also in favour of the battues. While gazing towards the great woods of Chedworth and Lord Eldon's estate, he commented:

> And here take place annually some of those big shoots which ignorant people are so fond of condemning as unsportsmanlike ... Why it should be cruel to kill a thousand head in a day instead of two hundred on five separate days, one fails to understand. As a matter of fact, the bigger the 'shoot' the less cruelty

takes place, because bad shots are not likely to be present on these occasions, whilst in small 'shoots' they are the rule rather than the exception. Instead of birds and ground game being wounded time after time, at big *battues* they are killed stone dead by some well-known and acknowledged good shot.[57]

Even before the rise of immense shooting estates, gamekeepers were obliged to ensure there were enough birds to shoot, and so they had long been ruthless in controlling anything they considered to be vermin, whether people, animals or birds. They also did whatever they could to control poaching, in order to preserve all the gamebirds, and the eighteenth century therefore saw mantraps and spring guns set up in the countryside, as well as rights of way blocked.[58] The traps had two strong, steel jaws, often with serrated edges, which snapped shut by means of powerful springs when someone stepped on the pressure plate between the jaws. Hidden in grass and undergrowth, they were designed to detain poachers and other intruders, but frequently resulted in a badly broken leg. Spring guns were mounted with a trip wire to the trigger, and the gun was fired when the trip wire was inadvertently pulled.[59]

Quite often, the signs alone and the warnings placed in newspapers proved a sufficient deterrent, as in one Norfolk newspaper: 'November 16th, 1782. MAN-TRAPS and SPRING-GUNS. NOTICE is hereby given, that Man-traps and Spring-guns will this Day be set in the Woods, Plantations, and Gardens, belonging and adjoining to BARNINGHAM-HALL, near Holt, in Norfolk.'[60] These devices were made illegal in 1827, and in the early twentieth century James Cash, who wrote about rural Cheshire for the *Manchester Evening News*, commented:

Let us recall that relic of olden days, the woodland sign to beware of Spring guns and man-traps. The latter was a horrible contrivance ... but to-day it is happily not seen outside museums. The threat, in the interests of game preservation, was often nothing more than the printed sign itself, and possibly even to-day these woodland signs, weather-worn and green with moss,

may linger. Certainly they were unpleasant warnings in the bird-
nesting days of our boyhood. In lighter vein, it is interesting to
recall the sagacious dog [Ponto] in *Pickwick Papers* – the pointer
of surprising instinct, which would not follow its master into an
enclosure, but stood transfixed, staring at a board: 'Gamekeeper
has orders to shoot all dogs found in this enclosure.'[61]

All the while they were controlling poachers, a number of gamekeepers illicitly shot and sold game to inns and other markets,
which William Daniel described in 1802:

> [They] are very apt to keep *their hands in* by killing Game
> unknown to their employers, and what perhaps is first sent as
> a present to a friend, soon becomes an object of *merchandize*,
> and readily finds its way to market, through the medium of
> the *Coachmen* and *Guards* to *Mail* and other coaches … These
> Gentlemen, in conjunction with the *Porters* of the different
> inns where they arrive at, carry on almost a *public traffic* in
> this article of Game, and at prices which render it astonishing
> how *purchasers* are met with, viz. *four* and *five* shillings (and
> sometimes as high as *eight*,) a brace for *Partridges*; [and] *twelve*
> to *sixteen* for *Pheasants*.[62]

The 1831 Game Act repealed and replaced most of the earlier
statutes relating to game and gave increased powers to gamekeepers. It also ended the property qualification and decreed that
all game belonged to the landowner. By the late Victorian era in
England and Wales, there were 17,000 gamekeepers and, by the
1911 census, twice as many gamekeepers than policemen in rural
areas.[63] As well as curbing poachers, gamekeepers had to preserve
game birds by killing vermin, resulting in a huge decline in the
number of birds of prey and an increase in animals such as rats,
which were a menace to young pheasants, along with pigeons,
rabbits and grey squirrels.[64]

William Hudson was scathing in his view of the large shooting
estates in an 1894 report for the Society for the Protection of Birds,

saying that landowners were the one class who had the power to preserve the bird population:

> The desire for a large head of game, a big autumnal 'shoot' – the ignoble ambition to transform a great estate into a kind of glorified poultry-farm, where you shoot young birds, instead of catching them in the usual way and wringing their necks – has overbalanced all other considerations. Hence the partridge and pheasant coddling policy, and the pitiless persecution of all birds whose presence is, or is ignorantly supposed to be, a check on the excessive multiplication of the one or two species chosen for preservation.[65]

Many birds were destroyed by gamekeepers through ignorance and custom. Lord Lilford was a leading ornithologist in the late nineteenth century who travelled widely, especially in the Mediterranean. He deplored the slaughter of barn owls, and on asking one gamekeeper why he did so, 'he assured me that though he could not say for certain that they killed the game, yet they were unlucky and *no good* to any one. I tried to convince him that their unluckiness was the fact of his destroying them, and that they did immense good to many people, but I feel sure that I only wasted my eloquence upon him, with the result of persuading him that I was a harmless lunatic.'[66]

John Atkinson said that not long after his arrival in the parish of Danby, he saw a beautiful spotted woodpecker, but it 'was shot by the gamekeeper, who showed his prize with great pride to "my lord." But he got such a wigging from "my lord" for the slaughter that I do not think he showed his next "rare bird" victim.' The gamekeeper was Robert Raw, who became a good friend: 'Although a gamekeeper, he was still in many particulars a bird-lover ... I was, however, unable to convert him to my view touching the harmlessness (in the game connection) of the kestrel. It was a "hawk," and the name was as fatal to the poor birds that bore it as the proverbial "bad name" to a dog.'[67] William Sterland related a similar attitude in the Sherwood Forest: 'The Jay ... is one of the most beautiful of our native birds; but he bears a bad character, from his predatory habits, and suffers accordingly. The keepers shoot every one they meet with, and one cannot go far in our woods without seeing their dead bodies dangling from the lower branches of a tree, and bleaching in the wind.'[68]

Even though rookeries were believed to bring good luck, there was a tendency, not just by gamekeepers, to want to reduce the number of young rooks, and while some used traditional crossbows, a day or two of rook shooting with guns was also deemed good entertainment. In about 1898 Arthur Gibbs took part in one at Ablington, though he failed to understand its attraction:

> Up to yesterday I had never shot a rook in my life. The accuracy with which some people can kill rooks with a rifle is very remarkable. I have seen my brother knock down five or six dozen without missing more than one or two birds the whole time. One would be thankful to die such an instantaneous death as these young rooks. They seem to drop to the shot without a flutter; down they come, as straight as a big stone dropped from a high wall. Like a lump of lead they fall into the nettles. They hardly ever move again. It is difficult work finding them in the thick undergrowth.[69]

That night he returned from a neighbouring farmhouse and heard the most lamentable sounds coming from the rookery: 'There seemed to be a funeral service going on in the big ash trees. Muffled cawings and piteous cries told me that the poor old rooks were mourning for their children. I cannot remember ever hearing rooks cawing at that time of night before ... I wonder if the poor rooks caw all night long after the "slaughter of the innocents?"'[70]

Before the early eighteenth century, wild pigeons were rare, but their numbers increased greatly for a variety of reasons, not least by the growing of crops like turnips and clover, to which they are partial, while the massacre of birds of prey by gamekeepers left few predators. Gibbs said that thousands of migrating wood pigeons came to the stubble fields each year from northern

Europe, and so pigeon shooting with the aid of decoys turned into a favourite sport for farmers:

> A few dead birds are placed on the stubble to attract the flocks, and a grand variety of flying shots may be obtained as the wood-pigeons fly over. The year 1897 was remarkable for this shooting. Between November 20th and 30th two of our farmers killed close on a thousand of these birds ... Tom Peregrine [the gamekeeper] remarked that 'he never saw such a sight of dead pigeons. The cheese-room up at the farm was full of them.' The vast flocks that blacken the skies for a few short weeks in November disappear as suddenly as they come.[71]

Around the same time, Charles Dixon in south Devon mentioned that immense numbers of pigeons roosted in the local woods, and so an annual shoot was held, which he described as a battue: 'The guns are in some cases stationed over an area of six or eight miles of country, and the birds are thus practically kept on the move all day. Upon the approach of dusk great slaughter takes place as the Pigeons seek their accustomed roosts.'[72]

Another new sport was trap shooting, in which live pigeons were released one by one from mechanical traps in order to be shot. First established in the eighteenth century, it was criticised by William Daniel in 1802:

> As a mode of shooting to bet large sums of money upon, it is perhaps the least objectionable, since every shooter has an equal chance as to the distance from whence the bird is sprung, but it certainly is not the *exact* shooting that a *sportsman* will ever try or fancy as an amusement; besides, the mind that thinks at all must feel a repugnance at the idea of first confining, and then setting at liberty, hundreds of domestic animals doomed to instant death, or, what is worse, to languish under wounds that in the end prove mortal ... there is no excuse for the wanton barbarity of it; for the shooting of *Pigeons* and of *Game* are so widely different.[73]

In spite of his misgivings, from the mid-nineteenth century the sport became hugely popular worldwide. Pigeon clubs were established, which put on competitions that attracted a good deal of betting. Considerable numbers of live pigeons were slaughtered, and of one match William Greener commented: 'One of the best scores first recorded is that of Captain A.H. Bogardus, who on July 2nd, 1880, succeeded in scoring 99 birds out of 100, the 47th bird falling dead out of bounds.'[74]

Pigeons were reared in dovecotes to supply trap shooting competitions, especially blue rock pigeons in Lincolnshire, while others were imported or stolen. In 1836 Charles Waterton at Walton Hall in Yorkshire said that dovecotes were at risk from a plundering set of vagabonds 'who attack the dovecots in the dead of the night, and sometimes actually rob them of their last remaining bird. The origin of this novel species of depradation can be clearly traced to the modern amusement, known by the name of a pigeon-shooting match.' He explained that when a club contracted somebody to supply pigeons, they turned to poachers and other criminals. The pigeons were stolen by climbing to the top of dovecotes at night and frightening the birds, who became entangled in nets when they tried to fly out: 'Thus the hopes of the farmer are utterly destroyed, and a supply of birds is procured for the shooting matches in a manner not over and above creditable to civilised society . . . The dovecots in this neighbourhood have been robbed repeatedly; and it is well known that the pigeons which have been stolen from them have fallen at shooting matches near forty miles distant.'[75] All the pigeons were old ones, he said, and so the clubs will have known that their owners would never have parted with them.

Towards the end of the nineteenth century, live pigeons were replaced by artificial ones, and clay pigeon shooting became especially popular, in which clay discs were catapulted from traps in order to simulate the flight of pigeons. At the International Pigeon-Shooting Competition held as part of the World Fair (l'Exposition Universelle) at Paris in the summer of 1900, clay pigeon shooting was held, as well as two live pigeon shooting events. The *Pall Mall*

Gazette reported that for the first live pigeon competition, nearly two hundred entries were received:

> The shooting for this prize commenced on Tuesday; however, the huge number of entries rendered it impossible for the contest to be finished off on the same day as had been arranged, so the final heat was left over till yesterday afternoon [Thursday]. The result was a win for Mr. Mackintosh [Australia], who brought down all his birds to the number of twenty-two. The Marquis de Villaviciosa [Spain] came next with twenty-one kills, and the third place was taken by Mr. Edgar Murphy [United States] with twenty birds. The winner receives the handsome prize of 5,000 f. (£200) . . . On Monday next the shooting for the Grand Prix de l'Exposition, to which a prize of 20,000 f. (£800) is attached, will commence.[76]

Donald Mackintosh of Australia was a highly successful competitive shooter, both in Australia and on the prestigious European circuit. He was the winner of the initial event and came joint third in the next. The World Fair also incorporated the second modern Olympic Games, with competitors taking part if they could afford to do so. Out of the many shooting events, it was never clear which ones formed part of the Olympics, but decades later Mackintosh was posthumously awarded gold and bronze medals. Hundreds of pigeons were shot in these two events, the only time live animals were intentionally killed at an Olympics, which may well explain why an attempt was made to rewrite history by rescinding these medals years later.[77]

The reasons for shooting any type of birds were various and complex, and those in the countryside often carried a gun through habit in case an opportunity arose to shoot a rare bird so it could be examined more closely and perhaps preserved. Alternatively, birds might be shot to be eaten or simply for some sport. To Arthur Gibbs at Ablington, shooting was part of his tranquil rural life at the close of the nineteenth century:

On a bright, warm day in October ... it is pleasant to take one's gun and, leaving behind the quiet, peaceful valley and the old-world houses of the Cotswold hamlet, to ascend the hill and seek the great, rolling downs, a couple of miles away from any sign of human habitation. You may get a shot at a partridge or a wood-pigeon as you go ... But on the other hand you will be equally pleased if your gun is not fired off, for it is peace and quiet that you are really in search of, – the noise of a shot and the jar of a gun do not suit your present mood. After walking for half an hour you come to a bit of high ground, where you have often stood before, and, resting your gun against a wall, you gaze at the view beyond.[78]

Around 1930 James Cash commented: 'Birds are of infinite value in the scheme of Nature. Who therefore causes their destruction? The ignorant type of both farmer, gardener, and gamekeeper; the random gunner; the bird-catcher; the trader; the feathered woman; the private collector. It is altogether wrong that supremely useful creatures such as kestrels and owls should be sacrificed to game or any other pursuit. Jays and magpies also justify their existence in many ways.' He then questioned the need to carry a gun when studying nature: 'Omit the gun, and study the living creatures in their haunts, with the aid of either field glass or camera. That is the happiest way of studying Nature.'[79]

Those involved in the scientific study of birds, almost always men, were frequently armed with a gun. In 1837 William MacGillivray imagined giving tips to a young man wanting to shoot and study birds:

Ah, there – a Sparrow-Hawk. You have hit. Down he comes whirling through the air. He has got on his back, the fierce little fellow! How he clutches the cane, and screams. Shall I kill him? Do: – he is doomed, and it is well to terminate his woes at once. Hit him on the back of the head. By the by, there are several ways of killing a bird ... the best of all, because the quickest and surest, is to thrust a pin between the occipital bone and first

cervical vertebra, and cut the spinal cord, which extinguishes
life in an instant. It is a barbarous business this practical ornithology of yours ... Ladies indeed cannot become practical
ornithologists ... Botany is the best study for ladies and other
gentle beings ... the study of Birds and Quadrupeds, with the
aid of powder and shot, is that which I prefer; and I know few
occupations more delightful than that of poring over the entrails
of a rare bird until you have satisfied yourself as to some minute
point of structure.[80]

He added some further reflections: 'Is it not cruel after all to shoot
birds, especially for mere sport? Perhaps it is. Some kill birds for
food; and I suppose they do right. Others slaughter them to make
money; and possibly they too are blameless. Some shoot for study,
and some to supply the naturalists and the museums with specimens ... As to the persons who shoot birds for mere sport, all I can
say of them is that they are mere sportsmen. I have known very
good men among them, and bad ones too.'[81]

Huge advances in knowledge were made possible by publishing
and sharing information, but whenever a rare bird was spotted, it
had to be shot, and the science was often reduced to a collecting mania. Books, journals, diaries and newspapers overflowed
with the slaughter of rare birds, with barely any twinge of regret
expressed. In a compilation of birds found in Devon, Edward
Moore, a physician in Plymouth, included a typical comment
when talking about a rare black stork: 'A beautiful specimen was
shot on the Tamar, November 5. 1831, and is now in Mr. Drew's
collection; I saw the bird while warm, and took note of it.' Later
on, a Penzance solicitor and ornithologist, Edward Hearle Rodd,
acquired the specimen for his collection, which passed on his death
to his nephew Francis Rashleigh Rodd.[82]

In Scotland, a Mr Angus described how the beautiful song of a
woodlark made him kill it: 'In the last week of March 1863, I shot
a male Wood Lark in the enclosure of Scotston House [Aberdeen].
Being the first time I had heard this pleasing songster, I was particularly struck by its mode of singing. It continued flying in circles,

trilling its sweet music without intermission for half-an-hour or longer, except once or twice when it alighted for a moment on the top bar of a wooden fence.'[83]

The following year, 1864, Robert Gray spent some of the winter near the mouth of the River Tyne in Scotland, with his friend David Robertson. They managed to speak to a party of men shooting on the shore, where hundreds of wigeon and mallards were present: 'On enquiry, we were told that each of these men earned as much as £2 per week by disposing of the birds they killed with ordinary fowling pieces. They were intelligent men in their way, and seemed tolerably well acquainted with the habits of wild fowl.'[84] When the tide receded, it was possible to cross the estuary, and Gray was pleased that one of the men evidently appreciated the scene:

> The sun was setting in a blaze of crimson and gold, throwing a deeply impressive shade over the stretch of sand and water. The man paused, as we imagined, for a little reflection on the singular sublimity of the scene, and stood nearly waist deep gazing cloudward a few seconds with no apparent intention. In another second, however, we were undeceived; there was a sudden upward movement of his arms, and, after a momentary flash of light, we heard a report such as a duck shooter's gun alone can make, and the two poor wanderers, part of a small flock we had not observed, fell into the river not far from where he stood. One of them – in my left hand as I write – is the most perfect male I have ever seen, but even now, on looking at the pair, I cannot help thinking that the glories of that sunset were dimmed by their death.[85]

The man had shot two wigeon, which are medium-sized ducks, and Gray obviously purchased the pair from him and had them stuffed, a reminder of an idyllic landscape, but tinged with sadness.

CHAPTER FIVE

———•◆•———

STUFFED

Mr. J. Gunn [a geologist] ... regretted so large
a space was occupied by birds in the museum.
There was hardly room for an insect.

Norwich Mercury August 1873,
on a new display at Norwich Castle Museum[1]

The practice of stuffing birds really has its roots in ancient
Egyptian mummification or embalming, by which birds such as
falcons, hawks and, above all else, ibises were preserved. The ibis
with its long, curved bill was the sacred bird of Thoth, who was
the god of knowledge, wisdom and scribes, as well as protector of
physicians, frequently depicted as an ibis or as a human with an
ibis head. The ibis was once a common sight in Egypt along the
swampy shores of the River Nile, and millions of these birds were
mummified from around 650 BC. At Saqqara, the vast cemetery
of the city of Memphis, roughly 1¾ million ibis mummies were
deposited as votive offerings to Thoth, but the largest number –
about 4 million – is from the necropolis of Tuna el-Gebel in
Middle Egypt.

The live birds were plunged into vats of liquid resin and then
neatly wrapped in layers of linen bandages to be sold to pilgrims
as offerings. Because so many were needed, even juvenile birds
were mummified, and some were fakes, containing little more than
a few feathers, random bones, reeds and sticks. The ibis offerings

were placed in pottery jars, thousands of which were moved each year into miles of underground catacombs. With such wholesale slaughter and subsequent changes to the Nile Valley environment, sacred ibises are no longer seen there.[2]

Imagine if archaeologists in the future were able to excavate the contents of a British stately home from the nineteenth century. They would unearth quantities of rectangular glass-fronted cases containing stuffed birds and might be tempted to interpret them as similar votive offerings. Instead, they represent the collecting mania for stuffed birds that developed in the eighteenth century, during the Enlightenment, when natural history was a popular topic of study. In that pre-photography era, methods of research were limited, and so birds were caught and often killed in order to be measured, drawn, dissected and possibly stuffed.

Information was freely shared between enthusiasts, as seen in the letters of the Reverend Gilbert White, an Oxford graduate who lived at Selborne in Hampshire. For many years he was a curate at nearby Farringdon and subsequently at Selborne, and as a well-educated clergyman he could read Latin, which was the common language of scholars across Europe, including early ornithological studies. From the late 1760s, White kept up a lengthy correspondence about birds with the Welsh naturalist Thomas Pennant, another Oxford graduate, now and again quoting from sources written in Latin. Pennant published numerous books, including his hugely successful volumes of *British Zoology*, filled with engraved illustrations, and these would have been in many naturalists' libraries.

Apart from acquiring the latest books, numerous naturalists collected dead birds, either as 'skins' that could be stored in cabinet drawers for reference, or as 'mounts' – birds preserved for display by taxidermy, which was literally 'the arranging of skin'. Some of them supplied each other with birds, and from his home on the island of Iona, Henry Graham frequently posted gifts of interesting birds and eggs to the naturalist Robert Gray. In early April 1852 he wrote: 'I have got two more Black Guillemots' skins for you which exhibit the changes the plumage undergoes between winter and summer.' Three weeks later he sent another letter: 'It is possible that I shall leave this note with my own hands at your address in Glasgow, accompanied by a small box containing a pair of Red-legged Crows [choughs], one skinned and the other simply disembowelled and embalmed in the style of an Egyptian mummy.'[3]

Graham enjoyed the stimulating correspondence with like-minded people, commenting a few months later to Gray: 'At the present day Ornithology takes its place as a respectable science, and especially as a popular and pleasing one. *Blackwood* [magazine] remarks: – "We remember the time when the very word *Ornithology* would have required interpretation in mixed company, and when a naturalist was looked on as a sort of out-of-the-way but amiable monster." Happily the case is very different at the present day.'[4] Serious naturalists had little need for an array of stuffed birds in glass cases, as the skins were preferable, which William MacGillivray recommended in 1837: 'Skins may be kept in drawers, or in chests having moveable trays of various depths. Two cabinets of moderate size are sufficient to contain specimens of all the British birds, of which, however, an individual can never make a complete collection ... skins thus preserved are preferable to stuffed or mounted specimens, and occupy comparatively little room.'[5]

Stuffed birds became fashionable items of interior decoration, from magnificent displays in country homes, especially sporting estates, to single birds for mantelpiece ornaments. There were no rules about what birds were considered attractive, though it surprised Charles Dixon in Devon that the superstitions attached to live magpies were heeded: 'We know people who would not have a stuffed Magpie in their home on any account, assuring us that bad fortune would be sure to follow.'[6]

Wealthy sportsmen in particular had taxidermists deal with fine specimens that they killed in order to display them as trophies. Peter Hawker's obsession was the shooting of birds and acquiring rare specimens, rather than scholarly study. He was frequently out in all weathers along the coast at Keyhaven, where he wrote in his diary for 28 January 1829: 'Saw 2 scoter ducks, birds I never met with before, except stuffed in museums; blew off a cartridge and floored them both.'[7] At the start of the following year, he noted: '3 brent geese and 1 old cock sheldrake [shelduck], one of the finest and cleanest specimens that any museum could have shown, and therefore I sent him to the prince of ornithology, Mr. Leadbeater, to stuff for me.' Four years later, he even sent a bird by coach to London for stuffing: 'Last night a stormy petrel flew against Hurst lighthouse, and was taken alive; I bought him for a shilling, and had him booked per coach to Leadbeater to stuff for my collection.'[8]

Hawker gave the renowned Benjamin Leadbeater of Brewer Street in London many commissions, considering him to have the finest knowledge of birds in Europe, as well as being the best at taxidermy. Taxidermists were also referred to as naturalists, and they worked on all types of animals, but those who specialised in birds tended to be known as bird stuffers or bird preservers. Countless people of varying skills practised as taxidermists, many in conjunction with other businesses or combining their skills with a keen interest in ornithology, and many shot their own birds. In mid-nineteenth-century Sussex, Arthur Knox, a former army officer living at Petworth, found Henry Swaysland highly reputable:

Mr. Swaysland, of Brighton ... has done much within the last few years to elevate the character of his art. From a correct knowledge of the proportions and attitudes of birds – the result of out-door observation – he succeeds in restoring to each its peculiar form and expression. Indeed, his specimens exhibit a life-like spirit which I have never seen surpassed, and contrast advantageously with those unhappy families of woodpeckers and kingfishers which one sometimes sees trying to stand in impossible attitudes within the shop-window of the ordinary bird-stuffer.[9]

Early attempts to preserve birds were rudimentary, with the skins simply being stuffed with rags, cotton, wool or clay, which too often ended in decay, particularly through attacks by insects. Possibly the oldest stuffed bird to survive is an African grey parrot that for forty years was the pet of Frances Teresa Stuart, Duchess of Richmond, who died in 1702. Not long afterwards her parrot died and was stuffed, and its survival is probably due to having been kept in a showcase. X-rays have shown that the skeleton is still intact, which would not be the case with later techniques. The parrot is now on display in Westminster Abbey next to the Duchess of Richmond's life-size wax effigy.[10]

In 1748 the Royal Society published a paper by the Frenchman René-Antoine Ferchault de Réaumur on several ways of preserving birds, especially those needing to be transported from distant countries for display in natural history cabinets back home. His first option literally involved skinning the creature, discarding the internal organs and bones and then stuffing it: 'take off their skin with all the feathers upon it, from the body and the thighs, leaving the legs, the wings, and . . . the whole neck with the bill sticking to it. Filling afterwards the skin thus taken off with some soft stuff, either straw, hay, wool, or flax &c. or even stretching it over a solid mould of the shape of the bird, you give to this skin, as near as possible, the form of the body of the bird, which it had when it covered its flesh and bones.'[11]

Another method resembled embalming, with the body of the bird being emptied out and refilled with spices and powdered alum, then placed in a box also filled with alum: 'The outward powder will make it dry the sooner, and keep off voracious insects, which will not care to attempt to pierce through it in order to come to the flesh they are fond of. During the first days, and even during the first weeks, the birds may cast a bad smell, which you need not be uneasy at, for it will lessen in proportion to the bird's drying.' Réaumur also advised attaching notes written on parchment, such as the bird's name in the country of capture, how many eggs were laid, whether or not it was resident, and if it was good to eat.[12]

In the late eighteenth century, arsenical soap was invented by the French pharmacist Jean-Baptiste Bécoeur, which was applied to the inside of skins to prevent decomposition and insect attack.

It revolutionised taxidermy and was used for well over a century and a half.[13] Bird specimens in museums can still retain traces of arsenic, which is now known to be toxic. One reader of a science magazine in 1865 warned that arsenical soap could lead to 'serious consequences ... owing to its getting between the nails or into any cut that the operator may have on his fingers when using it',[14] and he therefore suggested what he thought was a less dangerous formula, and yet it still contained arsenic. In complete ignorance, arsenical toilet soap was even sold to ladies as a method of ensuring a pallid complexion to prevent them looking like coarse labouring women. The use of arsenic may have been a reason behind the early deaths of so many Victorian naturalists.

William MacGillivray was just fifty-six years old when he died, though he himself did not approve of arsenic. To prevent skins being ravaged by moths, he advised, 'they ought to be frequently inspected. The application of a little rectified oil of turpentine I have always found a more effectual remedy than camphor, spices, old tallow, or solution of corrosive sublimate. As to the arsenical soap itself, it is not to be trusted to entirely, for being applied only to the inner surface of the skin, although it may preserve that part, it does not prevent moths from breeding among the down.'[15]

Not everyone could afford taxidermists, and so in 1865 James Gardner, who described himself as naturalist to the Royal Family, published a manual for amateurs, since he believed that taxidermy was an essential skill: 'All travellers should be able to shoot, and fish, and use the needle, and prepare the skins of animals. These arts in themselves go some way towards giving you what is called a polite education: and though, of course, I do not pretend to set them in competition with learning and languages, and science and the ologies, I humbly submit that the possessors of the last cannot but be rendered richer and happier by the acquisition of the first.'[16]

His manual included instructions on preserving and displaying birds, but his initial advice was how best to obtain them: 'It is not perhaps necessary to say, that to make a collection the amateur may either kill or purchase his specimens. If he kill them himself he should be careful to use only such small shot as will not greatly injure their skins.'[17] The importance of this point was demonstrated in a comment by Charles Stubbs of the Post Office at Henley-on-Thames: 'On the 12th of last August [1866] a flight of from forty to fifty crossbills . . . visited the grounds of Park Place, near here; they were seen about for two or three days, and then took their departure. The keeper, Mr. Hitchcock, shot three, but, his gun being loaded with large shot, they were so mutilated as to be quite unfit for mounting.'[18]

Gardner's method of stuffing was to skin the bird while carefully keeping the feathers, beak or bill, feet and skull attached. This was done by first of all placing the side of the bird to be displayed downwards on a table covered with a clean cloth: 'make an incision about three inches in length in the skin under the wing, taking especial care not to injure the feathers . . . Then dislocate the first joint of the wing, and gradually dress back the skin to the neck, which must also be dislocated. This is best done at the second or third joint of the vertebrae (or back bone).'[19]

A small wire hook was next inserted into the body so it could be suspended by string, enabling the remaining skin to be removed more easily:

> Now dislocate the joint of the other wing, and remove the skin down to the legs, which must be broken at the second joint. Great precaution is necessary in detaching the tail . . , Having so far freed the skin from the body, remove your hook, and fasten it to the neck bones which remain attached to the skull. You can now skin the neck to the ears, which operation requires great care . . . Then sever the neck from the head, clean out the brain from the skull, and remove the eyes and all the flesh, leaving the cranium as clean as possible. Skin the legs to the toe-nails, clean the leg and wing bones, and remove all fragments and particles of flesh and fat from the skin, which will then be ready for preserving.[20]

Most of the bones were discarded, along with anything that would putrefy, such as the eyes, flesh and inner organs – though not always wasted. In November 1892, the bones that had been removed from a MacQueen's bustard were presented to the museum of the Natural History Society at Newcastle by Pearce Coupe, a taxidermist at Marske-by-the-Sea on the Yorkshire coast. This was a large bird from Asia, rarely seen in Britain, and had been shot in a field near Marske. Richard Howse of the Natural History Society, who had negotiated to purchase the stuffed bird for the museum, said in his report that after skinning the bird, Coupe and a few friends had consumed the discarded flesh: 'The body of the bird was cooked and eaten; those who partook of it pronounce that the flesh was savoury but rather tough.'[21]

Not every bird was preserved skilfully, and one mishap happened to a large seabird known as a sooty tern, which is only rarely seen in Britain. According to Dr Henry Graves Bull from Hereford, 'a specimen was picked up dead at Marston, near Pembridge, after stormy weather, in May, 1885. It was unfortunately placed in the hands of a local bird-stuffer who, to make it fit a small case, carefully cropped the longest primary feathers of each wing.'[22] John Harvie-Brown in Scotland mentioned the fate of a great grey shrike: 'I have one specimen in my collection, which was shot in Dunipace parish, about seven years ago [1858]: it was stuffed by a man in the village, evidently with clay or stucco, by its weight, and spoiled accordingly.'[23]

The Reverend Henry Housman – uncle of the celebrated poet A.E. Housman – was unaware of Gardner's manual on taxidermy when, in 1881, he decided to publish his own book to show boys how they could create a museum (girls were rarely considered), just as he and his late brother Frank did as children. Their father had been a solicitor, and in 1840, when Henry was eight years old and his brother eleven, the family was living at Woodchester House in Gloucestershire.[24] The boys were given a room for their museum, and they embarked on hunting for specimens: 'On looking back ... I am sure neither of us has ever regretted the time thus spent. We may not have got through our Latin and Greek as fast as many other boys of our age ... but we were learning to observe, to use our eyes and ears, and to acquire that indefinable love for all living creatures.'25

When Frank shot a goldcrest, they paid a young village lad, Joe Wise, eighteen pence to stuff it. He worked in the local pin factory and was highly skilled at taxidermy, but rather than pay him again, the two brothers decided to stuff their own birds. The first victim was a wood pigeon, and it did not take them long to discover that taxidermy required immense skill:

First, we essayed to skin the creature by making an incision down the middle from stem to stern, beginning at the beak, then down the neck, along the breast and stomach, till we could get no further. So far, so good; but the feathers would come off at every turn, and those that were left soon got hopelessly sodden and caked together with the blood, fat, and contents of the stomach into which the knife had dug. Still, nothing daunted, we went on, pulling the skin away from the flesh until we came to the legs.26

Dealing with the legs was quite a performance, but the tail proved an even greater challenge:

At last, getting desperate, we cut right through the abdomen, as low down as possible, and a precious mess was the result. We then tied a string round the half-stripped carcass, and securing the other end of it to the door handle, we tugged the skin off the back, and so came to the wings. These we treated as we had done with the legs, only the difficulties were greater, owing to the number of joints and turns which had to be surmounted before we arrived at the meat-less region, and in accomplishing this some of the long feathers came out, the others resolutely fixing themselves the wrong way. Skinning the neck was easy, seeing that the whole length of the throat had been cut open; but the head was worse than all the rest put together.27

The wood pigeon had turned into quite a sight:

Melancholy and abject in the extreme was the spectacle which that skin presented! Without, however, stopping to cry over it, we entered upon the second stage of the process. How was the skin to be preserved? Pepper and salt had been the original idea, but on reflecting that insects abhorred turpentine, we inferred that this would prove still more effectual. Whereupon we began to lay this on and rub it in as thick as we could, until not our room only but the whole house reeked with the smell.[28]

Gardner's manual would later provide much information on stuffing and presentation. When a bird was skinned, he said, the skin and feathers turned inside out, and the exposed interior surface of the skin needed coating with a preservative paste to prevent it being infested with insects and rotting. He gave a recipe for a preservative paste manufactured from arsenic, white soap, camphor, salts of tartar, powdered chalk and spirits of wine: 'Thoroughly incorporate the whole of the materials, and keep it in jars or glass bottles, stoppered and marked POISON!'[29]

Afterwards, the skin and feathers were turned right side out, and any ruffled feathers straightened. Lengths of wire wrapped in fibrous hemp, jute or wool were positioned inside the skin and wings to create a wire framework, along with solid pieces of cork or soft wood. Cavities were filled with the same sort of soft materials until the body of the bird was re-created. Artificial eyes, Gardner said, were fixed in place with gum-arabic and 'can be purchased at any respectable bird-stuffer's . . . In choosing the eyes, you must be careful to put the right sort in the right bird.' Over the next few days, the bird needed to be dried in front of a fire, before being displayed in a glass case, under a circular or semi-circular glass dome or, he suggested, 'let him stand on top of your bookcase'.[30]

The Housman boys walked into Stroud to buy a pair of glass eyes, but found it difficult to position them: 'One would seem tolerably firm in its socket, but the instant we began to insert, the other seized the opportunity of tumbling out; but by dint of perseverance and gum, we at last persuaded both to stick in.' Even so, the mounted pigeon looked far from well:

we could not conceal from ourselves it was a pitiable object to behold. What few feathers had survived the operation, were either blood-stained, grease-stained, or turpentine-stained; the head would not stay at the end of its wire, but dropped down to the breast; no art could hide the gash from chin to tail; the skin of the thighs clung tightly to the wires, and no persuasion of ours had any effect in inducing the ruffled plumage to lie down, the quills of the wings setting the example by sticking straight out; the expression of the eyes was ghastly.[31]

Because only a small part of the original bird was actually preserved in taxidermy, errors by bird stuffers could cause serious issues when used as scientific specimens and as models for illustrations. The appearance of stuffed birds depended entirely on their skills and knowledge. It also took much artistic skill to set up a bird and display it in a case, and in London one street seller of birds' nests explained that he also collected materials to be sold for use in glass cases: 'some buys bulrushes for stuffing; they're the fairy rushes the small ones, and the big ones is bulrushes. The small ones is used for "stuffing," that is, for showing off the birds as is stuffed, and make 'em seem as if they was alive in their cases, and among the rushes; I sell them to the bird-stuffers at 1*d*. a dozen.'[32] Gardner advised making artificial branches from wire and covering them with glue, mahogany sawdust and moss:

Moss of the proper kind may be found in hedges and orchards all over the kingdom, or it may be purchased cheaply of the naturalist ... Artificial rock-work, &c., may be easily made by stretching calico over wooden frames and then painting it so as to imitate nature. By strewing sand and patches of moss here and there a very pretty effect is produced in rock-work and foliage ... As for the cases themselves: if you are not a pretty good carpenter you had better get them made by a cabinet-maker. A group of birds under a circular or semicircular glass shade, on a wooden stand, gilt round the edge, and a bit of chenille at the junction of the wood and glass, looks very pretty.[33]

One of the country's finest collections of properly preserved birds was that of Edward Lombe, a barrister from Melton Hall in Great Melton, near Norwich. From the early years of the nineteenth century he amassed his collection, which was the handiwork of Benjamin Leadbeater. Lombe died in 1847, his son died only a few years later, and in 1873 his collection was presented to the Norwich Castle Museum by his daughter, Julia Clarke. The glass display cases had to be moved from Melton Hall, as the *Norwich Mercury* described: 'The British Birds are in 36 uniform cases, and number 551 specimens, representative of 280 species ... There are several other smaller cases containing specimens of foreign birds ... The cases and their contents were removed in the furniture vans of Mr. A. S. Howard, Prince of Wales' Road, Norwich, and we are pleased to say that no glass was broken, nor a bird displaced.'[34]

Mrs Clarke was a remarkable woman, a highly esteemed bene-factor from the nearby town of Wymondham, but collectors tended to be male, as women had few opportunities to pursue such inter-ests. Mrs Caroline De Vitré was an exception. Born in 1825, her first husband died in 1847, and she married Matthew De Vitré a decade later, establishing a fine collection of stuffed birds at their home, Formosa Place, a substantial house on Odney Island in the River Thames at Cookham. Many significant birds found and shot in the locality went into her collection.[35] Now and again, women's names crop up as bird stuffers, so their presence may have been more common than the written evidence suggests.

More than a dozen notable private collections of local birds or ones collected from further afield were listed by William D'Urban and Murray Mathew in their 1892 book of birds from the county of Devon, including Francis Pershouse of Torquay, who had 'about 900 specimens of British Birds collected in Devon, Sussex, and the Orkney Islands, beautifully mounted and cased by himself. There are examples of the White-bellied Brent, Greater Shearwater, and Black Guillemot from Torbay.'[36] The collection of Sir Henry Peek was top of the range: 'A splendid collection of British Birds is arranged in a series of large wall-cases, placed around the corridor at Rousdon. It was formed by Messrs. Swaysland, the well-known taxidermists of Brighton. An elaborate catalogue of the species has been printed.'[37] A wealthy grocer, Peek had been a Member of Parliament and in 1868 purchased the entire parish of Combpyne Rousdon in Devon, where he commissioned estate buildings, a church and a mansion known then simply as 'Rousdon'.[38]

At the Great Exhibition in London in 1851, the Newcastle naturalist John Hancock amazed visitors with his lifelike displays of stuffed birds in glass cases. For two decades, he had been forming his own collection, shooting many of the specimens himself. Others were given to him by friends and correspondents or were acquired through purchase and exchange. He was meticulous in keeping notes for each individual bird, such as for one saker falcon: 'Caught in the rigging of a ship crossing the Mediterranean between Alexandria and Candia, Jan., 1845. Received this bird alive.'[39] Exhausted migrating birds often became entangled in rigging, as happened to a hoopoe near Newcastle: 'Caught upon the rigging of a ship coming into the Tyne about the year 1848.' These exotic-looking birds migrate from Africa to southern Europe in the spring and occasionally end up in Britain. One memorable day years earlier, a companion of Hancock killed another one: 'Shot on the coast between Cullercoats and Whitley on the coronation day of King William, Sept. 4th, 1831, by Thos. Harvey in company with J. Hancock.'[40]

The Victorian era was the peak of taxidermy, and Rowland Ward of Piccadilly in London was perhaps the finest practitioner. In 1880 he published *The Sportsman's Handbook*, which was a guide to preserving specimens for travellers. 'It is only in comparatively recent times that taxidermy has been elevated to claim any real art position,' he said. 'What has been gained for it has not been achieved by mere skill, but by extended and more accurate observation of nature in its living forms.'[41]

Serious ornithologists, collectors, naturalists and taxidermists frequently shot rare specimens and were rarely without a gun. They also obtained birds by any other means, and being well known in the community, local people brought them birds to buy. John Dutton was a chemist and druggist at Eastbourne in Sussex, and he reported being offered a great grey shrike: 'On Saturday last, the 30th of December [1865], Mr. Adams, fishmonger, of this place, sent me for inspection a fine male great gray shrike, in the flesh [not skinned], that had been shot at Pevensey a day or two before. It was a beautiful specimen, and, to show that people know what to ask for rare birds here, the price he put on it was ten shillings. I did not buy it, having a very beautiful one in my collection.'[42]

A range of outlets offered stuffed birds and skins for sale, from the premises of the finest London taxidermists to backstreet bird stuffers. Virtually every town and city had at least one bird stuffer, and in a directory of 1878, the county of Devon was typical in having several 'bird & animal preservers' listed for Honiton, Tiverton, Brixham, Newton Abbot, Plymouth, Kingsbridge, Exeter, Torquay and Teignmouth.[43] When visiting the village of Fortuneswell on the Isle of Portland in 1854, Walter White considered it to have a decayed, unkempt appearance, and yet even here 'you see a post-office, and shops with specimens of stone and fossils and stuffed birds'.[44]

Taxidermists were routinely offered birds for sale that had just been shot, which they could either prepare as skins to sell to customers, ready for stuffing if desired, or they could mount the bird and display it for sale in a glass case. They knew their customers and would contact them if something to their liking became available. They were also the first port of call in any town for those interested, who might first peer in their windows for the latest acquisitions before venturing inside. Green woodpeckers were so routinely killed in Norfolk, according to Henry Stevenson, that numbers 'pass into the hands of our bird preservers to make "show" cases for casual customers'.[45] In August 1863 William Boulton, a physician from Beverley in Yorkshire, was visiting Scarborough and went to Mr Roberts, a bird stuffer: 'I saw a fine female specimen of the Alpine accentor, which had been shot near Scarborough. Last winter (1862–3) a poor man offered for sale to Mr. Roberts a string of larks and small birds he had shot. Mr. R. bought them, and found this bird amongst them. I purchased it, and it is now in my collection.'[46]

Cecil Smith was a magistrate and keen ornithologist living at Bishops Lydeard, near Taunton, on the edge of the Somerset Levels. 'There have been several black terns seen in the marsh this year [1866],' he wrote, 'especially in the neighbourhood of Sedgemoor: I have heard of as many as thirty being seen at one time. I saw two specimens at Mrs. Turle's, the birdstuffer, at Taunton, on the 14th of April, and two more on the 28th, which latter are now in my possession. All ... had been killed on Sedgemoor a few days before I saw them.'[47] The taxidermist was forty-five-year-old Mary Turle, one of several birdstuffers in Taunton. She described herself as a naturalist and had premises in the town's High Street, where she ran the business herself. That same year her daughter, Jane Yandle, emigrated to Auckland in New Zealand with her husband William, where she became well known as a furrier and taxidermist, especially for birds.[48]

The fish and poultry markets were also scoured by taxidermists and collectors alike, where rare birds might be spotted. William Machin, a compositor by trade and a member of the London Natural History Society, was lucky: 'I obtained a remarkably fine specimen of this bird [a Richard's pipit] from among a number of larks in Leadenhall Market, on the 8th of March last [1866]: it is now preserved and in my collection.'[49] D'Urban and Mathew checked out the West Country markets:

> Some thirty years ago, before the present commodious market-house had been erected in Barnstaple [in 1855], the farmers' wives were wont on the market-day to set their panniers on either side of the High Street. It used to be a matter of great interest to us, in those days, especially in the winter-time, after a severe frost, to walk the whole length of the street inspecting the various stalls for the sake of discovering if any rare birds had been brought in amongst the numerous Snipe, Woodcocks, Wild Ducks, Wood-Pigeons, &c., which would be exposed for sale. In hard weather we should be certain to meet with three or four Bitterns; we have known of a dozen brought into the town in a single week ... Bitterns were almost equally numerous at the

time we refer to in the neighbourhood of Weston-super-Mare, in Somerset, where we have examined a great many which were brought into the local bird-stuffer from the adjacent levels.[50]

The pair complained that the barn owl was being senselessly slaughtered, so that 'In every little bird-stuffer's shop the Barn-Owl may be numbered by the half-dozen distorted and caricatured, his face and wings, perhaps, converted into fire-screens.'[51] Lord Lilford had seen skins used in a similar way:

> I am inclined here to enter my protest against a fashion which is only too prevalent of shooting birds for the purpose of making their skins into hand-screens [to protect ladies' faces from fires]. I have seen portions of Kestrels, Owls, and many other birds . . . stuck upon handles with the head, legs, and feet in impossible positions between the wings, and with staring glass eyes, almost always of the wrong colour, doing outrage to ornithological decency and all the best feelings of those who love birds.[52]

In 1903, Robert Service in Dumfries expressed his feelings about the treatment of barn owls: 'Perhaps no other bird figures so often and so prominently as an ornament for the parlour window-sill, or as one of the occupants that gaze with glassy eyes and a never ending stare from the case on the lobby table. Everyone to his, or her, taste. Barbarian habits die out slowly. The stuffing of Barn Owls is a barbarous taste, only less vile than that of using this poor bird's skin as the centre-piece of a drawing-room hand-screen.'[53]

Commercial dealings between taxidermists and collectors were based on trust and ornithological knowledge. Collections tended to be formed from birds obtained in a specific area, such as a country or county. Conscientious collectors only acquired birds that they had personally seen and shot, but some wealthy individuals had an obsession for rarities – birds from abroad that had lost their way. The rivalry between collectors made prices rise, and so fraud was inevitable. In 1962 research finally proved (after being long suspected) that George Bristow, a taxidermist and gunsmith from St Leonards-on-Sea, had carried out a massive hoax on the bird world from 1892 to 1930, apparently by importing skins from abroad that were preserved on ice, using the nearby port of Hastings.[54] He passed them off as being acquired locally and over the decades gave vague details of provenance of these extreme rarities, such as shot by a market gardener or picked up on the beach, and they were also added to the official list of birds found in England. Similar fraud was also likely elsewhere, leaving rare birds in some collections suspect.[55]

Bristow was never prosecuted, but one court case showed how the trade purchased goods from each other. William Sear was obviously familiar with how it worked and saw an opportunity to deceive two London naturalists, George Ashmead and a woman, Ellen Harper. At the Old Bailey in 1896, he was accused of obtaining feathers and skins from them worth several pounds, but he claimed in his defence to be acting for other taxidermists, including James Gardner, who was called as a witness: 'I am a taxidermist, of Oxford Street – the prisoner has come in to make inquiries for several years, but I have had no business with him, and never authorised him to represent me, or act for me.'[56] Sear was found guilty and sentenced to eight months' hard labour.

In his manual, Gardner talked about purchasing skins: 'Groups of British birds form very pretty drawing-room or library ornaments, the variety in their colours, sizes, and forms contrasting well, and providing a good study for the young naturalist. Or a group of foreign birds might consist of humming-birds, golden orioles, and finches, with a bird of paradise or a cock of the rock. The skins of these, properly prepared for stuffing vary in price from 1s. to 25s., the cost of a good bird of paradise.'[57]

Birds of paradise were widely considered to be the most beautiful birds in existence. Found mainly in New Guinea, their skins were originally traded between the native tribes of neighbouring islands. They were also purchased by early European explorers, who assumed them to be legless birds that spent their lifetime floating in the sky, held aloft by their immensely long plumes of feathers. At their death, these birds supposedly fell to the ground and were gathered up and sold. The renowned Swedish scientist Carl Linnaeus aptly – but mistakenly – named the species *Paradisaea apoda*, 'without feet', as related by the editor of an 1827 catalogue of the collection of the Newcastle Museum: 'Linnaeus ... has recorded, by his trivial name *Apoda*, the vulgar error of this bird being produced without feet, which had at first been propagated from the specimens being usually brought to Europe in that state, for the purpose of sale as head-dresses.'[58]

In 1854, a young officer on board a trading vessel from Canton (now Guangzhou) in China was allowed to go ashore with his captain while moored in a small bay on the north coast of New Guinea:

> As we were drawing near a small grove of teak-trees, our eyes were dazzled with a sight more beautiful than any I had yet beheld. It was that of a bird of paradise moving through the bright light of the morning sun. I had seen the stuffed skins of these birds for sale in Amboyna and other places of the China sea, and had gazed upon the combination and arrangement of colours with much wonder and more admiration; but I now saw that the birds must be seen alive in their native forests, in order to fully comprehend the poetic beauty of the words 'birds of paradise'.[59]

He shot one bird, very pleased that he had brought along a musket: 'I have but one poor and selfish excuse for the crime of killing a thing so beautiful as a bird of paradise, and that was the desire of possessing a thing so exceedingly lovely – to hold it in my hands, examine its plumage, and feast the admiring vision on all the astonishing charms of colour.' He described the skill of the native taxidermists: 'The birds prepared for the foreign market are killed in the night by the natives, who climb the trees in which they are perched, and shoot them with small arrows made from the stems of palm leaves. The skin is taken off the birds in a very artistic manner, and dried in smoke. It is then stuffed, and the feathers cleaned.'[60]

These stuffed skins, the young man reported, proved so popular that they were faked: 'the Chinese have made artificial ones of the feathers of parrakeets and other birds. These imitations could not possibly rival the beauty of the real birds; and yet so little was known till lately of these birds that a celebrated ornithologist has been deceived, and given illustrations of them from the artificial skins.' He said that the idea of these birds spending their lives in the sky was totally wrong and that they actually had strong legs and large feet: 'So ugly are the feet and legs, that the natives of New Guinea, in preparing the stuffed skins of these birds for a foreign market, always cut them off.'[61]

In an era before any kind of colour photography, the brightest colours were flowers or paints used by artists, and so the brightly coloured plumage of many stuffed birds, especially exotic ones imported from afar, caused astonishment. In 1858 John Keast Lord was appointed as naturalist to the British North American boundary commission, and in the Rocky Mountains he shot several hummingbirds. Back in England a few years later, he was well aware that he might be criticised:

I dare say hard epithets will be heaped upon me, – cruel man, hard-hearted savage, miserable destroyer, and such like, – when I confess to sitting and shooting numbers of these burnished beauties. Some of them are at this moment before me as I write; but what miserable things are these stuffed remains, as compared to the living bird! The brilliant crests are rigid and immovable, the throat feathers, that open and shut with a flash like coloured light, lose in the stillness of death all those charms so beautiful in life ... It is useless pleading excuses; two long days were occupied in shooting and skinning.[62]

His remorse was unusual at this date, because books and journals of the eighteenth and nineteenth centuries usually record the slaughter of rare birds without any cares, but in 1860 Cornwall Simeon also expressed his disapproval of the pointless massacre of birds: 'There appears to exist too often an insane desire to kill rare birds, for no other reason than because they are rare, not with a view to add to the stock of knowledge already possessed with regard to the birds, but from a morbid wish to gratify the vanity of the person who kills them.' He devised an imaginary local newspaper piece entitled 'Rare Bird':

> On Saturday last that enthusiastic and accomplished ornithologist, Mr Snooks, was so fortunate as to obtain two specimens, male and female, of that rare bird the *Peregrinus fidens* ['trusting Peregrine']. They had been for some time observed in the neighbourhood, and many of our naturalists had been eagerly on the watch to secure them. We heartily congratulate our esteemed fellow-townsman on the attainment of this trophy, which will serve to add new lustre to his already celebrated name. From the fact that the female bird had a feather in her bill, when she was shot, there can be no doubt that these interesting visitors were in the act of constructing a nest when they fell before Mr Snook's unerring tube.[63]

Over a decade later, Simeon went on a walking tour of the Cornish Lizard coastline, accompanied by his friend Charles Johns, a Victorian schoolmaster and author of popular books on natural history. Johns subsequently updated his own guidebook with a list of birds known from the area, ironically including real examples of such slaughter: 'GREAT BUSTARD (*Otis tarda*). – The only specimen procured in this district was a female, which was shot, in February 1843, in a turnip-field belonging to Joseph Lyle, Esq., of Bonythin, where it had been observed for several days.'[64]

Some of the targeted rare birds were struggling to survive, on the point of extinction, like the great bustard. Others were rare visitors to Britain, having flown off-course, but it was not enough for everyone to observe them – they had to be shot and stuffed. Henry Stevenson in Norwich was saddened by the treatment meted out to the white stork, which was a rare visitor that occasionally ended up on England's east coast, rather than in Holland: 'But here, unhappily, instead of wooden boxes being erected in our towns and villages for their nesting accommodation, the only box provided is the birdstuffer's case, wherein the victim of misplaced confidence inevitably finds its last home.'[65]

In early May 1866, George Harding from Stapleton, now a suburb of Bristol, was fascinated by the behaviour of several exotic bee-eaters, as he informed *The Zoologist*: 'I have to record the occurrence of this beautiful and rare visitor ... The birds, when first observed, were hawking for bees round a number of fruit trees in blossom, and in the neighbourhood of a number of bee-hives. Their flight is most graceful and beautiful; at one time soaring in graceful circles at a great height, and darting with great velocity after their prey.'[66] Over eighteen months later, his local newspaper featured a report of a meeting of the Bristol Naturalists' Society, during which he gave a talk: 'On the 5th of May, 1866, the author observed a small flock of these birds, whose movements he had the opportunity of watching for six or seven hours; four of them were shot, three of which were now in his possession. Mr Harding exhibited stuffed specimens of the birds obtained by him on that occasion.'[67]

Henry Graham in Iona told Robert Gray that he would rather own engravings than stuffed birds: 'I envy you very much the fine engravings that you are procuring. I prefer a really good portrait of a bird to the finest mounted specimen, and a collection of the first has such great advantages of portability and durability that I would sooner possess it than a collection of real specimens. However, this is a matter of taste, and depends very much whether one remains in a permanent home or not.'[68]

Over three decades later, William Warde Fowler in Oxford emphasised the desirability of studying birds in their natural habitat, rather than 'the best plates in a book, or the faded and lifeless figures in a museum. You may shoot and dissect them, and study them as you would study and label a set of fossils: but a bird is a living thing, and you will never really know him, till you understand *how* he lives.'[69] James Cash in rural Cheshire made a similar point: 'To a great many people the kingfisher is little more than a name. Stuffed specimens are seen in museums, or in glass cases in private houses where the colours fade and the birds are often stiff and unnatural. The kingfisher must be seen alive and alert in its own natural haunts, where its brilliant plumage can be enjoyed to perfection.'[70]

Concerns increased in late Victorian times about the persecution of birds. The excuse that they were vermin was no longer supported, and legislation was passed to protect birds, most notably the Act for the Preservation of Sea Birds in 1869 and the Wild Birds Protection Act of 1880. The unintended consequence was that it was now much more difficult to obtain birds legally for stuffing. Existing collections had no protection, and gradually numerous private collections disappeared without trace, as tastes changed and owners died. Some collections were sold off, often ending up in the hands of anonymous collectors and dealers, while others were obtained by museums.

In 1854, Dr Francis Plomley, a physician at Maidstone in Kent, donated his own natural history collection, notably stuffed birds, to the museum at Dover. It was celebrated by a dinner in his honour, during which he was presented with a vote of thanks 'written on vellum, and superbly emblazoned with the Borough Arms, &c., and having a fac-simile of the Corporation seal in gutta percha, tastefully suspended . . . executed by the Messrs. G. and C. Smith, of the Heralds' College, Doctors' Commons'. After enthusiastic applause, the newspapers reported Dr Plomley's response:

> The occasion was one, not only of mutual thanks, but in which Dover was united with him in commemorating the formation of the first Local Natural History Museum in England . . . The worthy Doctor proceeded to remark on the cause of Museums failing to accomplish the great objects of their formation, and that collections illustrating localities would inspire interest otherwise dormant, lead to the study of Natural History, elevate the mind, and result in the development of industry, pleasure, and wealth, to a degree at present inconceivable . . . he hoped they would all exert themselves in promoting the important purpose for which museums were established . . . He thanked

them honestly and sincerely for giving him the opportunity of placing his collection in a position where, if no pecuniary advantage resulted, it would prove an attraction to the people, especially to the poor.[71]

A final vote of thanks was to the generosity of the South-Eastern Railway Company for transporting the collection from Maidstone to Dover free of charge.

Countless collections were discarded after succumbing to decay and infestation, and D'Urban and Mathew recorded one example: 'In the autumn of 1859, a great multitude of Terns of various species frequented the Barnstaple river, and among those which came into our hands were two in immature plumage ... Unfortunately they were but indifferently preserved, and, becoming infested by moth, had to be destroyed.'[72] A few years later Cecil Smith spent six weeks in the Channel Islands: 'There is a tolerably good Museum in St. Peter's Port, Guernsey, but it is now unfortunately much neglected, many of the best specimens being moth-eaten to such an extent as to be scarcely recognizable.'[73] Smith's own collection was dispersed at a sale in London in 1890, with some specimens being purchased by Mathew.[74]

When Edward Hearle Rodd died at Penzance at the age of sixty-nine in 1860, his collection of birds was bequeathed to his nephew Francis Rodd, who was also a knowledgeable ornithologist and owned Trebartha Hall, near Launceston. There were at least forty-five cases containing 270 specimens, all mounted by William Vingoe, the best taxidermist in Cornwall. In 1949, the local newspaper reported on their fate: 'For 60 years a fine collection of Cornish birds, considered the best in the West of England, was on view in the Western or Museum Room and galleries of Trebartha Hall. Many specimens were extremely rare and all were beautifully mounted ... In 1940 the collection was presented to the Royal Institution of Cornwall and is now exhibited at the County Museum, Truro.'[75] It was later suggested that with wartime conditions, the collection was never accessioned, leading to its dispersal.[76] The tragedy continued with Trebartha Hall, which was said to be the most significant country house in Cornwall, hundreds of years old, with an estate that for generations was a bird-lover's paradise. The house was occupied by the military in World War Two and deliberately demolished in 1949, because the repair costs were too great.

After the Reverend Charles Penrice of Plumstead Hall in Norfolk died in 1853, Henry Stevenson purchased much of his collection of birds, though 'some of the rarer local specimens had been given away'. That was the least of the problems:

> Amongst those purchased by myself were specimens of the osprey, goshawk (adult), hen harrier (adult male), marsh harrier, buzzards ... golden oriole (female), roller, eared grebe (summer plumage), long-tailed ducks (immature), smew (adult male), white-eyed pochard, &c., &c., with most of the common British birds; but beyond half-dozen of the best and rarest, I found it useless to attempt preserving them, having been badly stuffed in the first instance, and sadly injured by neglect.[77]

Other collections suffered from a variety of other damage, even bombing, as happened on 21 October 1940, when the museum in the Market Square of Dover in Kent was struck: 'The bomb passed through one of the skylights in the Long Room ... and exploded on the paved floor of the Covered Market below. The explosion ... completely wrecked a 40ft. long section of the range of mahogany wall casework for ornithological specimens ... birds and other specimens were hurled about, adding considerably to the debris in the Market below.'[78] Shelling of the town in 1942 and 1943 did further damage, tragically destroying much of the celebrated Plomley collection of birds.

Even when museums managed to keep open, their displays were not always admired. In October 1942 May Smith, a schoolteacher, recorded in her diary that she went with a friend for the day into wartime Nottingham. Too early for the cinema, they looked at an art exhibition at the nearby museum, then wandered through the natural history galleries, with 'rooms full of stuffed birds in all attitudes, colours and shapes – all looking very stuffed and very depressing, and rather moth-eaten, some of them very supercilious. The last room of these creatures proves too much for us. We collapse on the hard bench and giggle at the straits to which we have been reduced.'[79]

After the war, cases of stuffed birds fell out of favour and therefore out of fashion, and collections were dismantled, broken up and sold off, even by museums. Instead of admiring these stuffed birds as pieces of history, remnants of the past, they were treated with disapproval. Yet today, that attitude is changing again, and such museum displays form the only way many of us will ever get to see such wonderful birds, and they are now conserved as historic artefacts, as well as valuable natural history specimens that still have much to offer.

One curious but significant example of taxidermy was carried out by Walter Potter at Bramber in Sussex. Born in 1835, nature was his passion, and over several decades he created displays of stuffed animals, including birds, to illustrate stories. The most famous was based on the nursery rhyme 'Who killed Cock Robin?' These whimsical displays proved to be a huge tourist attraction long after his death in 1918, but eventually the collection changed hands and finally ended up in Cornwall. As Victorian art, these displays were of national importance, but in 2003 the collection was allowed to be broken up and sold, an act of cultural vandalism.[80]

CHAPTER SIX

·◆·

CAGED

The bullfinch ... is generally content to live in a
small cage, and does not care so much to fly about.

The Young Woman's Book 1877[1]

Keeping birds as pets indoors in cages or tethered to perches, as
well as outside in aviaries where they were free to fly, was a pastime
enjoyed by all levels of society. Even ornithologists were able to
breed birds and study them as living creatures close-up and in all
weathers, which was a very different experience to studying shot
specimens. When the Reverend Edmund Dixon in Norfolk was
compiling a book on dovecotes and aviaries, published in 1851, he
was given information by William Rayner, a surgeon and naturalist
from Uxbridge in Middlesex, who told him that he had ninety-four
species of birds in captivity in an aviary made of iron wire that was
33 feet long, 10 feet wide and 17 feet high. Inside was a stone water
fountain, and many trees were planted, and amongst his birds were
robins, wrens, cuckoos, green woodpeckers, skylarks, magpies and
kingfishers.[2]

Rayner's aviary was of similar size to one constructed three
centuries earlier at Kenilworth Castle in Warwickshire as part of
the lavish entertainment and hospitality that Robert Dudley, Earl
of Leicester, had to provide in 1575 for a visit by Queen Elizabeth
I. That aviary was described as 'a square cage, sumptuous and
beautiful', measuring 30 feet long, 14 feet wide and 20 feet high,

filled 'with lively birds, *English*, *French*, *Spanish*, *Canarian*, and ... *African*'.[3]

Collecting exotic animals including birds became a pastime of royal courts from the twelfth century, evolving into Royal Menageries open to the public and scientific community. Entrepreneurs also had collections that they took round the country as travelling menageries, and zoological gardens were opened for the paying public. In London, James I had a royal menagerie and aviary, which was expanded by Charles II when the extensive St James's Park was remodelled, with large cages ranged along the carriage road that is still known as Birdcage Walk.[4]

Wild birds were ideal as pets for ordinary people as well, since they took up little space and needed minimal care. The main expense was a supply of food and a cage, and in return birds might learn to recognise their owners and even show them affection, which were precious qualities for the majority of people who were leading difficult lives. They also provided entertainment with their antics, song and mimicry, a few could be taught tricks, while some successfully bred in captivity.

The most common birds for indoor cages were bullfinches, goldfinches, skylarks, linnets, thrushes, nightingales and canaries. Songbirds were preferred, but anything curious or unusual might be tried, such as cuckoos. One morning in July 1802 at Over Stowey in Somerset, William Holland went to his church where his manservant Morris had placed a young cuckoo in a cage that had hatched in a wagtail's nest in the tower: 'I have had the cuckoo now brought down into the garden where the cage hangs under the bough of a tree, & two water wagtails are continually feeding him.' That evening Morris took the cage indoors, but the bird did not survive: 'Alas! the poor cuckoo, with all its feeding, is dead; perhaps overfed. Ann [a maidservant] says that one Gookoo kill'd the other, meaning Morris.'[5]

Edwin Waugh was known especially as a dialect poet of Lancashire life, and in 1862 he visited a poor cottage rented by an elderly couple in Preston. On a table, he recalled, 'there was a rude cage, containing three young throstles [thrushes]', and after he had chatted to the couple for a while, the young birds began to chirp. '"Yer yo at that! [You look at that!]" said the old man, turning round to the cage; "yer yo at that! Nobbut three week owd!" "Yes," replied the old woman; "they belong to my grandson theer. He brought 'em in one day—neest an' all; an' poor nake't crayters they were. He's a great lad for birds." "He's no warse nor me for that," answered the old man; "aw use't to be terrible fond o' birds when aw wur yung." '[6]

Those who kept birds invariably considered themselves to be bird lovers, but over two decades earlier, William MacGillivray gave his reasons for excluding the sensitive topic of avian pets when writing about birds:

> It is not my purpose to treat of caged birds, so that we shall escape a great deal of unnecessary sentimentalism, and save the time that might otherwise be lost in learning how dear a pet was Matilda's Goldfinch, how sweetly it sung, how neatly it preened its plumage, how delightedly it nibbled the nice bit of white sugar presented to it, and how excruciating were its agonies, as well as those of its gentle mistress, when its dear little bones were crunched by the serried teeth of cruel grimalkin [cat], that sworn foe of all birdlets.[7]

When investigating the lives of London's labouring classes in 1849–50 for the *Morning Chronicle*, the journalist and social reformer Henry Mayhew found that most songbirds were purchased by the working people and tradesmen: 'Grooms and coachmen are frequently fond of birds; many are kept in the several mews, and often the larger singing-birds, such as blackbirds and thrushes.' He thought that anyone who delighted in birds and flowers had a certain intelligence: 'I have seen and heard birds in the rooms of tailors, shoemakers, coopers, cabinet-makers, hatters, dressmakers, curriers, and street-sellers,—all people of the best class ... The bird-lover ... is generally a more domestic, and, perhaps consequently, a more prosperous and contented man.'[8]

Mayhew reckoned that London had about two hundred

birdcatchers, who were 'all labouring to give to city-pent men of humble means one of the peculiar pleasures of the country – the song of the birds'. They took all manner of live birds from the nearby commons and open fields, using call birds that were known as mules if they were bred from two different species of bird. Large clap-nets were laid out on the ground and secured by iron pins, leaving two flaps: 'In the middle of the net is a cage with ... the "call-bird." This bird is trained to sing loudly and cheerily ... and its song attracts the wild birds. Sometimes a few stuffed birds are spread about the cage as if a flock were already assembling there. The bird-catcher lies flat and motionless on the ground, 20 or 30 yards distant from the edge of the net.'[9] At night, birdcatchers mimicked the birds instead or else used a whistle.

After several birds had been lured to the net, the birdcatcher would pull on a line, causing the flaps to envelope the victims. One dealer told Mayhew that the greatest number of birds captured by one 'pull' was almost two hundred, and at the height of the season the numbers varied from fifty to one hundred and fifty. The live birds were normally carried to London in wickerwork hampers, though some birdcatchers hired a donkey cart or pony if the catch was substantial. Mayhew also learned that birdcatchers worked all their lives at this trade, and 'last winter two men died in the parish of Clerkenwell, both turned seventy, and both bird-catchers – a profession they had followed from the age of six'.[10]

By the end of the nineteenth century, in Headington Quarry village near Oxford, many a poor villager supplemented their earnings from September to March by catching birds, usually at night. The Buckland family roamed the rough woodland of nearby Shotover:

> Most that I can remember is people going down there catching birds, like bullfinches. They used to go down there and set these traps. Song birds. Sell them – linnets – quite a few of 'em used to do it. Flog 'em down the town or around – everybody kep' a bird in a cage ... My two brothers ... was devils at it ... Another thing they used to do was to go with nets at night, and wait till it was

dark, and then go up and bash the tree ... and the birds would come out and go into the net ... There was a greater variety of song-birds in them days than what there is now.[11]

The most prized songbird was the nightingale, which performs not only in the daytime but all through the night, when most other birds are silent. In the sixteenth century Pierre Belon, a French traveller and naturalist, declared that nightingales, being so melodious, could be trapped by discordant sounds, which they hated: 'For if someone ties up a cat at the foot of a tree and attaches a line to it, with the end far enough away, in some place where there are nightingales, and pulling the line, makes the cat screech (or some other animal whose cry is unpleasant and piercing), then the nightingales, showing signs of being alarmed, will fly around the cat, as if indignant.'[12] At that moment, it was possible to take the nightingales with birdlime.

After spending the winter in Africa, nightingales fly northwards to Europe and Asia to breed. For several years from 1740, Alexander Russell was a physician to the English factory (trading centre) in the city of Aleppo in Syria, where, he said, the abundant nightingales were greatly admired and 'not only afford much pleasure by their song in the garden, but are also kept tame in the houses, and let out at a small rate to divert such as chuse it in the city; so that no entertainments are made in the spring without a concert of these birds'.[13] A century later, in 1847, *Chambers's Edinburgh Journal* reported: 'In Moscow, the bird-fanciers keep large numbers of nightingales for sale; the average price is fifteen roubles. They are so abundant in Warsaw, that the streets are filled with their music.'[14]

At that time the nightingale was equally common in Britain, but is now a scarce summer visitor. Few people today have ever heard its glorious and powerful song, and except for the occasional owl, hardly any other birds are heard at night. And yet throughout history, the nocturnal song of the nightingale formed part of the fabric of British life and was taken for granted for a few weeks each year. Their decline began in the nineteenth century, when

huge numbers were captured as cage-birds. Although that trade
has largely ceased, environmental factors have exacerbated this
shocking situation.

It was difficult to resist nightingales, whose attraction was
described in 1847 by *Chambers's Edinburgh Journal*:

> The nightingale's voice may be heard over a circle of a mile
> in diameter, nearly the distance at which the human voice is
> audible. Possessed of such exquisite powers, they are much
> prized as cage-birds, but are not easily domesticated, owing to
> their delicate and sensitive nature ... But when accustomed
> to their captivity, they sing all the year through, except in the
> moulting season, and their music is then said, by a strange
> contradiction, to surpass that of their wild state.[15]

The Reverend Murray Mathew was one of many ornithologists
who kept birds in cages and aviaries, and on one occasion he took
several nightingales from Surrey to north Devon by steam train:

> We carried them down by the night mail in the guard's van, and
> in their cages the birds sang throughout the journey. Several of
> them lived for three years at Barnstaple, and treated us to rich
> concerts. They would begin to sing just at the early dawn of a
> summer's morning; first one would warble a few notes, and the
> others, in emulation, would tune up as well, until we had the full
> choir. During the middle of the day they were generally mute,
> nor did they sing much of evenings, with the exception of the
> winter time. They were then brought into the dining-room for
> the sake of the warmth of the fire, and when the lamps were
> lighted they often commenced to sing, and would provide us
> with a concert during dinner.[16]

Feeding caged birds like nightingales was onerous, though made
easier by unhygienic Victorian kitchens. Mathew said that 'their
favourite *bonne bouche* was a fat cockroach, and great was the excitement in all the cages of a morning when a pie-dish full of writhing

monsters from the kitchen regions was brought into the room, and the birds were fed in turn by means of a quill-pen'.[17]

In a late Victorian manual that aimed to give young women information to enable them to add some pleasure and usefulness to their lives, Miss Mary Anne Dyson, a Christian educator of girls, provided chapters on caring for pets. 'The nightingale's cage,' she advised,

> should be from twelve to eighteen inches long, ten or twelve broad, and twelve high, with wooden or wire bars on the front and sides, the top made of green baize or cloth, and green gauze curtains to hang round the sides, to shut out the strong light and let in fresh air. There should be three perches, two near the bottom of the cage, and one higher up; and these must be covered with green baize or cloth, because this bird's feet are so tender. He should have a bath every day, but this must not be left in the cage, and the floor must be covered with dry sand.[18]

In the 1840s, the naturalist Edward Jesse, who was deputy surveyor of royal parks and palaces, said that he did not object to the trade in cage-birds, but he did take exception to nightingales being caught and had a poor opinion of the nightingale catcher, who 'is generally a stealthy, downcast vagabond, most justly detested by all owners of groves, plantations, and hedge-rows, possessing any good taste, within twenty miles of the metropolis. I knew one of these men, who passed much of his time in the pretty lanes of Buckinghamshire ... He was a hard-featured, uneducated man, looking very like a veteran poacher.' He added: 'The nightingale catcher's season is very short, but he makes the most of it; and it is greatly to be regretted, that in the exercise of his craft, he deprives so many persons of those exquisite cadences which are justly appreciated by all lovers of harmony and nature ... if there were no purchasers, there would be no nightingale-catchers.'[19]

James Edmund Harting knew a gamekeeper who was extremely skilled at catching nightingales: 'In one season alone he caught fifteen dozen, receiving eighteen shillings [about £50 today] a dozen for them in London. He told me also that, on one occasion, he caught no less than nineteen nightingales before breakfast in the grounds of one gentleman, and in sight of the windows; for which, as I told him, he ought to have been transported.'[20] Jesse's vagabond nightingale catcher also took other songbirds, 'such as thrushes, blackbirds, woodlarks, and blackcaps; and it was extraordinary in how short a time he tamed them and brought them to resume their song'.[21] Usually it was only male songbirds that were selected, as females were thought not to sing much, something that recent research has questioned.

The fervently held belief was that birds were a resource provided by God, which the English naturalist Eleazar Albin stated in his 1737 book on songbirds: 'SINGING Birds are so pleasant a Part of the Creation ... They were undoubtedly designed by the Great Author of Nature, on Purpose to entertain and delight Mankind, who, for the Generality, are well pleased with these pretty innocent Creatures.'[22] Being entertained by birds kept in cages was therefore considered perfectly acceptable. He gave advice on how to look after and train songbirds, but disapproved of the treatment of chaffinches: "'Tis a custom among the Bird-men, when they want to learn the *Chaffinch* a song, to blind him when he is about three or four months old; which is done by closing up his eyes with a wier [wire] made almost red-hot, because, as they say, he will be more attentive, and learn the better; but I am sure it would be much better never to confine them in cages, than purchase their harmony by such usage.'[23]

Miss Dyson in *The Young Woman's Book* discussed various birds – the goldfinch 'has a merry little song, and is very pretty and lively', while 'The blackbird does not sing so well as the thrush, but he will learn to whistle tunes, and to imitate the songs of other birds, and to say some words.' Jackdaws and magpies, she said, 'will learn to repeat words, and even sentences, and so will the starling, which is a very clever, lively bird, and may be taught to whistle the tunes also'.[24]

In Shakespeare's play *Henry IV, Part I*, which was first performed around 1596–7, Hotspur declares that he will drive Henry IV mad by having a starling constantly utter the word 'Mortimer' – referring to Edmund Mortimer, another claimant to the throne, who has been captured by Owen Glendower:

> *He said, he would not ransom Mortimer;*
> *Forbade my tongue to speak of Mortimer;*
> *But I will find him when he lies asleep,*
> *And in his ear I'll holloa, 'Mortimer!'*
> *Nay, I'll have a starling shall be taught to speak*
> *Nothing but 'Mortimer,' and give it him,*
> *To keep his anger still in motion.*[25]

Thomas Durham Weir of Boghead House, about 20 miles south-east of Glasgow, was a very keen curler and ornithologist, and he corresponded regularly with MacGillivray, who valued the information he provided. In the 1830s, Weir told him a story about Hugh Paton, a carver and gilder of picture frames in Horse Wynd, Edinburgh, who had a pet starling, 'which I have heard pronounce most distinctly the following sentences. When I entered the shop, he said to me, "Come in, Sir, and take a seat – I see by your face that you are fond of the lasses – George, send for a coach and six for pretty Charlie – Be clever, George, I want it immediately;" and many other sentences to that purport. He was taught by a shoe-maker in Stewarton, Ayrshire.'[26]

According to Albin, 'The *Starling* ... does not sing naturally, but has a wild, screaming, uncouth note; yet for his aptness in imitating man's voice, and speaking articulately, and his learning to whistle divers tunes, is highly valued as a very pleasant bird; and when well taught, will sell for a great deal of money, five guineas [over £600 today] or more.'[27] His advice was to take the young birds from their nest when about ten days old and constantly talk while feeding them. The custom was to slit their tongues with a silver sixpence in order to encourage them to speak rather than sing, but Albin disagreed: 'To slit their tongues, as many people advise and practice, that the birds, as they say, may speak the plainer, is of no service, they will talk as well without, as I have found by experience; as will likewise *Magpies*, and other talking birds.'[28]

The starling was referred to as 'the poor man's and the peasant's parrot,'[29] and the custom of slitting tongues was long-lasting. William Howitt encountered the practice when staying with his grandfather near Heanor in Derbyshire in the early nineteenth century: 'there was a starling that had his tongue slit with an old sixpence, and had been taught to talk; and he used to sit in a sunny gutter over the kitchen door, watching the maid-servants going in and out about their work, and would say – "Molly Gibson, why don't you milk the cows?" or "Molly Gibson, you've left the gate open." '[30]

In 1846, the Scottish poet Thomas Aird wrote about magpies and tongue slitting in 'A Summer Day'.

> With all the short thick rowing of her wings
> The Magpie makes slow way. But her glib tongue
> Goes chattering fast enough. In yonder fir,
> The summer solstice cannot keep her mute.
> Surely the bird should speak: Take the young pie,
> And with a silver sixpence split its tongue,
> 'Twill speak incontinent; thus the notion runs
> From simple father down to simple son,
> In many parts. Oft in our boyhood's days
> We've seen it tried; but somehow, by bad luck,
> It always happened that the poor bird died,
> When, doubtless, just upon the eve of speech.
> Sore was the splitting then, but far worse now;
> The sixpence then, worn till it lost the head
> Of George the Third, was thin as a knife's edge,
> And fitly sharp; the coin's now thick and dull,
> And makes the clumsier cleaving full of pain.[31]

George III issued relatively few sixpences, so Aird probably meant the thin silver sixpences of George II, worn away by being in circulation for so long. The newer ones of Queen Victoria subsequently made this practice more difficult to carry out.

Towards the end of the nineteenth century, William Greene, a

Fellow of the Zoological Society of London, warned against such cruelty when discussing military starlings (long-tailed meadow-larks) that were imported from South America:

> some people affirm that their tongues must be split, or they will neither talk nor sing ... It is a monstrous assertion, handed down from the barbarous old times ... Like other errors, it dies hard; and I am grieved to find it from time to time crop up in the most unexpected places, where one would have naturally looked for more enlightenment. No; no bird – starling or what not – should have its tongue slit or tampered with in any way; and if anyone makes the suggestion, advise the ignoramus to try the experiment on himself first.[32]

Some people used mechanical music organs to train birds, which Mary Russell Mitford included in her authentic but fictionalised 'Village Sketches' that were published in the *Monthly Magazine* and later in book form. In the 1820s, she said, 'Old Robin' the birdcatcher bred birds in an attic room and trained them to sing:

> The din is really astounding. To say nothing of the twitter of whole legions of linnets, goldfinches, and canaries ... the chatter-ing and piping of magpies, parrots, jackdaws, and bullfinches, in every stage of their education; the deeper tones of blackbirds, thrushes, larks, and nightingales, never fail to swell the chorus, aided by the cooing of doves, the screechings of owls ... and the eternal grinding of a barrel-organ, which a little damsel of eight years old, who officiates under Robin as feeder and cleaner, turns round, with melancholy monotony, to the loyal and patriotic tunes of *Rule Britannia* and *God Save the King* – the only airs, as her master observes, which are sure not to go out of fashion.[33]

Mayhew discovered that bullfinches were taught to pipe when very young by depriving them of light and food: 'The bird is wakeful and attentive from the want of his food, and the tune he is to learn is played several times on an instrument made for the purpose, and known as a bird-organ, its notes resembling those of the bullfinch. For an hour or two the young pupils mope silently, but they gradually begin to imitate the notes of the music played to them.'[34] The training lasted several weeks, with light and food given as a reward, and each bird was then sold for the equivalent of £250 today.

A bird-organ or serinette (from the French word *serin*, meaning a canary) was a portable barrel organ that was used from the early eighteenth century in eastern France for teaching tunes to canaries. They also became fashionable musical instruments for the amusement of refined ladies, and Jesse suggested that young ladies who spent much time caressing their favourite birds might be better employed teaching their canary an air that would after a few months prove a delight, a reflection of many pointless female pursuits in Victorian times: 'This is done by means of a very small organ pitched very high, termed by the French a "serinette," many of which they send to London, and they are procurable there at a trifling price. The bird should begin his lesson when quite young, as soon as or before he can feed himself. It is better ... that it should be brought up by hand, taken at a fortnight old, and played to from the beginning.'[35]

Much time was spent as well in teaching birds tricks, and Albin said that goldfinches could be taught to draw up water in a little ivory bucket fastened to a small chain: "Tis a pretty sight to see with what dexterity these little creatures will pull up their bucket, drink, and throw it down again; and lift the lid of a small box, or bin, with their bill, to come at their meat.'[36] Patrick Syme, an artist of natural history in Edinburgh, gave further details in a book on songbirds that he published in 1823: 'They may ... be taught to draw up little buckets or cups with food and water. To teach them this, there must be put round them a narrow soft leather belt, in which there must be four holes, – two for the wings, and two for the feet. The belt is joined a little below the breast, where there is a ring to which the chain is attached that supports the little bucket or cup: We have seen both the goldfinch and lesser redpole perform this action.'[37] He had observed bullfinches do tricks as well, including one that a lady had purchased from a French prisoner-of-war: 'The poor exile had painted the cage of his little captive like a prison, and the bird drew up two little buckets suspended by a gilt-chain, one containing seed, and the other water. This bullfinch was extremely tame, and, though bred in the woods of Greenlaw, (near Edinburgh,) it whistled a variety of troubadour songs.'[38]

Birds were owned by all classes of society, depending on what they could afford, and ones like sparrows were the cheapest, as Mayhew described: '*Sparrows* ... are netted in quantities in every open place near London, and in many places in London. It is common enough for a bird-catcher to obtain leave to catch sparrows in a wood-yard, a brick-field, or places where is an open space certain to be frequented by these bold and familiar birds.' They were not even intended for cages, which reduced their cost:

> The sparrows are sold in the streets generally at 1*d*. each, sometimes halfpenny, and sometimes 1½*d*. .... merely as playthings for children; in other words, for creatures wilfully or ignorantly to be tortured. Strings are tied to their legs and so they have a certain degree of freedom, but when they offer to fly away they are checked, and kept fluttering in the air as a child will flutter a kite. One man told me that he had sometimes sold as many as 200 sparrows in the back streets about Smithfield on a fine Sunday. These birds are not kept in cages, and they can only be bought for a plaything.[39]

Most birds, especially the more expensive songbirds, were purchased in the shops of specialist dealers, or else in outlets such as markets, curiosity shops, barbers and pubs. In towns and cities, birdcatchers also operated as street-sellers or street-hawkers from a single pitch, while some were itinerant street sellers who, Mayhew said, could be seen 'carrying the birds in cages, holding them up to tempt the notice of people whom they see at the windows, or calling at the houses ... Sometimes the cages with their inmates are fastened to any contiguous rail; sometimes they are placed on a bench or stall; and occasionally in cages on the ground.'[40]

Young linnets were the cheapest songbirds in London, 'sold in the streets at 3*d*. and 4*d*. each; the older birds, which are accustomed to sing in their cages, from 1*s*. to 2*s*. 6*d*.' Goldfinches were also popular: 'if any one casts his eye over the stock of hopping, chirping little creatures in the window of a bird-shop, or in the close array of small cages hung outside, or at the stock of a street seller, he will be struck by the preponderating number of goldfinches ... The demand for the goldfinch, especially among women, is steady and regular.'[41]

Old Billy, who was disabled, told Mayhew that he had been selling birds in the streets of the east end of London for twenty-six years, buying them from birdcatchers: 'It's all small birds I sell in the street now ... I sell more goldfinches than anything else ... One of my favouritest birds is redpoles, but they're only sold in the season. I think it's one of the most knowingest little birds that is; more knowing than the goldfinch, in my opinion. My customers are all working people, all of them. I sell to nobody else.'[42] Very often, newly purchased birds were taken home in a temporary wrapper, paper bag or other container, though a songbird might be sold with a small cage.

The favourite songbird was the linnet, but Mayhew discovered that their mortality rate was high: 'About one half of those birds die after having been caged a few days. The other evening a bird-catcher supplied 26 fine linnets to a shopkeeper in Pentonville, and the next morning ten were dead ... The "catch" of linnets – none being imported – may be estimated, for London alone, at 70,000 yearly.'[43] The Victorian poet Christina Rossetti compared a caged linnet to a free one in a short poem for children that she published in 1872:

> A *linnet in a gilded cage*, –
> A *linnet on a bough*, –
> *In frosty winter one might doubt*
> *Which bird is luckier now.*

> *But let the trees burst out in leaf,*
> *And nests be on the bough,*
> *Which linnet is the luckier bird,*
> *Oh who could doubt it now?*[44]

Although Rossetti hinted at the hardships of free linnets in winter, her summer scene shows that freedom was infinitely preferable to comfortable captivity. Verses about caged birds provided a message about lack of freedom for various sectors of society, with Rossetti's poem alluding to girls in Victorian society.

Paul Laurence Dunbar was one of the first nationally recognised African-American poets, and in his poem 'Sympathy' of 1899 he gave a powerful message about slavery and oppression:

> *I know why the caged bird sings, ah me,*
> *When his wing is bruised and his bosom sore, –*
> *When he beats his bars and he would be free;*
> *It is not a carol of joy or glee,*
> *But a prayer that he sends from his heart's deep core,*
> *But a plea, that upward to Heaven he flings –*
> *I know why the caged bird sings!*[45]

This popular sentiment was repeated in the music-hall ballad 'A bird in a gilded cage' that was published in 1900 as sheet music and sold hugely before World War One.

One of the most iconic cage-birds is the canary. When Spanish seafarers visited the Canary Islands in the North Atlantic as early as the fifteenth century, they took away small exotic birds called canaries, and Europeans became captivated by their song. Italian, Dutch and German breeders proved particularly successful in rearing, training and selling them, and canaries were brought to London by pedlars from the Swiss Alps, though the trade was disrupted by the Napoleonic Wars. According to Mayhew, 'they brought over about 2000 birds yearly. They travelled the whole way on foot, carrying the birds in cages on their backs ... the principal open-air sale for canaries ... was in Whitechapel and

Bethnal-green. All who are familiar with those localities may smile to think that the birds chirping and singing in these especially urban places were bred for such street-traffic in the valleys of the Rhaetian Alps!'[46]

In the 1840s, Charles Reiche emigrated from Germany to New York and ran a successful bird business, which led him to publish a book on looking after cage-birds. He described the canary trade:

> The best singers have been raised, within the last century, on the Harz Mountain, in the kingdom of Hanover, and in Thuringia, in Saxony, and have become quite an article of merchandise. They raise annually no less than *fifty thousand* [birds] in that country which are disposed of over the greater part of the earth, and sometimes at extraordinary prices. In the present year (1853), there were no less than ten thousand canaries brought over to the United States.[47]

Almost two decades later, he added: 'Since the above, the fancy has increased so much in this country, that in the last year, 1871, sixty thousand canary birds found their way to this market, of which forty-eight thousand were imported by ourselves.'[48]

Canaries became the most popular cage-bird in the United States and beyond, and Mary Dyson in *The Young Woman's Book* said: 'The canary seems the best bird to keep in the house; born and brought up in a cage, he knows no other home. He is easily tamed, will sing merrily all day long, and may safely be kept in a room without a fire all the winter, if his cage is put out of the way of draughts, and warmly covered at night ... Canaries will live ten or twelve years in confinement.'[49]

Miss Margaret Courtney, a poet and folklorist who lived at Penzance in Cornwall, reported a superstitious practice to prevent the death of cage-birds: 'At a recent meeting of the Penzance Natural History and Antiquarian Society a gentleman mentioned that ... when, some years since, the landlady of the "First and Last" Inn, at the Land's End, died, the bird-cages and flower-pots were also tied with crape, to prevent the birds and plants from dying.'[50] Grantley Berkeley, a sportsman and politician, mentioned another belief: 'I have never been able to make up my mind as to the well-known legend of the goldfinch. Cottagers and all country people believe that, if they go too often near the young goldfinches in their cage, they will one day find them all suddenly dead.'[51]

Seamen often had cage-birds on board ship, providing company and a reminder of home. On 19 November 1846, the smack *Commerce* of Portsmouth was heading towards Lymington in Hampshire from Poole in Dorset with a cargo of railway sleepers, but sank in the night during severe winds. Peter Hawker was staying at his cottage at Keyhaven, along this stretch of coast. 'A wreck off our north shore,' he noted in his diary. '"Hands" saved. Cargoes of railway timber rescued by our coastguard, who got a poor little goldfinch, drowned in his cage.'[52] As with all pets, their owners could become deeply attached, and during World War One, when the hospital ship *Llandovery Castle* was torpedoed in June 1918, the second officer returned to the sinking vessel in order to save his pet canary, which was reported in the *Daily Sketch*:

'George,' a pet canary, was saved from the Llandovery Castle. He is a bright and perky yellow songster, and looked very chirpy yesterday when he had his portrait taken for the *Daily Sketch*. Just as the last boat was leaving, Mr. Leslie Chapman, second officer, suddenly dashed back to rescue 'George.' He put 'George' in a cigarette tin, and there 'George' stayed until a biscuit tin could be found. But 'George' didn't mind. He had his regular ration of bird-seed in the open boat, drank water from the second officer's hand, and seemed reconciled to the fact that during war-time one must bear a few inconveniences. Mr. Chapman, whose home is at Romford, has now bought 'George' a new cage. 'I'm very fond of George,' he told the *Daily Sketch*. 'That's why I went back for him. Everybody in the boat regarded him as a mascot.'[53]

If a vessel did not sink but was deserted by the crew and passengers (a 'derelict'), the owner could pay salvage money to recover it within a year and a day if a live domestic animal, such as a caged bird, was still on board.[54]

Edmund Dixon, who lived near Norwich, gave an insight into the avian pets of the merchant seamen from the nearby port of Great Yarmouth: 'The Yarmouth sailors are very fond of buying Pigeons in the Mediterranean ports, and they are great pets on board ship. They breed there in lockers and hen-coops, and are sometimes allowed their liberty, and permitted to fly round about the vessel, while she is pursuing her course on a fine day. If the breeze is but steady they get on very well, and enjoy themselves as much as they would in calm weather on shore.'[55]

In the nineteenth century, the British Empire was expanding, and thousands of emigrants sought a better life in distant countries. Many settlers longed to see if the familiar birds of their childhood could thrive in their new country, either in the wild or as cage-birds. In about 1849, Old Billy the bird-seller told Henry Mayhew that he once worked in Poplar, Limehouse and Blackwall, along-side the River Thames in east London: 'I've sold larks, linnets, and goldfinches, to captains of ships to take to the West Indies. I've sold them, too, to go to Fort Philip [Melbourne, Australia]. Oh, and almost all those foreign parts. They bring foreign birds here, and take back London birds.'[56]

When cargoes of birds and animals reached their destination, they were regularly kept in captivity for a period to acclimatise. Acclimatisation societies were established around the world, initially in France in 1854, devoted to organising exchanges of flora and fauna between distant continents, made possible by faster sea journeys using steamships. All kinds of birds were sent from Britain as far as North America, Australia and New Zealand, some being intended to eradicate pests in agriculture. In 1867 *The Zoologist* magazine carried a notice from the Wanganui Acclimatisation Society in New Zealand, 'offering a premium of £1 per pair for any number of English house sparrows, not exceeding one hundred, delivered alive and in healthy condition'.[57]

Songbirds were particularly popular, and newspapers regu-larly reported on progress. A few years later the *Brisbane Courier* described the situation in New Zealand: 'The attempt to acclimatise the English skylark ... has been thoroughly successful, and these charming songsters are now quite numerous in the district around Auckland. Epsom and its neighbourhood is a favourite resort of theirs, and numbers of them are there to be seen soaring skyward in the early morning, making the atmosphere ring with their charming melody.'[58]

Not all attempts at acclimatisation were successful, and newspapers carried reports and letters about problems, while the *Melbourne Punch*, a satirical magazine modelling itself on the British *Punch*, published a mocking poem called 'The Acclimatised Lark', describing its nostalgia for its own homeland, mirroring the nostalgia of the emigrants that had prompted its transportation. The poem began:

> *I cannot sing, I cannot sing,*
> *My* heart *is* far *away,*
> *Where the wild bird flits on restless wing*
> *Through the merry summer's day.*[59]

The dangers of the uncontrolled introduction of foreign species soon became apparent, such as starlings and sparrows taken to North America that now compete with native species and destroy crops. The scientific basis for the exchange of foreign species was shown to be seriously flawed, and the acclimatisation societies disappeared.[60]

The foreign birds that were brought back to Britain were exotic, colourful species, often the result of private trade amongst merchant and naval seamen, which Mayhew described in the mid-nineteenth century: 'The commanders and mates of merchant vessels bring over large quantities; and often enough the seamen are allowed to bring parrots or cockatoos in the homeward-bound ship from the Indies or the African coast, or from other tropical countries, either to beguile the tedium of the voyage, for presents to their friends, or ... for sale on their reaching an English port.'[61] One merchant seaman told him that his last voyage had been to the Gold Coast of Africa (Ghana): 'I would never go to that African coast again, only I make a pound or two in birds. We buy parrots – grey parrots chiefly – of the natives, who come aboard in their canoes. We sometimes pay 6s. or 7s. in Africa for a fine bird. I have known 200 parrots on board; they made a precious noise; but half the birds die before they get to England. Some captains won't allow parrots.'[62]

Parrots were favoured by Royal Navy seamen and officers, both for their striking colours and their mimicry. In 1794, Midshipman Thomas Cochrane was serving with a frigate stationed off Norway: 'On board most ships there is a pet animal of some kind. Ours was a parrot, which . . . had learned to imitate the calls of the boatswain's whistle. Sometimes the parrot would pipe an order so correctly as to throw the ship into momentary confusion, and the first lieutenant into a volley of imprecations, consigning Poll to a warmer latitude than his native tropical forests.' One day a party of ladies was allowed to visit: 'several had been hoisted on deck by the usual means of a "whip" on the mainyard. The chair had been descended for another "whip", but scarcely had its fair freight been lifted out of the boat alongside, than the unlucky parrot piped "*Let go!*" The order being instantly obeyed, the unfortunate lady, instead of being comfortably seated on deck, as had been those whom preceded her, was soused overhead in the Sea!'[63]

The Cochrane family had close ties with the Hay family in Scotland, and four decades later, when he was thirteen years old, John Dalrymple Hay joined the Royal Navy as a first-class volunteer. By 1835 he was serving on board the brig *Trinculo*, which was involved in anti-slavery patrols off the west coast of Africa. The next year, Hay recorded, the ship set sail for England 'with three turtles and 240 parrots. The latter learned a good deal of the English language of a particular character, and a large proportion arrived safe at home. Mine reached my father's house in Scotland and lived for more than twenty years.'[64]

An old naval officer informed Mayhew that when he was ordered back to England from the west coast of Africa, he allowed his men to bring home any parrots and other birds they wished. Native boats filled with birds rowed out to the warship, and they were sold to the seamen:

> Before the ship took her final departure, however, she was reported as utterly uninhabitable below, from the incessant din and clamour: 'We might as well have a pack of women aboard, sir,' was the ungallant remark of one of the petty officers to his commander. Orders were then given that the parrots, &c., should be 'thinned' . . . something like a thousand were released; and even after that, and after the mortality which takes place among those birds in the course of a long voyage, a very great number were brought to Plymouth.[65]

Another favoured source for parrots was South America, and towards the end of 1806, when the seaman George Watson was at Rio de Janeiro during a disastrous naval and army expedition, he remarked: 'Parrots, and paroquets, are numerous here, also starlings, cardinals, and several other of the feathered creation, of variously coloured plumage, and may be shot or caught at sea many leagues off.'[66] Before colour photography, even the best artists could seldom capture the vivid colours of birds such as parrots. In December 1809, in a letter to his wife Sally, William Wilkinson, who was sailing master of the *Christian VII*, described in detail the colours of his new pet that he had acquired from a prize – a captured enemy ship:

> The parrot is at present doing pretty well. I was fearful at first it would not live, the weather being so cold, and it just coming from a hot climate. Besides, an hour after I had it on board the prize some sailors frightened it overboard which hurt it very much. But it is now recovered and by keeping it warm I hope to preserve it. They are very likely to die coming suddenly from Rio de Janeiro in South America. I never saw a bird so handsome. Sir J [his captain] says it is worth ten or fifteen guineas in England. I think it is one of last year's birds as it is scarcely tame

yet. It is now sitting on its perch close to me. The whole of its head, neck and breast is scarlet, its wings of different shades of green, with some yellow feathers. Its tail is black and between its wings is a most beautiful purple. I don't think it is the kind of parrot that talks, but its plumage makes up for more than that.[67]

The folklore surrounding imported parrots is scanty, but in one folksong the bird takes on a supernatural role, questioning the heroine when she tries to creep back home in the early morning. Usually referred to today as 'The Outlandish Knight', this song has been found in many countries under different titles. In Britain the earliest known version was probably composed in the latter half of the eighteenth century and was first published in 1776 with the title 'May Colvin',[68] a version that relates the story of the heroine, the king's daughter, eloping with her lover during the night. He attempts to drown her, after confessing to have murdered other young women in this way, but she manages to drown him instead and arrives home before daybreak, hoping that nobody will find out. However, the parrot notices her return and calls out, as in this verse:

Up then spake the pretty parrot,
May Colvin where have you been?
What has become of false Sir John,
That woo'd you so late the streen?[69]

She tells the parrot to be quiet, bribing it with luxuries such as a gold cage and an ivory perch, but by now the king has woken and asks the parrot why it is prattling. The parrot makes an excuse about being threatened by a cat, and so May Colvin escapes discovery.

Although the chances of birds surviving long voyages improved with steamships, many still died, as Mayhew reported: 'Java sparrows, from the East Indies, and from the Islands of the Archipelago, are brought to London, but considerable quantities die during the voyage and in this country; for, though hardy enough, not one in three survives ... About 10,000, however, are sold annually in London ... chiefly in demand for the aviaries of the rich in town and country.'[70]

Old Billy in London told Mayhew that he knew nothing about foreign birds: 'I know there's men dressed as sailors going about selling them; they're duffers – I mean the men.' Some bird duffers or swindlers sold fake foreign birds in the streets, claiming that they needed to sell their birds as they were about to set sail once more. The birds were in fact most likely to be female greenfinches that had been painted and their beaks and claws varnished. 'They study the birds in the window of the naturalists' [taxidermists'] shops for this purpose,' Mayhew explained. 'Sometimes they declare these painted birds are young Java sparrows ... or St Helena birds, or French or Italian finches.' The fakes were not restricted to copying foreign birds, because native birds were also enhanced, including turning blackbirds a deeper shade of black, 'the "grit" off a frying pan being used for the purpose'.[71]

Other deceptions, Mayhew said, involved starlings, which 'may be seen carried on sticks in the street as the tamest of the tame, but they are "braced." Tapes are passed round their bodies, and so managed that the bird cannot escape from the stick, while the fetters are concealed by his feathers, the street-seller of course objecting to allow his birds to be handled ... After having been braced, or ill-used, the starling, if kept as a solitary bird, will often mope and die.'[72]

Such huge quantities of birds being captured for the pet market led to numbers in the wild noticeably decreasing, though this was not the only cause, as James Cash in Cheshire pointed out in 1905: 'Rural folk have no special regard for the sneaking bird-catcher ... who, to a large degree, are responsible for the loss of the goldfinch. But the arts of the bird-catcher are not wholly to blame. The steady march of agriculture has had a great deal to do with it: waste land is constantly being reclaimed, and thus the natural haunts of the finch are lost for ever.'[73]

The popular *News of the World Household Guide and Almanac* in 1952 featured four birds, including the goldfinch:

Not very long ago goldfinches were a favourite cage bird, large numbers of these beautiful and striking birds being taken annually in nets to supply the markets of this country. Many were offered for sale under appalling conditions in Petticoat Lane and other open-air markets, but eventually public opinion was brought to bear on the matter and the goldfinch, along with other wild birds, was given some measure of protection by an Act of Parliament prohibiting its capture and sale.[74]

The almanac was referring to the passing of the Protection of Birds Act 1933, which aimed to stop the trade in wild birds unless they were born in captivity. John Buchan, better known as the spy novelist, was the Member of Parliament who introduced this private member's bill after Eric Parker, the editor of *The Field* magazine, had exposed the cruel trade in the capture and sale of wild birds.

The second reading of the bill was supported by Lord Buckmaster, who had been responsible for the passing of the Protection of Lapwings Bill in 1928. He argued that wild birds were not protected from August to March, something that was apparent to anyone visiting Club Row in London, which Parker had done: 'He found a barrow with wooden cages containing linnets, goldfinches, and a variety of other birds, and as buying was brisk he described exactly what took place. The woman in charge of the birds took linnet after linnet from a tiny cage. She held it up by its tail and legs, dipped its beak into a little glass of dirty water, poked it beak first into a cardboard box, closed the box and pocketed a shilling. She took and dipped a goldfinch and sold that.'[75]

Lord Buckmaster described how competitions were driving the capture of wild birds:

Bird competitions are established all over England and are conducted in almost every place you can think of, from the Crystal Palace to a public-house, and men are encouraged to take and keep these wild birds in the hope of getting some of the very valuable prizes that are provided ... Here are some of the lists of birds in the recent exhibition given at the Crystal Palace [in

January 1933]. I take one class alone: 'Hedge, house and tree sparrows; long-tailed, great, coal, marsh, blue and crested tits; bearded reedling; common, golden-crested and fire-crested wrens; pied and spotted fly-catchers; tree-creeper, wry-neck and nuthatch.' That is one class, and the first prize for a tree-creeper is £50. All down this catalogue you will find prizes that I estimate amounted to several thousand pounds.[76]

After giving further evidence of cruelty, Lord Buckmaster added: 'Surely we do owe something to these creatures. Those of us who take pleasure in enjoying the sight of them must feel that we owe them something ... I want to have the whole of this ended.'[77] The bill was passed by an overwhelming majority, but two decades later, in 1953, Lady Priscilla Tweedsmuir, the conservative Member of Parliament for Aberdeen, began to successfully steer her own private member's bill through the Commons, while her husband, Lord Tweedsmuir, the son of John Buchan, did the same in the Lords. The Protection of Birds Act 1954 protected wild birds, their nests and eggs all year round and was a highly influential and important piece of legislation that replaced all earlier laws.[78]

Until this awareness spread of the cruelty towards wild birds, the morality of confining birds in cages was rarely considered. It was not just keeping them in cages that was questionable, but also the decline of birds in the countryside, the manner of their capture, the mortality rate of those taken and the way they were trained, including blinding and tongue slitting. Perhaps the most poignant aspect was keeping migratory birds in cages, such as nightingales, since they became distressed when it was time to leave Britain for a warmer winter climate. Henry Mayhew in the mid-nineteenth century expressed his concern:

Like all migratory birds, when the season for migration approaches, the caged nightingale shows symptoms of great uneasiness, dashing himself against the wires of his cage or his aviary, and sometimes dying in a few days ... I am inclined to believe that the mortality among nightingales, before they are reconciled to their new life, is higher than that of any other bird, and much exceeding one-half. The dealers may be unwilling to admit this; but such mortality is, I have been assured on good authority, the case; besides that, the habits of the nightingale unfit him for a cage existence.[79]

Two decades later, Miss Anne Dyson admitted to her readership of young ladies that she was not keen on keeping nightingales, because of their agitation at migration time:

They will frequently start up suddenly and flutter their wings, and try to fly up ... Beautiful as their song is, I have never liked to keep them, for I think they must suffer from this restlessness. I should still less like to keep a skylark in a cage, for this bird seems so unfit for such a life ... He wants plenty of fresh air, and should be put outside the window whenever the sun shines warmly; but I think it would make me sad to see him looking up in vain to the sky, into which he must be longing to fly, and I would much rather hear his song while he is at liberty.[80]

CHAPTER SEVEN

———— ·◆· ————

MIGRATION

In April, the cuckoo show his bill;
In May, he sing both night and day;
In June, he change his tune;
In July, away he fly;
But in August, away he must.

Traditional Suffolk chant[1]

For thousands of years, the high point of the traditional agricultural year in rural communities was the bringing in of the crops. In September, after weeks of gruelling summer labour, the successful harvest was duly celebrated before the cycle of the year drew to a close. At this time of year, the daylight hours start to decrease and the birdsong diminishes. No longer do the birds need to draw attention to themselves and their territory – they have mated, made nests, demarcated territories, laid eggs, raised their young and moulted in preparation for the autumn. Many of them now migrate before the winter arrives.

It is known that numerous birds fly considerable distances to overwinter in a warm climate. Even some so-called native birds migrate, such as blackbirds and robins, while birds of the same species come in from colder countries, creating the impression that they are resident and loyal all year round. Amongst the multitude of temporary summer residents are house martins, swifts, yellow wagtails, blackcaps, willow warblers, nightjars, nightingales and wheatears, but two birds, the cuckoo and swallow, hold particular significance in relation to the changing seasons.

Before much of the world was explored, what actually happened to any of these birds was unfathomable. Migration is so extraordinary that it was difficult for our ancestors to comprehend, and it took a long time to be understood and accepted. Instead, outlandish ideas circulated to explain why so many birds disappeared for months on end. One theory involved the moon, while another was that the arrival of winter visitors such as fieldfares was nothing more than the summer birds changing their plumage in readiness for the inclement weather ahead. A further idea was that birds were transformed towards the end of summer into different birds or into creatures such as mice.

Peter Hawker was shooting on his Longparish estate in mid-September 1837: 'I killed a cuckoo that flew past me like an arrow, and I took him for a hawk, as I never saw a cuckoo so late before.'[2] Being similar to kestrels or sparrowhawks, the hawk-like flight of cuckoos gave rise to the idea that they did not migrate, but turned into hawks for the winter season, and in 1912, Geoffrey Egerton-Warburton, who was the rector of the village of Warburton in Cheshire, commented: 'The widespread belief that cuckoos turn into hawks in winter is still seriously held in Cheshire to-day, even by farmers.'[3]

A sign of the cuckoo's impending migration to Africa in July or August is the fading of its distinctive call, which John Nicholson in Hull spoke of in the late nineteenth century: 'before the bird emigrates, its call is less full and more indistinct, sometimes failing to give utterance to one distinct "cuckoo." Hence the saying, "Cuckoo'll seean be gannin; she chatters rarely".'[4] Just before World War One, Edward Scobell, the rector of Upton St Leonard's near Gloucester, recorded some of his parishioners' beliefs: 'The cuckoo had its stories. It was not believed to be migratory: it was remarked, "He go abroad, not he; he be too lazy to fly from one parish to another." The bird was supposed to hide in granaries and hollow trees.'[5] As proof, Scobell mentioned a story he had heard of a family sitting round a fire one cold night, putting a log on the fire, only to hear the distinct sound of a cuckoo as it blazed, causing the sheepdog to bark. Before it could be rescued, the bird was burnt to ashes.

This was actually a traditional story, no doubt passed down through the generations across Europe. When Francis Willughby was in Italy in 1664, he visited the museum at Bologna created by the naturalist Ulisse Aldrovandi, who had died in 1605. He learned that Aldrovandi had heard of a similar story: 'Some (saith he) tell a story of a certain country-man of *Zurich* in *Switzerland*, who having laid a log on the fire in winter, heard a Cuckow cry in it.' On this, Willughby commented: 'We have also heard of the like stories in England, and have known some who have affirmed themselves in the middle of Winter, in a more than usually mild and warm season, to have heard the voice of the *Cuckow* . . . For my part I never yet met with any credible person that dared affirm, that himself had found or seen a *Cuckow* in Winter-time taken out of a hollow tree, or any other lurking-place.'[6]

Willughby was one of the earliest English naturalists, and a few years after his death in 1672 his work on birds was published by his colleague John Ray, initially in Latin, as was customary for scholars, but two years later in English, showing that Willughby was undecided on migration: 'What becomes of the *Cuckow* in the Winter-time, whether hiding herself in hollow trees or other holes and caverns, she lies torpid, and at the return of Spring revives again; or rather at the approach of Winter, being impatient of cold, shifts place and departs into hot Countrys, is not as yet to me certainly known.'[7] Because cuckoos lay eggs in the nests of other birds and do not raise their young, they tend to slip away somewhat earlier than other birds, unnoticed, leading to the idea that they went into hiding for the winter.

By contrast, swallows gather in highly visible numbers just before migrating, as witnessed by James Harley, a Leicestershire naturalist: 'On the evening of 10th September 1839, just as our chimes were going six, myriads upon myriads of swallows assembled in the air over the south-eastern part of the town [Leicester], until the whole face of the heavens became literally peopled with them.'[8]

The departure of swallows and other migrant birds was considered as an especially melancholy event, not least because it emphasised the cruel winter that lay ahead and a sense of abandonment. Archibald Hepburn felt sadness as he watched the swallows leave his village of Linton in East Lothian: 'On the morning of the 24th September last [1839], a flock passed over our reapers, casting no looks behind on the rural homesteads, which they once held so dear, but rushing onwards in an undeviating course and maintaining a sullen silence. I marked their flight until they blended with the blue ether, and thought of the bright days which were gone, and the storms which were soon to come.'[9]

William Fowler expressed his sense of loss at the copious numbers of birds leaving Oxford each autumn:

It is at first rather sad to find silence reigning in the thickets and reed-beds that were alive with songsters during the summer term. The familiar pollards and thorn-bushes, where the Willow-warblers and Whitethroats were every morning to be seen or heard, are like so many desolate College rooms in the heart of the Long Vacation. Deserted nests, black and mouldy, come to light as the leaves drop from the trees – nurseries whose children have gone forth to try their fortune in distant countries.[10]

There was widespread obsession about the autumnal departure of swallows, and William Holland at Over Stowey in Somerset was typical. 'I have not seen any swallows for four or five days past,' he wrote in his diary in mid-October 1809. 'I suppose they are gone, yet I saw them much later last year. I should [think] they must be gone about the tenth or twelfth of this month. They do not, as some suppose, go to a day, but much depends on the season.' Less than a week later, he was uneasy that the swallows had gone, considering that the weather in Somerset remained mild: 'I do not see a swallow moving. They have clearly left us these eight or nine days ago. I think they were after this last year, when the season was more harsh than it is just at this time, though we had some cold, uncomfortable weather [at] the beginning of October, which carried them all off, I presume.'[11]

All manner of ideas once circulated about where swallows went. In 1602 Sir Richard Carew, an antiquary from Antony in Cornwall, published some thoughts in his *Survey of Cornwall*: 'In the West parts of *Cornwall*, during the Winter season, Swallowes are found sitting in old deepe Tynne-workes [tin mines], and holes of the sea Cliffes.'[12] Views changed little there in two and a half centuries, with Robert Hunt, a chemist and antiquarian, saying in 1865: 'I find a belief still prevalent amongst the people in the outlying districts of Cornwall, that such birds as the cuckoo and the swallow remain through the winter in deep caves, cracks in the earth, and in hollow trees; and instances have been cited of those birds having been found in a torpid state in the mines and in hollow pieces of wood.'[13]

William Holland certainly assumed that the swallows over-wintered close by, perhaps in a state of hibernation like hedgehogs or bats, and to him the proof was in the way they seemed to return in the spring: 'I cannot think they emigrate far, but creep out of some hole or crevice in the neighbourhood, for they do not come all in a body, but two or three at a time.'[14] The following autumn, 1810, he returned to the same topic:

to my great surprise, I saw a great many swallows again after I thought they had been gone. I have been on the watch ever since the first of October, but saw none till this day, when they came out in a kind of a cluster. I do not well comprehend this matter. They do not travel abroad for the winter season as some suppose, because in that case they could not be seen and then retire again all of a sudden, and so, on the return of summer they do not appear all at once, but first one or two, and then a few more, as if from some lurking place at hand, and not from a distance.[15]

The possibility that they might hibernate in his own church at Over Stowey led him to declare: 'I will have the inside of the Church Tower searched this winter.' Unfortunately, some of his later diary entries are missing, so we have no idea if this futile task took place.

Holland would have been amazed to learn that swallows actually embark on a hazardous journey of over 6,000 miles across France, eastern Spain, Gibraltar, Morocco and the Sahara and then to South Africa, with others taking a route down the west coast of Africa. The movement of migratory birds between continents is phenomenal, and birds have an uncanny ability to navigate the globe, following the same route every year, possibly using as a guide the earth's magnetic field and other influences including the moon and stars.[16]

In the mid-eighteenth century, bird migration had become a lively topic of discussion, in which the London merchant and botanist Peter Collinson was involved. He had an extensive circle of scientists and collectors as correspondents, including Benjamin Franklin, Hans Sloane, Carl Linnaeus and the Danzig naturalist Jacob Theodor Klein, who took the widely held view that swallows spent the winter underwater. Some scientists even postulated that they turned into lumps of mud or clay.[17] This underwater theory was not new, and in his 1602 *Survey of Cornwall*, Sir Richard Carew mentioned what the Swedish author Olaus Magnus had said about swallows in northern Europe in a book on folklore and history that was printed in 1555:

> For he saith, that in the North parts of the world, as summer weareth out, they clap mouth to mouth, wing to wing, and legge in legge, and so after a sweete singing, fall downe into certaine great lakes or pooles amongst the canes [reeds], from whence at the next spring, they receive a new resurrection; and he addeth for proofe hereof, that the fishermen, who make holes in the ice ... do sometimes light on these swallowes, congealed in clods, of a slymie substance, and that carrying them home to their stoves, the warmth restoreth them to life and flight.[18]

In 1865 Robert Hunt mentioned that the same belief persisted in Cornwall: 'A man employed in the granite quarries near

Penryn, informed me that he found such a "slymie substance" in one of the pools in the quarry where he was working, that he took it home; warmth proved it to be a bird, but when it began to move it was seized by the cat, who ran out on the downs and devoured it.'[19]

On 6 March 1758 Collinson wrote a letter to Klein, part of which he read out three days later at a meeting of the Royal Society in London, taking issue with the theory of swallows disappearing underwater in the autumn:

> This is not to be comprehended, being so contrary to nature and reason; for as they cannot live in that state without some degree of breathing, this requires a circulation of the blood, however weak and languid. Now as respiration is absolutely necessary for circulation, how is it possible to be carried on, for so many months, under water, without the risque of suffocation? Besides, if so remarkable a change was intended, the great wisdom of the Almighty Creator would undoubtedly be seen in some particular contrivance in the structure of the organs of the heart of this bird to enable it to undergo so very remarkable a change of elements.[20]

He suggested to Klein that when swallows started to disappear, one of them should be caught and kept in a container of water. If it adapted to the aquatic surroundings, that would furnish reasonable proof of Klein's theory. Collinson was nevertheless sceptical, having often seen multitudes of swallows gather on the reeds of the islands in the River Thames towards the end of September, with no proof of them slipping into the water:

> I never heard or read of any fisherman or other person that has ever found in the winter months a swallow under water in a torpid living state, for if such a very marvellous thing had ever happened, it would have been soon communicated to the nation. Besides, as these islands of reeds and willows are annually cut down for several uses and yet not a swallow has been discover'd

in his aquatic abode, and considering the multitudes I have seen on these reeds and willows in the autumn, if they took their winter's residence under water it is most reasonable to think in a river so frequented, and in so long a course of years, some swallows would have been found in that situation.[21]

Instead, he was sure that swallows spent the winter in some distant country, and the clinching evidence for him came from the French naturalist Michel Adanson, who had spent five years exploring Senegal. This west African state was a focus for trade, including the Atlantic slave trade, by the rival nations of France and England, and in October 1749 Adanson was on board a vessel off the Senegal coast when he spotted a few swallows that were undoubtedly migrating from Europe to South Africa:

at half past 6 in the evening, we were about 50 leagues from the coast, when four swallows came to look for shelter on the ship and alighted side-by-side on the shrouds. I easily caught all four and recognised them as true swallows from Europe. This fortunate encounter confirmed the suspicion I had formed that at the approach of winter these birds pass over seas to get to the countries of the torrid zone as soon as winter approaches. Indeed, I have since remarked that you only see them in Senegal in this season.[22]

Carl Linnaeus in Sweden remained unshakeable in the belief that swallows spent the winter underwater, and so Collinson urged him to offer a reward to any fisherman who found swallows in the water or beneath the ice in Sweden's lakes. He also suggested another experiment: 'At the time the Swallows are nearest absconding, they resort in vast numbers to the reeds and bushes on the sides of rivers and lakes, so may be easily taken in the night with a net . . . take five or six swallows and tie a weight to their legs and sink them under water. If they survive after laying there in seven days, who will doubt their living in the lakes.'[23]

To reinforce his own argument that swallows migrated elsewhere rather than hibernated, Collinson made the point that he had observed masses of them high in the sky in September:

> I have for many years been very watchful in taking notice of the times when the swallows leave us, and have twice seen them undoubtedly taking their flight at two different years on the 27th and 29th of September. Walking in my garden about noon on a bright clear day looking up to the sky at a very great height, I distinctly saw innumerable numbers of swallows soaring round and round, higher and higher, until my eyes was so pained with looking. I could no longer discern them.[24]

In 1686, only a few decades before Collinson's letter writing, Charles Morton, a nonconformist minister and teacher, went to America and forged strong connections with Harvard. With the development of telescopes in the seventeenth century, increased knowledge about the moon and the heavens promoted scientific and religious debate. By focusing on the teachings of the Bible, Morton put forward the idea that while the winter whereabouts of animals was known, the disappearance of birds was a mystery, though he did not believe that swallows and other birds spent time underwater. One reason was that they were too cheerful at the point of their departure, singing and chattering. Furthermore, he said, 'their flights are also high, but never over any sea-water, that I can hear of; therefore, I conceive, they leave not the land to go beyond sea ... their cheerfulness seems to intimate, that they have some noble design in hand, and some great attempt to set presently upon; namely, to get above the atmosphere, hie and fly away to the other world.'[25]

   To Morton, the moon was their obvious destination, which also explained why swallows returned so suddenly in the spring, as if dropping down from the heavens. In response to the immense distance involved, he said: 'I know not better to answer it, than ... by dividing the year into three parts; allow one third for staying here, another one third there [on the moon], and the remaining one third for their coming and going.'[26] He also wondered whether

So-called vermin, mainly birds, suspended as a deterrent by a gamekeeper

Andrew Carnegie's coach tour to the West Country in the summer of 1884

Catching birds with birdlime on a bush (right) and a call bird in a cage (left)

An idyllic Victorian view of swallows migrating over Gibraltar to their old home at a Surrey cottage

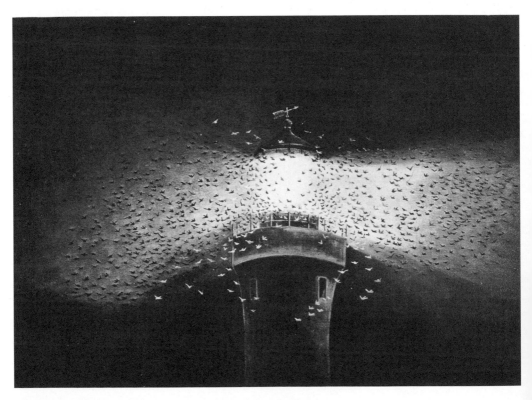

Migrating birds at night round the Eddystone lighthouse, 12 October 1901

Retrieving birds caught in nets on the mudflats of the Wash, Lincolnshire

Sparrow catchers in winter, hunting with nets for sparrows roosting in hayricks

A duck decoy pipe with a detachable net at the end

A decoyman killing wild ducks trapped in netting at an East Anglian duck decoy

Hoodwinked falcons being carried by a cadger on a wooden frame (cadge)

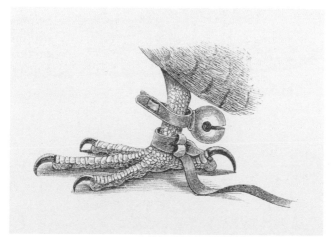

A jesse with a bell

A pigeon badly stuffed by the Housman boys

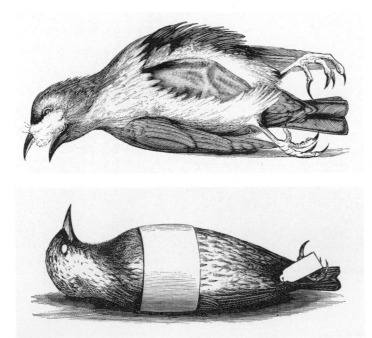

Starting to remove the skin of a starling (left), with cotton wool placed in its mouth, and (below), the starling skin with internal organs and bones removed, prepared for a collection

A Yorkshire 'climmer' preparing to
descend the cliffs to collect eggs

'Climmer' William Wilkinson descending
the cliffs at Jubilee Corner, Bempton,
Yorkshire, to collect eggs

Wren, nightingale, thrush and blackbird (left to right)

Dicky Bird Society medal of 1886 with 'Uncle Toby' (left) and the members' pledge (right)

Dovecote of circular, beehive shape, sixteenth century, at Aberdour, Scotland

Nesting boxes ('pigeon-holes') at a dovecote, possibly fourteenth century, with roof missing, Stoke-sub-Hamdon, Somerset

there might be ethereal islands between the earth and the moon, like the rocky islands in the sea on which birds congregated.

The involvement of the moon seems nonsensical today, but anyone looking through a telescope or binoculars by night may glimpse migrating birds silhouetted against a full moon. Robert Service of Dumfries was not just interested in birds, but was also a very keen astronomer. Gazing through a powerful telescope on the evening of 12 September 1902, he spotted migrating wild ducks against the moon:

> I occasionally see birds pass the disc of the sun, or the moon, when I happen to be looking at these bodies at certain times of the year ... I was looking at the sunrise upon the mountains extending out into the small unilluminated portion of the moon's eastern hemisphere, when a flock of birds passed across the moon in an instant. They were wild ducks, and nine, per-haps ten, were upon the disc at once. The direction of flight was S.S.W. by S. The instrument I was using is an 8½-inch reflector, and the power I had on was 60. I had about two-thirds of the moon's image in view, and it filled the whole field. A calculation of the distance and height of the birds was in the circumstances rather difficult, depending almost entirely upon estimation. Upon the whole I think they would be nearly five miles distant, and about 3000 feet high.[27]

Although so many cherished birds abandoned the country in the autumn, William Fowler had a more positive outook: 'things are not so bad as they seem. The silence is not quite unbroken: winter visitors arrive, and the novelty of their voices is cheering, even if they do not break into song; some kinds are here in greater numbers than in the hot weather, and others show themselves more boldly, emerging from leafy recesses in search of food and sun-shine.'[28] This new set of winter arrivals, from Scandinavia, Russia, Siberia, central Europe, Iceland and the Arctic, includes fieldfares, redwings, bramblings, woodcocks, waxwings, short-eared owls, Bewick's swans, whooper swans and many ducks and geese, which all contribute to the constantly changing kaleidoscope of different types of bird. In addition, various birds move from the north to the south of the country so as to spend winter in a less hostile climate and with more daylight hours for feeding.

Birds of passage – migrating birds – rest and feed at various stopping-off points before resuming their journey, so in times past they risked being shot by ornithologists and others, especially if they were a rare sighting. Some migrating birds arrive in Britain in error, perhaps blown off course by stormy weather. These are categorised as rare bird sightings, but are rare only in the sense of being accidental visitors. Irruptions are mass movements of birds far beyond their normal range, usually caused by food shortages or adverse weather conditions in their traditional territories. Such events are still not well understood, but in the past were inexplicable. Pallas's sandgrouse breeds over a vast area of central Asia, and major irruptions occurred in 1863 and 1888, with sightings as far as the British Isles, where every opportunity was taken to shoot these unusual visitors. Some were discovered after flying into telegraph wires, and John Mansel-Pleydell, a founder of the Dorset Natural History and Antiquarian Field Club, remarked: 'I am glad to say that the recent migration of this bird into England ... has reached Dorsetshire. Six of these birds were picked up dead, or dying, last week at Stoborough, Wareham, on May 28th [1888], under the telegraph-wires – a proof, if any were needed, that the steppes of Tartary are not yet furnished with this higher state of civilisation.'[29]

Four years later, William D'Urban and Murray Mathew wondered what happened to the birds that managed to survive all winter:

> The Sand-Grouse, which arrived in the United Kingdom in May and June 1888, continued with us until the beginning of the following spring, when they gradually disappeared. In many places great care was taken to protect them, in the hopes that they might remain and nest; but when the restless spirit of migration came upon them, they wandered off – and whither?

Did they press westwards, to perish on the Atlantic, or did any of them safely return to their distant Eastern home by the shores of Lake Baikal?[30]

Woodcocks migrate to Britain in huge numbers each autumn after breeding in northern Europe, and their return was eagerly anticipated, because they were considered excellent to eat. Where they came from was a fitting subject for discussion over dinner, and in December 1717 John Thomlinson, a curate at Rothbury in Northumberland, noted in his diary: 'Dined with Robert Snowdon ... Great dispute about woodcocks, what becomes of them in summer – some say they go to Sweden, Norway, etc., and come here in winter ... All agreed that swallows lye in a state of death.'[31]

Charles Morton assumed that, like swallows, woodcocks arrived from the moon – and he thought he had proof. In the mid-seventeenth century, his friend Thomas Travers, rector of St Columb Major in Cornwall, was told by a reliable sea captain that one woodcock overshot its passage to earth from the moon and became stranded above the sea: 'A ship out at sea, farther from land than any birds used to be found, discovered a bird aloft in the air, hovering over them, as high as they could discern; which bird descended towards them ... and at last lighted on the deck. It was a woodcock, so wearied that they took it up with their hands ... It came not from any coast, but down right from above.'[32]

Winter visitors such as fieldfares and woodcocks return to their breeding grounds in March and April, and birds such as swallows and cuckoos arrive back in Britain soon afterwards. With increased worldwide travel in the eighteenth century on board merchant and naval ships, scientific-thinking seafarers had an opportunity to observe the natural world, and before Admiral Sir Charles Wager's death in 1743, Peter Collinson often heard his old naval friend relate what had happened one spring as his ship entered the Channel on the homeward journey to England: 'a great flock of swallows came and settled on all his rigging. Every rope was covered. They hung on one another like a swarm of bees. The decks and carving about the ship was filled with them. They seemed almost spent and famished, was only feathers and bones, but being recruited with a night's rest they took their flight in the morning.'[33] Because another sea captain reported a similar experience with swallows in the Channel when sailing from Philadelphia to London, Collinson was rightly convinced that swallows were birds of passage, returning from a long journey. In fact, these exhausted birds had already flown thousands of miles from South Africa, and ships provided a useful overnight resting place before they embarked on the final leg of the journey to their breeding grounds in Britain.

After so many wretched winter months, this reappearance of swallows in the spring, usually in April, was a joyous sign of hope, something that was widely anticipated and mentioned time and again in diaries and newspapers. On 7 April 1704, Nicholas Blundell, a landowner from the village of Little Crosby near Liverpool, wrote: 'I heard the Cookow and saw one Swallow,'[34] while on 20 April 1764, Richard Hayes of Cobham in Kent noted: 'I have seen 1 swallow but have not heard of any body's seeing two.'[35] On 19 April 1810, in the vicarage at Over Stowey, William Holland and his wife Mary were about to sit down to eat, when he excitedly stopped: 'I thought I saw something fly about like a swallow, which took my attention so much that I was on the watch, and at last saw it again, and alight on Mr. Frost's chimney. I called out [to] my wife to view the bird, and [it] shewed its white breast and black head. "But one swallow," said she "makes no summer." '[36]

Mary Holland was using the old proverb, common across all Europe, of 'One swallow does not a summer make', which has its origins with the fourth-century BC Greek philosopher Aristotle. In defining happiness, he said that it is not a fleeting experience, but takes a lifetime to achieve: 'One swallow does not make a summer; neither does one fine day. And one day, or indeed any brief period of felicity, does not make a man entirely and perfectly happy.'[37] Some translations say that 'one swallow does not make spring' – the ancient Greek word actually means 'spring', which makes better sense because meteorological summer does not start in

Britain until the month of June. Mary's husband then saw a second swallow: 'While we were talking I saw another, and they flew off together, and I pointed them out and cried, "There are two now." "I agree," said my wife. "It is summer." This being fairly early in the season, I wrote this down to be remembered.'[38] According to Margaret Courtney in Cornwall, 'It is the custom here to jump on seeing the first [swallow] in the spring.'[39]

Other birds also gave hope that winter was at an end. On 20 February 1774, Richard Hayes recorded: 'The sky lark now begins to usher in the spring, by piping his melodious note aloft in the sky.'[40] Many more birds than now were seen arriving in the spring, often exhausted and emaciated and in numbers that are today unimaginable, making the changing seasons feel distinctive. In Oxford in the 1870s, William Fowler said that the number of yellow wagtails was not particularly significant, but over the following decade they increased greatly, and on 28 April 1887 he witnessed a throng of them at Port Meadow, alongside the River Thames:

> The river's brink was studded for a full mile with their bright yellow breasts ... it was cold, windy, and pelting with rain ... As I walked along the bank ... they got up from beneath my very feet, for they were sheltering themselves from wind and rain just under the lip of the bank. Then they would settle on the turf, all turning towards me and showing like brilliant spots of yellow ... They had just arrived, and seemed to be as yet unpaired, but in a few days they were nearly all distributed over the country side in pairs, and the great assemblage had vanished.[41]

Occasional setbacks seemed to occur when birds disappeared again. William Roberts, the rector of Whittington in Shropshire, slipped a note between baptism records for 1778: 'April 8th to the 12th. The weather was so very warm and inviting that the swallow, the bat, and the wasp left their dormitories, and the cuckow proclaimed her joy.' He obviously believed that the birds had been hibernating during the winter, because he then wrote: 'April 13 & 14: A sudden return of winterly weather, frost and snow and cold to excess, the sleepers are retired to rest, and the song of the cuckow no where heard.[42]

The arrival of the cuckoo was highly significant as an audible marker of spring, on which Geoffrey Egerton-Warburton in Cheshire commented in 1912:

> The coming of the cuckoo seems to be of more interest to people here than any event in natural history, and cuckoos are, I should say, more plentiful with us than in many places, and are nearly as often seen as heard. I must have seen a dozen one day in May from the high road during a short drive of a few miles, and generally speaking, in May not a day (I should not be far out if I said not an hour of the day) goes by without our knowing by sight as well as sound that there are cuckoos in the garden.[43]

There was always an expectation of hearing several cuckoos, and in 1871 their arrival in south London on 10 April triggered a typical note in the *English Mechanic*, a weekly science magazine: 'A bevy of these harbingers of spring made their appearance on Streatham and Tooting-commons on Easter Monday and poured forth a welcome chorus of "cuckoo" for several hours, which echoed for miles around.'[44] In many areas of Britain today, the call of the cuckoo has completely disappeared owing to their catastrophic decline, which is not just a tragedy for nature, but also for the country's heritage. Although the bird was rarely visible, the distinctive call – literally 'Coo-koo' – was anxiously awaited, and more superstitions are attached to this bird than any other.[45]

When visiting different localities, the agricultural writer William Marshall kept meticulous records of significant aspects of the year, and during freezing weather in the Midlands in 1784, he noted: 'notwithstanding the backwardness of spring, the cuckoo began to call the 26th April, in a cold sharp white-frosty morning'.[46] Others claimed that the 21st was the arrival date of the cuckoo, but Worcestershire people reckoned the 20th, because that coincided with the fair at Tenbury Wells. In Shropshire in 1866, it was pointed out that cuckoos 'appear in this part every spring with the greatest regularity; Orelton Fair, which usually falls on or about the 24th of April, being the day on which we look for them, and they are, generally speaking, true to a day'.[47]

Everyone in Sussex was certain that the cuckoo returned on the 14th, and that was the day of the annual fair at Heathfield, when a mythical old woman supposedly released the first cuckoo of the season from a basket or bag. According to the *Sussex Advertiser* in 1825, it is 'sometimes called The Cuckoo Fair, from tradition, which has impressed a belief on many, that the cuckoo, as the harbinger of summer, is always first heard in the morning of the day in which Heathfield Fair is held'.[48] Curiously, reports in newspapers over the years related that cuckoos were often heard for the first time at Heathfield on that particular day, as in 1854: 'The "Old Woman," so long famed from "John-o'-Groats to Land's End" for her punctual attendance at this fair to "turn out the cuckoo," was this year again at her post especially strict to time, on Wednesday last, the 14th, and her cheering bird of monotonous song was freely dispensing her notes in the neighbouring coverts early in the evening of the day.'[49]

Charlotte Latham lived for years at Fittleworth in West Sussex, where her husband Henry was vicar until his death in 1866. Her experience of cuckoos was obviously related to Heathfield, some miles to the east:

> There is a childish legend current with us ... that a certain old woman of irascible temper has charge of all the cuckoos, and that in the spring she fills her apron with them, and, if she is in a good humour, allows several to take flight, but only permits one or two to escape if anything has happened to sour her temper. This spring a woman of the village complained quite pathetically of the bad humour of the cuckoo-keeper, who had only let one bird fly out of her apron, and 'that 'ere bird is nothing to call a singer.'[50]

Norman Wymer, a native of Sussex, wrote in 1950 that wherever they were, anyone from Sussex would hear the cuckoo on the 14th.

On that day, he said, many would also sing 'Sumer is icumen in', the earliest surviving complete English secular song. The manuscript from Reading Abbey dates to the mid-thirteenth century, and the song was composed for several voices:

> *Sumer is icumen in,*
> *Lhude sing cuccu.*
> *Groweth sed and bloweth med*
> *And springth the wde nu,*
> *Sing cuccu.*[51]

The original Middle English version can be rendered as:

> *Summer has arrived,*
> *Loudly sing cuckoo.*
> *The seed grows and the meadow blooms*
> *And the wood springs anew,*
> *Sing cuckoo.*

Another summer visitor was the wryneck, a type of woodpecker that migrates and was once common in southern England. In the spring it usually arrived just before the cuckoo and so was widely known as the 'cuckoo's mate'.[52] A correspondent to a Hastings newspaper in April 1895 wrote: 'On Monday last [8th] I heard the Wryneck's note (locally termed the Cuckoo's waiting-maid) in Shorndon Wood, Alexandra Park. This morning, Wednesday April 10th, I heard the cuckoo for the first time this year at 6.30 A.M. in Hoad's Wood, St. Helen's Down. This event ... seems to be a harbinger of milder and more seasonable weather for our district.'[53]

It became an annual tradition for readers to write to *The Times* announcing when the first cuckoo had been heard. In 1913 Richard Lydekker did just that, saying that he had definitely heard the cuckoo when he was in his garden at Harpenden in Hertfordshire on 4 February, an extremely early date. He was a respected source, being a palaeontologist and zoologist at the

British Museum (Natural History) in London, but a week later he was apologetic:

> Sir, – I regret to say that, in common with many other persons, I have been completely deceived in the manner of the supposed cuckoo of February 4. The note was uttered by a bricklayer's labourer at work on a house in the neighbourhood of the spot whence the note appeared to come. I have interviewed the man, who tells me that he is able to draw cuckoos from considerable distances by the exactness of his imitation of their notes, which he produces without the aid of any instrument.[54]

When the first cuckoo was heard, it was unlucky not to have any coins, but for those who had some, good luck could be secured. In 1851 Thomas Sternberg, a librarian, published *The Dialect and Folk-Lore of Northamptonshire*, the first person to popularise the word 'folklore' and use it in a book title. Of this superstition, he said: 'When the cry of the cuckoo is heard for the first time in the season, it is customary to turn the money in the pocket and wish. If within the bounds of reason, it is sure to be fulfilled.'[55] Slightly different versions were common across the country, including Yorkshire: 'It is fortunate if you have money in your pocket when you first hear the cuckoo in any year. If you turn it then, you will have money in your purse until the cuckoo comes again'[56] – giving an entire year of financial security.

In Somerset, Charles Henry Poole wrote: 'when the cry of one is heard for the first time, it is usual to turn the money in the pocket and wish: children often sing –

> *"Cuckoo, cuckoo, cherry tree,*
> *Catch a penny and give it me".*'[57]

The cherry tree itself was generally regarded as lucky and could be used to foretell the future. It was also thought possible on hearing the first cuckoo to discover when somebody would be married by turning coins and chanting a similar rhyme:

Cuckoo, cherry tree,
Good bird, tell me
How many years I shall be
Before I get married.[58]

The tree was then shaken, and the number of times the cuckoo called indicated the number of years. Countless superstitions concerned marriage partners, because it was crucial for a young woman to marry in order to have some sort of financial security, including children to care for her in old age. Another belief was that if a young woman ran into a field on hearing the first cuckoo and removed her left shoe, she would see a hair that matched the colour of her future husband's.

Although birds such as the cuckoo and swallow suddenly reappeared in the spring, with scant indication of where they had been, it is now known that many birds worldwide follow specific flight paths, known as migratory flyways. Those coming from Africa to Europe tend to avoid the great expanse of the Mediterranean Sea, but use the shortest crossings, and so massive numbers of birds fly over the narrow Strait of Gibraltar.[59] White storks (which are actually white and black) are large birds that spend the winter in Africa, and in the spring they return to their breeding grounds in Europe either by the Gibraltar route or else by an eastern migration route, with the Bosphorus as their sea-crossing. They rarely come to Britain, but the folklore surrounding white storks is extensive, and children were often told that storks brought babies to new parents, suspended in cloth from their large bills.

A white stork was killed on the Bothmer Castle estate near the town of Klütz in northern Germany in May 1822. Incredibly, stuck right through its neck was an iron-tipped wooden spear 2½ feet long, of a type known from equatorial Africa, beyond the Sahara desert. The stork had obviously been attacked by a hunter there, but took off and migrated in an injured state all the way to Germany. Another two dozen impaled storks were subsequently found, providing incontrovertible proof that they overwintered far away in Africa and that birds really did migrate. The original

Pfeilstorch ('arrow stork', a misnomer, as it should be 'spear stork') was preserved and is on display at the University of Rostock.[60]

Even so, uncertainty about migration persisted, with *The Sporting Magazine* noting in April 1830: 'On the subject of the emigration of birds ... it is probable, that if England continues to be at peace with, and to trade in all parts of the world, as she does at present, our natural historians will be eventually able to come to some more categorical conclusions.'[61] What really increased knowledge of migration was the ringing (or banding) of birds – putting rings round the legs of captured birds before releasing them, though this only works if a bird is recaptured and if the information on the ring is recorded and shared.

William Daniel, in his 1802 book on countryside sports, said that Edmund Morton Pleydell of Whatcombe House in Dorset told him that in February 1798 his gamekeeper caught a live woodcock in nearby woodland. A marked brass ring was put on its left leg and the woodcock was released. Most likely, it returned to northern Europe a few weeks later to breed and came back in the autumn. On 13 December 1798, the actual bird with its brass ring was shot by Mr Pleydell.[62] Peter Hawker, in his instruction manual for young sportsmen, confirmed this story, because Mr Pleydell had related it to him when he was at Whatcombe House, where the bird had been stuffed and preserved.[63]

In an updated edition of his book, Daniel included another incident: 'A second instance occurred in February 1802, when a woodcock was taken alive in the same wood; and, after a tin ring, with the date, was affixed round its leg, the bird was liberated from the front of the house; its flight was very high in the air, and towards the sea. Upon the eleventh of the following *December* the bird was shot in the same *wood*, where it was captured the preceding *February*.'[64] Decades later, John Mansel-Pleydell inherited the Whatcombe estate and heard a different version of the story:

Three Woodcocks were caught in the Whatcombe coverts the same year [1798] in the month of February. My grandfather, after placing a brass ring on a leg of *each*, let them go, and *all three* were killed the following winter. Two were preserved, and are still here at Whatcombe; the other escaped notice until after it had been cooked and sent to table, when the discoloured ring attracted attention. I think it may be inferred that the three birds had remained here the whole year.[65]

Most woodcocks were then winter visitors, with only a few remaining all year, so these three are more likely to have migrated and then returned to the same woodland.

House martins were another migrating species that presented possibilities for tracking by ringing, as they appeared to return to the same nesting site each year, and one story was told to Thomas Durham Weir:

The regularity of the arrival of these 'Joyous prophets of the year,' at their breeding places, is truly astonishing. David Falconar, Esq. [a botanist], told me that for a very long period of forty successive years, a pair of them had come to Carlowrie [near Edinburgh], either upon the 22d or 23d of April. On the forenoon of the 23d of April 1837, he asked his gardener if they had made their appearance? 'Not yet,' he replied. About four o'clock however, in the afternoon, he entered the house, in a great hurry, and with ecstatic delight announced to his master that they had just now alighted upon the top of the stable.[66]

A few years earlier, in about 1825, when the surgeon and naturalist Richard King was living in Kent, he was intrigued by house martins and tried to discover if they returned to the same place: 'Having selected a detached nest, I fastened a small piece of silk round one of the legs of its inmates, then sitting upon eggs. The following season the bird returned, and, with the garter still affixed, was secured in the same nest: a convincing proof of the instinctive knowledge attributed to it.'[67] In 1893 William Storey of Fewston in the West Riding of Yorkshire caught a pair of house martins that had built a nest under the eaves of his house. He put a split ring on one leg of each bird and then released them. They returned the following year, on 20 June, with the rings intact, and again in 1895, 'but the male was, unfortunately, killed by flying against the telegraph wires, the ring still remaining on its leg, and so proving its identity'.[68]

There were several early attempts like this at ringing birds, all small-scale experiments, but in 1890 Hans Mortensen in Denmark, a teacher in natural history at Viborg, began attaching aluminium bands to the legs of starlings, inscribed with contact details, in order to try to discover migration routes of birds. He subsequently turned to white storks, gulls, herons and ducks, using volunteers for this work. Gradually, the ringed birds were found across Europe after being shot.[69] Mortensen's work even continued throughout World War One, but in wartime Britain ringed birds were treated with suspicion. When a ringed black-headed gull was shot in Suffolk in 1916, it was forwarded to the military authorities. The ring, marked 'M – VIBORG: DENMARK', was passed to the Admiralty with a note that demonstrated their lack of knowledge about bird ringing: 'This looks as if the addressee had been trying to train a GULL as a carrier to England: this has often been tried, but is generally considered unreliable.'[70] Mortensen died in 1921, by which time ringing was being carried out in Britain and North America. Inevitably, the numbers of reported ringed birds were extremely small, but became significant over time.[71]

Another method of researching bird migration that had exceptional results was through lighthouse and lightship keepers, once ornithologists realised their potential. In 1879 John Harvie-Brown from Dunipace in Scotland teamed up with John Cordeaux who lived at Great Coates near Grimsby in Lincolnshire, and they circulated questionnaires to many keepers about the birds that had been observed. The results were so remarkable that a committee on migration was established under the British Association, and observations were collected over the next few years.

In 1883 William Eagle Clarke became involved. He was a museum curator at Leeds, but shortly afterwards moved to the Royal Scottish Museum in Edinburgh. As part of his research, he spent a month in the autumn of 1901 at the new Eddystone lighthouse that had been completed in 1882, some 14 miles south-west of Plymouth. It was hoped to observe summer visitor birds leaving Britain on their way to warmer winter retreats, and his first night-time sighting was in the early morning of 23 September. For two hours he stood outside, on the gallery round the light, 130 feet above the sea. Although the wind had moderated, heavy rain was falling:

> It was the first movement I had ever witnessed under such advantageous conditions and will ever remain ineffaceably impressed on my memory. I was aroused from my sleep by the keeper on duty with the words 'Birds, sir,' and was on the gallery a few moments later, when the scene which presented itself was very remarkable. The birds were flying around on all sides, and those illumined by the slowly revolving beams from the lantern had the appearance of brilliant glittering objects, while the rain shot past on either hand, as I stood on the lee side, like streams of silver beads. I was not a little surprised to discover how extremely difficult it was to identify the birds seen under such novel and peculiar conditions. Even the conspicuous spots on the breasts of the Song-Thrushes were entirely effaced as they fluttered in the beams towards the lantern, by the dazzling brilliancy of the light shining directly upon them.[72]

In the autumn of 1903, Eagle Clarke spent nearly five weeks on board the Kentish Knock lightship, about 33 miles from the Essex coast. This was quite a different experience, as the vessel, with a white revolving light and a powerful siren in times of fog, was close to the surface of the sea. The lightship was located beneath an array of migration routes, and he witnessed streams of migrants moving from east to west, coming to spend winter in Britain. A considerable drop in temperature on 10 October was followed by the greatest daytime movement of birds that he had ever seen: 'It set in at 8 A.M. with a marked passage of Starlings, Skylarks, and Tree-Sparrows. By midday it had assumed the nature of a "rush," which was maintained without a break until 4 P.M. It was a remarkable movement in many ways. Skylarks, Starlings, Chaffinches, and Tree-Sparrows not only passed westwards in continuous flocks, but many of these companies consisted of hundreds of individuals.'[73]

On the 18th, Eagle Clarke was taken off the lightship, and during the 4½-hour sailing journey to Southend, he saw continuous flocks of birds, mainly starlings and skylarks: 'we must have encountered tens of thousands of these birds during the passage. It was a revelation even to one – shall I admit it? – painfully familiar with the voluminous records of such movements chronicled in the migration schedules; but it is one thing to study in cold blood, as it were, masses of statistics, and quite another to witness these bird-streams actually flowing unceasingly before one, hour after hour.'[74]

CHAPTER EIGHT

———— ·◆· ————

WEATHER

A very large number of seabirds, recently dead,
were observed on the beach ... lying near the
edge of the water in considerable numbers, so
much so that a lady counted 240 in the space of
not more than two miles: many were gathered for
manure, one man collecting four cart-loads, partly
composed of sea-weed, but principally of dead
birds ... they extended along the beach in the
neighbourhood of Cromer for full six miles.

JOHN GURNEY in Norfolk, 11 May 1856[1]

Britain's changeable weather has always been an obsession and
a cause for immense anxiety. The research at the lighthouses
and lightships revealed that migration was precarious for birds,
something that William Eagle Clarke witnessed when he was on
board the Kentish Knock lightship in the autumn of 1903. He
recorded one particularly bad day, with downpours of rain and a
fierce wind:

There were squalls at intervals, which lashed the rain against
my face with such violence as to cause the skin to tingle for a
considerable time. How the migrants braved such a passage
was truly surprising. How they escaped becoming waterlogged
in such a deluge of wind-driven rain was a mystery. Yet on they

sped, hour after hour, never deviating for a moment from their course, and hugging the very surface of the waves, as if to avoid as much as possible the effects of the high beam wind. It was surely migration under the maximum of discomfort and hardship, indeed under conditions that approached the very verge of disaster for the voyagers.[2]

Storms, freak events or even prolonged winter weather in the past are known to have killed multitudes of birds. Migrating birds in particular risked being displaced or overwhelmed by storms. Thomas Coward, a Cheshire ornithologist, noted in the early twentieth century: 'I have seen the east coast tidal litter full of Goldcrests which had failed to make land.'[3] Formerly (and misleadingly) known as golden crested wrens, these are Britain's smallest birds, and in the autumn many arrive on the east coast from Scandinavia.

A mass of dead and dying birds washed ashore is usually termed a 'bird wreck'. It is a sinister and tragic event, and in more recent times, in February 1983, one such wreck was studied in detail, after a deep low depression from Greenland to the North Sea was followed by westerly gales that led to 34,000 seabirds washing up on the north and east coasts of Scotland and England.[4] Huge numbers were sometimes killed in these bird wrecks, but these fatalities possibly passed unnoticed on more remote coastlines where the human population was much smaller. Cornwall Simeon heard about one of the most disastrous wrecks ever recorded:

A very remarkable and extensive mortality was observed to prevail this autumn, 1859 ... amongst several species of seabirds along the west coasts of Scotland, Ireland, and England, the species more particularly affected by it appearing to be Guillemots, Razor-bills, Puffins, and Gulls, the greater part dead, but some still alive, though so reduced and helpless that they could be taken up by hand. A friend of mine in the Isle of Arran picked up in the course of one morning's walk upwards of fifty dead and dying, mostly Guillemots and Razor-bills.[5]

A local Scottish newspaper, the *Greenock Advertiser*, carried the story:

> During the gale of east wind which prevailed on Tuesday the 13th inst. [September], and for some days thereafter, great numbers of dead and dying razor-bills, puffins, guillemots, and gulls were washed ashore at Arran ... They are cast up dead on the beach, in great numbers, along the shores of Arran, more especially that part of the coast between Brodick and Lamlash. Within a few paces were found forty or fifty birds, many of them lying in heaps, and in some places the poor and emaciated auks might be seen floating, a short distance out at sea, in a dying state. All are wasted to mere shadows of their once plump bodies, and present a pitiful appearance of starvation. Much farther up the Clyde, large numbers of the birds are floating on the water dead.[6]

In his role as an inspector of the branches of the City of Glasgow bank, Robert Gray was able to travel throughout Scotland, taking the opportunity to study birds. He described how razorbills in this same bird wreck had perished in extraordinary numbers, along with other species: 'They were all in a wasted condition, being reduced almost to skin and feathers, and were found dead or dying in thousands over a wide extent of the sea, from the mouth of the river Clyde to the Irish coasts.' Royal Mail steam packets sailed daily, except Sundays, between Glasgow and Belfast, and Gray said that 'the master of one of the steam packets ... reported that he sailed his ship through miles of floating carcasses'.[7]

Terrible scenes also occurred on the Irish coast, which the *Belfast Daily Mercury* mentioned: 'Several masters of coasting vessels which have arrived in Belfast within the last few days report having seen in the [North] Channel, and in the Belfast Lough, thousands of sea gulls and other sea birds dead, floating on the water. The extraordinary circumstance of seeing so many dead birds was a matter of wonder and great mystery to the seamen.'[8]

All too often, bird wrecks on many coasts went unnoticed,

because the corpses were quickly dispersed by tides and predatory creatures, but Charles Johns once spotted the last traces of numerous puffins: 'I have seen a portion of the sea-shore in Cornwall strewed for the distance of more than a mile with hundreds of their remains. All the softer parts had been apparently devoured by fishes and crustaceous animals, and nothing was left but the unmistakeable parrot-like beaks.'[9]

Birds inland were also at the mercy of severe weather, and Murray Mathew described what happened to the rooks in his first winter at Wolf's Castle in Pembrokeshire: 'In the severe winter of 1880 thousands perished. Their dead bodies were to be seen high up in the trees suspended frozen among the branches, and when the deep snow disappeared, hundreds were discovered to have been buried beneath it, especially in the vicinity of small splashlets, where the birds had sought in vain for food.'[10] During such suffering, birds were liable to persecution, as William Holland in Somerset noted in his diary in January 1802: 'Still snow, a vast quantity fell last night and it now continues to snow ... What terrible weather this is for all kind of birds, no food to be found any where. And man, cruel man adding to their calamity by hunting after their lives in every quarter, the whole regions resound with pops and explosions.'[11]

On the Hampshire coast in February 1831, Peter Hawker was at his shooting cottage in Keyhaven. Undeterred by the weather, he went out for some sport: '*2nd*. – A tremendous hurricane, with an overwhelming fall of snow, and with the wind south-west. An extraordinary influx of fieldfares, not less than 20,000 dispersed round Keyhaven and Westover, and so tame that you might have kept firing from morning till night.'[12] Some years later, in mid-March 1845, he observed: 'A Siberian gale the whole day. Small birds starved to death with cold', while the next day: 'Heavy snow all day, and intense frost. Redwings and thrushes half starved, and rabbles of boys knocking them down with sticks in a bitter north-east gale.'[13]

Such cruelty by children when birds were vulnerable was commonplace and usually unchecked, as was witnessed in

mid-summer 1856 by Frederick Smith, an entomologist in the
Zoology Department of the British Museum. He was staying
at the coastal town of Deal in Kent when a plunge in temperature occurred:

> the early part of the day was warm, but a continued drizzling
> rain fell; this, however, did not prevent swallows and swifts
> from hunting after their prey much as usual: towards evening
> a sudden atmospheric change took place, the thermometer
> fell rapidly; it became so cold that an overcoat was not
> uncomfortable. Sitting at the window, and amusing myself by
> watching the swifts, which were very numerous, I was struck
> by observing that their flight was unsteady; they fluttered up
> against the walls of houses, and I saw several even fly into
> open windows.[14]

A girl came to the door, saying that her mother thought he might
like to buy a bat, as they knew he bought all sorts of bugs:

> On her producing it, I was astonished to find it was a poor
> benumbed swift. The girl told me that they were dropping down
> in the streets, and the boys were killing all the bats; the church
> [St George's], she said, was covered with them. Off I started
> to witness this strange sight and slaughter. True enough; the
> children were charging them everywhere, and on arriving at the
> church in Lower Street, I was astonished to see the poor birds
> hanging in clusters from the eaves and cornices; some clusters
> were at least two feet in length, and at intervals benumbed
> individuals dropped from the outside of the clusters. Many
> hundreds of the poor birds fell victims to the ruthless ignorance
> of the children.[15]

Bad weather affects people as well as birds, and so meteorological knowledge has always been of vital importance, especially
for those working on the land or at sea. Folklore once played a
significant part in attempts at forecasting weather, along with the

observation of natural indicators, such as cloud formations and the behaviour of animals, most notably birds.

The wind direction could be easily seen from weathervanes set up on lofty buildings, such as church spires, many of which incorporated a cockerel emblem. In the Bible story of the Last Supper, Jesus foretold that Peter would deny him three times before cock-crow (dawn). This did indeed happen, and after repenting for his actions, Peter took the cockerel (or rooster) as his symbol. In the sixth century Pope Gregory I declared the cockerel to be a fitting Christian emblem, which some churches incorporated in weather-vanes – or weathercocks as they are frequently called. In the ninth century Pope Nicholas I decreed that every church should display a cockerel emblem, and weathercocks became commonplace. They still adorn numerous church spires, steeples and towers in Britain, and because churches are aligned east–west, they had no need to display the compass points.

Weathercocks came to be relied on for predicting changes in the weather by the wind direction. On 9 August 1843, the inhabitants of parts of central England were terrified by a particularly violent storm, and John Jordan watched as it approached the village of Enstone in Oxfordshire, where he was vicar, relying for information on the weathercock: 'The storm rose very slowly and majestically, and directly in the teeth of the wind, which, though imperceptible in the slightest degree, was indicated by the weathercock to be in the east, while the storm came from W.N.W.' As it came closer, 'The weathercock still indicated the wind to be with us due east ... At about a quarter past one, I observed that the weather-cock had just shifted from due east to due north. I was at once satisfied that we were in the wind of the storm, and must expect its fury.'[16]

Peter Hawker expressed exasperation many times when travel-ling to Keyhaven for some sport, only to pass weathercocks showing that the wind direction had altered, making conditions unfavour-able. In early December 1829, he was frustrated at being stuck in London with a fine easterly wind, when shooting conditions would have been perfect, and related how, on the evening of the

11th, 'I mounted the "Telegraph" [stagecoach] at eight, and was in Keyhaven Cottage (just one hundred miles) within the twelve hours. But most wonderful and remarkably unfortunate, I had no sooner got about halfway on my journey down, than my implacable enemy the weathercock flew into the west, and it began a hurricane and rain just as I entered the gig which took me from the coach at Lymington.'[17] Unable to go shooting, he considered himself a 'gaol-bird', a term that he often used.

The parish church at Ottery St Mary in Devon has what is probably the oldest weathercock in use in the country, dating to about 1340. It was constructed with two tubes through its body, evidently intended to produce a crowing noise in windy conditions.[18] At Shrewsbury on 6 June 1593, the weathercock was taken down from the top of the spire of their medieval church of St Mary the Virgin, before repointing took place on what was almost the highest spire in England. The occasion was recorded in a local diary: 'The weathercock was seen and shown to many, which was, and is, of brass weighing 12 lbs. Half the length of the said cock from the bill to the end of his tail is a yard lacking 2 inches [2 feet 10 inches], and the depth from the top of the comb to the bottom of the breast is half a yard and one inch [19 inches], and [it] was put on the steeple top again the 14th day of June by one John Richmond of Acton Reynold [Reynald], mason.'[19]

The spire of the medieval church of St Mary the Virgin in the village of Hemingbrough in Yorkshire was, in August 1762, likewise repointed, giving the opportunity to replace the weathercock. A few weeks later, the churchwarden described the final stages of the ceremony: 'Oct. 7, 1762, the new weathercock was put up, music played at the top, and ale was drawn up from the windows of public houses by a rope.'[20]

Kingfishers and woodpeckers were also relied on to show the wind direction. In Christopher Marlowe's play *The Jew of Malta*, first performed at the Rose Theatre in London in 1592, the Jewish merchant in his counting house on Malta asks which way the wind is blowing, because an easterly is needed to bring his ships from the eastern Mediterranean into port safely:

But now how stands the wind?
Into what corner peers my halcyon's bill?
Ha! to the east? yes. See how stand the vanes –
East and by south.[21]

A halcyon was a kingfisher, and a dead specimen suspended by the neck or breast, indoors or outside, was believed to act as a weathervane, pointing out the direction of the wind with its long bill. In classical mythology, different stories existed that involved Alcyone and her husband Ceyx. In one, they are both changed into kingfishers by the gods, who ensure that the sea remains calm for seven days either side of the winter solstice, the 'halcyon days', enabling them to mate, make a nest and lay and hatch eggs.[22]

The ancient Greeks and Romans liked to stay in sight of land and avoided sailing in winter, and these myths developed to explain the few calm days in winter when it was possible to sail in the Mediterranean. In Britain, a spell of fine weather can occur in early November, also known as the 'halcyon days' or St Martin's Little Summer. The earliest English dictionaries, such as that compiled by Samuel Johnson, define 'halcyon' as peaceful, placid, quiet and still, derived from a bird called the halcyon 'that breeds in the sea; there is always a calm during her incubation'.[23] In reality, kingfishers dig tunnels in banks overlooking water in which to lay eggs. They do not make nests on the sea, nor do they lay eggs at the winter solstice.

Half a century after Marlowe, the physician and naturalist Sir Thomas Browne included the kingfisher in his *Vulgar Errors*, which attempted to prove or disprove hundreds of popular beliefs or misconceptions. According to him, kingfishers were suspended by the bill: 'That a kingfisher, hanged by the bill, sheweth in what quarter the wind is, by an occult and secret propriety, converting the breast to that point of the horizon from whence the wind doth blow, is a received opinion, and very strange – introducing natural weather-cocks, and extending magnetical positions as far as animal natures.'[24] His experiments showed that the theory was false: 'we cannot make it out by any we have attempted; for if a single kingfisher be hanged up with untwisted silk in an open room, and where the air is free, it observes not a constant respect unto the mouth of the wind, but, variously converting, doth seldom breast it aright. If two be suspended in the same room, they will not regularly conform their breasts, but oftimes respect the opposite points of heaven.'[25]

Even so, the belief endured. The poet and novelist Charlotte Smith (whose work Jane Austen knew well) wrote a book on birds for children, which was published in 1807, the year after her death. Of the kingfisher, she said:

> I have once or twice seen a stuffed bird of this species hung up to the beam of a cottage ceiling. I imagined that the beauty of the feathers had recommended it to this sad preeminence, till on enquiry I was assured, that it served the purpose of a weather-vane; and though sheltered from the immediate influence of the wind, never failed to show every change by turning its beak from the quarter whence the wind blew. So that some superstition as to the connection between the wind and the Halcyon seems, like many other relicts of almost forgotten prejudices, to linger still in our cottages.[26]

Dead green woodpeckers received similar treatment, something that Thomas Sternberg had observed in mid-nineteenth-century Northamptonshire (where it was known as the eekle): 'This bird may be said to be the countryman's barometer: when dead, he hangs it up by the legs, and judges of the weather by the state of its tongue; before rain it expands so much that it protrudes from the mouth, while in mild weather it remains shrivelled up in the head.'[27] In Surrey, the loud, laughing cry of the yaffle, as the green woodpecker was called there, was supposed to portend rain, with another of its names being the rainbird.[28]

Such lore was widespread, and at Wreyland in Devon, John Torr kept records of the weather, noting on 1 September 1847: 'Woodpecker called aloud for wet: wish he may be true, the turnips want it.'[29] The Reverend Edward Scobell in Gloucestershire, at the start of the twentienth century, said that a friend mentioned wood-peckers calling for rain: 'I've allers noticed that when the "Ayquils" hollohs "weet, weet," we gets rine. If you listen to them you can hear them speck quite plain: "wet, wet." They've been hollohing very loud this last d'y or two, and see what rine we've got. They hollohs as they flies along.'[30] Scobell thought that a lesser spotted woodpecker was meant, but the green woodpecker is more likely.

Countless other birds across the British Isles were associated with foretelling the weather, in particular rain and storms. If swallows touched the water while flying, then rain was supposedly approach-ing. At Gairloch in Scotland in late Victorian times, the red-throated diver 'when flying frequently utters a loud wailing cry, which is said to prognosticate rain', while Gilbert White mentioned in a letter of 1769 that the mistle thrush 'is called in Hampshire and Sussex the storm-cock, because its song is supposed to forebode windy wet weather'.[31]

In East Yorkshire, John Nicholson related that a whole array of birds performed such a weather-forecasting function – the frequent harsh cry of the corncrake (or landrail) meant rain, as did the con-tinual calling of a cuckoo, while pigeons congregating on the ridge of a house roof signified a storm of wind or rain. As for the robin, he said, 'Should this bird go about the hedge chirping mournfully, though the day be bright and the sky cloudless, it will rain ere long; and when you see him singing cheerfully, on some topmost twig, it will soon be fine, though the rain be pouring down.'[32] Robert Gray in Glasgow certainly believed that robins could foretell weather: 'I have frequently observed that in the evenings, when the robin sings from a high perch, such as the top of a tree, or the edge of a chimney-can on the house top, one may calculate with certainty on fine weather for the next day.'[33]

At Wells in Somerset, one inhabitant reported that jackdaws appropriately gave forecasts from the cathedral's weathervane:

Time out of mind the citizens of Wells, whenever a jackdaw has been seen standing on one of the vanes of the cathedral tower, have often been heard to say 'We shall have rain soon.' I have closely observed the habits of these cunning birds for nearly twenty years [from about 1842], and particularly with respect to the old saying about the weather; and as sure as I have seen one or more of them on the cathedral vanes, so sure has rain followed – generally within twenty-four hours. I have mentioned these facts to many persons, and from several have learnt that the same circumstances have been a 'household tale' in different localities for many years past. Two places I may mention: Croscombe, near Wells; and Romsey, Hants.[34]

A similar forecast was known at Norwich:

When three daws are seen on S. Peter's vane together,
Then we're sure to have bad weather.[35]

The evening movement of gulls was another weather indicator, according to Cornwall Simeon:

Changes of weather may be foretold with considerable accuracy by observing the flight of Gulls, as, after feeding inland, they, according to their invariable custom, wing their way homewards towards evening to their roosting-places in the cliffs; making this transit in fine weather high and in comparative silence, but in bad blustery weather, and before rain, much more noisily and nearer to the ground, merely skirting the tops of the coverts which lie in their course.[36]

A particular flight pattern of rooks was described by Charles Johns as a sign of rain: 'A flock will suddenly rise into the air almost perpendicularly, with great cawing and curious antics, until they have reached a great elevation, and then, having attained their object, whatever that may be, drop with their wings almost folded till within a short distance of the ground, when they recover their propriety, and alight either on trees or on the ground with their customary grave demeanour.'[37] An additional forecast by rooks was mentioned by a York correspondent to *Notes and Queries* in 1880:

'An old man has told me that he observed whenever the rooks congregated on the dead branches of trees there was sure to be rain before night; but that if they stood on the live branches the effect would be *vice versa*.'[38] Thomas Nelson, also in Yorkshire, observed that if anyone destroyed or robbed a swallow's nest, then terrible penalties would ensue, including rainfall on the farmer's crops for a month.[39]

Writing in 1845, Richard Lubbock of Eccles in Norfolk considered the golden plover was truly called *pluvialis*, from its restlessness before bad weather. He referred to this species as *Charadrius pluvialis*, though *Pluvialis apricaria* is used today – *pluvialis* is the Latin word for 'rainy' or 'rain-bringing'. He recalled one memorable incident:

> A few years back, one day in the end of December, I stood upon an eminence overlooking a level of marshes; the day was beautifully mild and bright. I was struck by the perpetual wheelings, now high now low, of large flocks of this bird and the peewit [lapwing]. They were not still for a moment, and yet I could discover no cause of disturbance. Some hours afterwards I went again to the same hill, and found them in the same perturbed state. I was so persuaded that this restlessness was the harbinger of stormy weather, that I wrote a letter excusing myself on that plea from fulfilling an engagement at a distance.[40]

The next morning, the weather was as calm as the day before, though the plovers had all departed, and then, Lubbock said, 'About five P.M. the wind began to howl, signs of tempest came on, and before morning so much snow fell, that in the lanes were drifts six and seven feet in depth.' Imminent severe weather, Simeon said, could also be sensed by sparrows:

> It is curious to observe how the approach of hard weather is heralded by the flocking together of House Sparrows in rick-yards. This was very noticeable in the spring of 1853 [on the Isle

of Wight], when, after the severe frost and snow of the winter had passed away, and given place to more genial weather, scarcely a sparrow was to be seen in the homesteads. Suddenly however they were again filled with large flocks of them, and within two days after, on the 19th of March, came a biting easterly wind and heavy fall of snow, accompanied by a frost from 4° to 6°, which lasted several days.[41]

Without any scientific forecasting beyond readings from barometers and thermometers, it is not surprising that people used observations of birds as real or imagined signs. Bad weather could have drastic effects on livelihoods and health, and therefore intense interest was taken in any animal behaviour that might give warnings. Birds certainly do have the ability to sense changes in the weather, though people do not always have the ability to interpret the signs properly, except in hindsight.

While most observations of birds took place in daylight, some flights of birds by night were noticeable for the noise they made. For more than three decades, until 1906, Arthur Savory was a gentleman farmer at Aldington, near Evesham in Worcestershire. 'We could always tell,' he later wrote, 'when really severe winter weather was coming, by the flocks of wild geese that passed overhead in a V-shaped formation ... we often heard them calling at night as they passed.'[42]

In the mid-nineteenth century at Folkestone in Kent, one of the oldest fishermen there, by the name of Smith, told the surgeon and naturalist Frank Buckland about the sinister 'Seven Whistlers':

I never thinks any good of them, there's always an accident when they comes. I heard 'em once one dark night last winter. They come over our heads all of a sudden, singing 'ewe-ewe,' and the men in the boat wanted to go back. It came on to rain and blow soon afterwards, and was an awful night, sir; and, sure enough, before morning, a boat was upset, and seven poor fellows drowned. I knows what makes the noise, sir; it's them long-bill'd curlews; but I never likes to hear them.[43]

Seamen and fishermen were highly superstitious, believing in storm-spirits and paying attention to apparently supernatural occurrences. Many myths are attached to the idea of the Seven Whistlers, which are sometimes identified as curlews, but are more likely to be whimbrels, whose call often has seven notes. While some people appreciated the sound of them flying overhead at night, others were terrified that they were Gabriel's hounds – or corpse hounds – in search of unsaved souls, including those of unbaptised children.[44] Robert Hunt in Cornwall added: 'To whistle by night is one of the unpardonable sins amongst the fishermen of St Ives.'[45]

The storm petrel was also feared by fishermen, who believed it to cause or foretell turbulent weather. Mainly black in colour, this small ocean seabird of ill-omen lives most of the time far out in the Atlantic or Mediterranean, coming ashore only to nest and only at night. Flocks of storm petrels around a ship were taken as a sign of a brewing storm, though they were more likely to be after the small fish and plankton disturbed by the ship's wake or waiting for food scraps to be thrown overboard. The appearance of these birds ashore, though, is usually a result of them being driven inland by strong winds, sometimes ahead of a big storm.

From the eighteenth century, seamen used the term 'Mother Cary's chickens' for storm petrels. One possible reason is because the Latin *Mater Cara* ('dear mother') is a name given to the Virgin Mary, even though storm petrels were more commonly considered to be the souls of dead seamen or else witches, as in John Masefield's poem 'Mother Carey'. First published in 1902, it was written as if narrated by an elderly boatswain who has seen far too many storms. He explains that Mother Carey is a cruel old woman and the mother of the witches, luring young seamen to their death:

> *She's the mother o' the wrecks, 'n' the mother*
> *Of all big winds as blows;*
> *She's up to some deviltry or other*
> *When it storms, or sleets, or snows.*[46]

Being skilled at sailing, Henry Graham often ventured from Iona to other islands, and on 8 September 1852 he removed two young storm petrels from the island of Soay, which he tried to rear. In a letter to Robert Gray in Glasgow on the 20th he said: 'On the whole they were very amusing and interesting pets ... Last Saturday night, however, was a wintry night; it hailed, and the north wind blew hard; the high hills were covered with snow, and the spirits of the Stormy Petrels departed amidst the roaring of the equinoctial storm.' He told Gray that he had posted him both dead birds, but a few weeks later was apologetic: 'I was disappointed to hear of the Petrels arriving in such bad condition, as I hoped that the post would have taken them quickly enough to prevent their being spoiled.'[47]

Mariners often relied on birds to reveal when land was near. In the biblical story of the Great Flood in the Book of Genesis, Noah sent out a raven from the Ark to look for land, and it flew back and forth. When he sent out a dove, it returned with a sprig of an olive tree, which was a sign that the water was receding. This was a mirror image of the flood story in the earlier Mesopotamian poem *Epic of Gilgamesh*, which related that a dove and a swallow were released, but finding nowhere dry to rest, they both returned. Then a raven was sent, which found food to eat and never returned, proving that it had reached land.

The raucous cries of gulls warned seafarers of approaching land, saving them from possible disaster, especially in poor weather when fog and mist enveloped landmarks and lighthouses. At Flamborough Head in Yorkshire, the gulls certainly kept vessels out of danger, guiding them to safety. They remained invaluable even after Flamborough Head's lighthouse was built in 1806, but in the 1860s so many gulls were being shot for sport that an increased number of shipwrecks occurred. One local newspaper in November 1861 related a typical scene: 'A heavy gale from the north-west, accompanied by rain, snow, and hail, commenced off Flamborough, at one A.M. on Saturday morning. One vessel, the Harbinger, of Scarborough, went down off the sands, and all hands were lost on the south side of Flamborough Head.'[48]

Francis Morris wrote to *The Times* in January 1869 about the terrible slaughter of gulls around the coast, particularly at Flamborough. His prolific literary output included books on birds, and he was rector of Nunburnholme and before that of Nafferton, close to Flamborough Head, where he had witnessed numerous shipwrecks: 'when the sea roke (the Yorkshire word for sea-fog or mist) closes in round the becalmed vessel which is helplessly drifting on towards the dangerous shore, and even the bright Flamborough revolving light cannot be seen, the harsh screams of ten thousand sea-gulls are as useful to the crew as the far-famed Inchcape bell, and they may well bless these "Flamborough Pilots" and hail their cry.'[49]

In his letter, Morris called for the massacre of gulls all round the coasts to be stopped, and the following month the Act for the Preservation of Sea Birds was introduced to the House of Commons by Christopher Sykes, who was the recently elected Member of Parliament for the East Riding of Yorkshire:

> The grounds on which he brought the measure forward were no mere sentimental or humanitarian grounds, though these were strong enough ... He appealed to the House also in the interest of our merchant sailors, for in foggy weather those birds, by their cry, afforded warning of the proximity of a rocky shore, when neither a beacon-light could be seen nor a signal-gun heard. He held in his hand a paper proving that with the decrease of those birds the number of vessels which had gone ashore at Flamborough Head had steadily increased. For the services they rendered to the mariner those birds had earned for themselves the name of the 'Flamborough pilots.'[50]

The Act was passed later that year, making it illegal to wound, kill or attempt to kill various named seabirds, including fulmars, gannets, petrels, puffins, guillemots and kittiwakes, though only during the breeding season between 1 April and 1 August and excluding St Kilda, a remote group of islands in the Atlantic, 100 miles west of the Scottish mainland. This was one of the earliest pieces of legislation to protect wild birds in the British Isles.[51]

While birds were relied on for revealing imminent changes in the weather, some beliefs were connected to long-range forecasts, such as in Yorkshire, where it was said that 'Many cuckoos make a fine summer'.[52] More curious long-range forecasts were obtained by reading breast-bones of ducks and geese. Ethel Rudkin was from the village of Willoughton in Lincolnshire and collected material on local history and folklore. She said that a forecast for three winter months, from 21 December to 21 March, was visible on the breast-bone of a cooked Christmas goose, because after being picked clean of meat, the bone's discolouration indicated any bad weather. She had a bone from Christmas 1939 that revealed deepening storms from the beginning of February 1940, continuing into March. It had been interpreted for her by the estate carpenter, George Hopkin, who had learned the art 'at Hemingby, Lincs., where he was apprenticed, and his "boss" there has learnt it from the old man to whom he was apprenticed, and so on, back through generations. It has always been accounted the most correct forecast.' A less complex forecasting method was reported a few decades earlier from Yorkshire: 'In parts of Richmondshire some persons say that the breast-bones of ducks and geese, after being cooked, are observed to be dark coloured before a severe winter, and much lighter before a mild winter.'[53]

Rooks supposedly had powers of knowing when storms were likely to endanger trees. On 10 September 1816 a terrific storm hit north-east England, which was described by local newspapers: 'In Newcastle many buildings were injured, and the large ash tree, which had for many generations stood in the west corner of the vicarage garden in Westgate-street, was blown down, to the great regret of the inhabitants. It was somewhat remarkable that this tree, which, for many years, had been frequented by rooks was this year deserted by them, a solitary pair only making their nest in it.'[54] Something similar occurred a few decades later on the coast at nearby Marsden: 'a tree was forsaken by rooks with no apparent cause. Their landlord, an ardent naturalist, regretted their flight, and perplexed himself for an explanation. It came to him after not many days. A storm broke over his domain; the forsaken tree was thrown down by the wind; it was rotten at the core, and could not withstand the blast. The branches on which the birds had built were found to be untrustworthy – their strength and stability gone.'[55]

A resident of Woodbury in Devon, Eric Ware, said that long-range forecasts were given by the position of rooks' nests, as in 1939: 'The rooks were shot in the elms [in May] around Oakhayes Lane in the usual noisy way. If nests were very high up it would be a sign of a dry summer; if low, wet. They must have been high that year, because certainly August and September were long and hot.'[56] In Sussex Mrs Latham recorded another belief: 'They say that a tree with a magpie's nest in it was never known to fall. The instinct of the bird may lead it to fix upon a firmly-rooted tree, but the assumed fact is stated as a proof of its supernatural knowledge and of its being in league with the powers of darkness.'[57]

CHAPTER NINE

⸻ ·❖· ⸻

NESTS AND EGGS

The robin of the red breast,
Martin and swallow –
If e'er ye steal one egg of theirs,
Bad luck'll follow.

Anonymous rhyme[1]

Tales about the nests of birds can shine a light in the shadows of social history. Henry Stevenson described how, in mid-nineteenth-century Norfolk, he watched house martins collecting mud from the sides of roads and watercourses 'or settling on the roads in busy groups to avail themselves of the temporary moisture afforded by the water-carts in dry seasons'.[2] Mud is a common lining of nests, and birds such as swallows and house martins make their nests almost entirely from it. In an era before piped water supplies, horse-drawn watering-carts went round the unsurfaced dirt streets, damping down the choking dust, and also delivered water to customers, spilling some in the process and creating ideal sources of mud for birds.

When Frederick Kirkman, a school certificate examiner, edited a series of books on British birds in 1910, he included gannets, which laid eggs in colonies, with their nests located on rocky ledges overlooking the sea: 'In some places the usual material is seaweed, but in others grasses, rushes, etc., while all kinds of articles, such as paper, rags, straw from wine bottles, bits of cork, old clothes,

and even the remains of a parasol have been found in the nests ...
[and] golf-balls, toys, and candle-ends.'[3] The candle ends were
discarded remnants in an era when candles were still widely used
for lighting, though nowadays the nests tell a different story as they
incorporate much plastic debris.

Suitable material for nests was varied, and in Victorian Scotland
Robert Gray said that the nests of the chaffinch (which they called
a shilfa) depended on the locality:

> In rural places, away from the dust and smoke prevailing near
> cities and large towns, the nest is a perfect model in its way for
> neatness and compactness ... [but] those obtained in the out-
> skirts of Glasgow are built of dirty straws, pieces of paper, and
> bits of blackened moss intermixed, forming as a whole such a
> cradle as a country shilfa might feel ashamed of. I once took one
> from a smoke begrimed hedge within the city boundaries which
> had, among other odd things adhering to it, three or four postage
> stamps exhibiting various effigies that a juvenile collector would
> have prized.[4]

In order to lay eggs, many birds make nests, using twigs,
feathers, moss, lichen, hair and mud, but others lay camouflaged
eggs in a hollow in the ground. Birds such as puffins and kingfishers dig or reuse tunnels or burrows, while seabirds tend to lay a
single egg on a rocky ledge. Hair for nests was obtained from the
ubiquitous horses, cattle and sheep, even in urban areas where
for centuries horses were an everyday sight. In mid-nineteenth-
century Sussex, Charlotte Latham said that the villagers were
afraid of human hair being used:

> No hair either cut or combed from your head must be allowed to
> be thrown carelessly away, lest some bird should find it and carry
> it off, in which case your head would ache during all the time
> that the bird was busy working the spoil into its nest. 'I knew
> how it would be,' exclaimed a servant to me one day, 'when I
> saw that bird fly off with a bit of my hair in its beak that blew

out of the window this morning when I was dressing; I knew I should have a clapping headache, and so I have.'[5]

At the end of the century, Margaret Courtney in Cornwall gave a warning when cutting hair: 'locks shorn off must be always burnt, it is unlucky to throw them away; then birds might use them in their nests and weave them in so firmly that there would be a difficulty in your rising at the last day [at the resurrection].'[6]

In Derby, a newspaper revealed how lace was used in a nest, after being washed and left to dry outside: 'Two years ago [1849], some Brussels lace collars belonging to two ladies residing in Friargate, Derby, were laid to dry upon a grass plot, whence they disappeared, and they were supposed to have been stolen. Yesterday week, however, a man employed in pruning a tree in a neighbouring garden found an old bird's nest completely lined with the lost lace.'[7]

James Edmund Harting made a detailed study of Shakespeare's numerous references to birds and pointed out one line in *The Winter's Tale*: 'When the kite builds, look to lesser linen.' It is spoken by Autolycus, a petty thief who boasts that his trade involves sheets, rather than the small pieces of linen that the red kite steals, which were frequently laid out on hedges to dry. Harting explained this episode with a real example of a nest:

This line may perhaps be best illustrated by a description of a kite's nest, which we have seen, that was taken in Huntingdonshire, and which is still in the possession of a friend [John Hancock] in Newcastle. The outside of the nest was composed of strong sticks; the lining consisted of small pieces of linen, part of a saddle-girth, a bit of a harvest glove, part of a straw bonnet, pieces of paper, and a worsted garter; and in the midst of this singular collection of materials were deposited two eggs.[8]

Nearly thirty years later, at the end of the nineteenth century, Lord Lilford repeated this story, having heard it from Hancock, though he was told that the nest came from Rockingham Forest in Northamptonshire. The red kite, he said, was now a rare sight in England, but on his frequent travels to Spain, the black kite was still common,

> constantly to be seen circling alone or in pairs about the villages, on the look-out for chickens, refuse, or materials for its nest, which is often built of somewhat curious materials . . . A Spaniard who accompanied me in my bird-collecting rambles in Central Spain, in 1865, assured me that he had once taken a purse containing nine dollars from a Kite's nest, and I first learned the news of the murder of President A. Lincoln from a scrap of Spanish newspaper found in a nest of this bird near Aranjuez.[9]

Another famous author used a nest analogy. When writing to her nephew Edward from her Chawton cottage in mid-December 1816, Jane Austen sympathised with him for having lost two and a half chapters of a novel he had written: 'It is well that I have not been at Steventon lately, & therefore cannot be suspected of purloining them; – two strong twigs & a half towards a Nest of my own, would have been something.'[10] She went on to say that his strong, spirited, manly sketches would actually have been impossible for her to join to her own 'little bit (two Inches) wide of Ivory on which I work with so fine a Brush as produces little effect after much labour'.

The words 'nest' and 'egg' have ended up in all sorts of ways in everyday language, such as nest eggs, crow's nests and walking on eggshells. 'Nest egg' can mean a useful sum of money put aside, but can also refer to the artificial china eggs to encourage hens to lay, and a dictionary of 1690 defined the term as: '*To leave a Nest-egg*, to have alwaies a Reserve to come again.'[11]

The colour and size of an egg depend on the species of bird, as does the number of eggs produced in a single clutch. Some land birds are capable of laying numerous eggs and having more than one clutch in a year, while seabirds tend to produce just one egg each breeding season. They were all at risk of having their eggs plundered by people, for a whole host of reasons. Some were procured for hatching, like those of the bullfinch, in order to train birds from a very young age, though any unfertilised eggs (like most chicken eggs today) would not have developed into an embryo. Many more eggs were removed from nests for collectors, for industrial processes and for eating, not just from domestic hens and other fowl, but from wild birds. While collectors were eager to acquire examples of eggs from as many species as possible, few birds were safe from the collecting mania that took off in the Victorian era. Oology (the study of eggs, from the ancient Greek 'oon', meaning 'an egg') became a branch of ornithology, but what seems more curious is that nests were also in demand.

In his 1849–50 research of London's labouring classes, Henry Mayhew interviewed street sellers of nests and eggs and learned that they tended to take all the eggs with its nest, since that was more profitable. His main informant was a 'gypsy-looking lad', who in the summer would gather nests of more than thirty species from the surrounding countryside, setting out by night on a Monday and returning on Wednesday morning:

> I gets the eggs mostly from Witham and Chelmsford, in Essex; Chelmsford is 20 [actually 28] mile from Whitechapel Church, and Witham, 8 mile further. I know more about them parts than anywhere else ... Sometimes I go to Shirley Common and Shirley Wood, that's three miles from Croydon, and Croydon is ten from Westminster-bridge. When I'm out bird-nesting I take all the cross country roads across fields and into the woods. I begin bird-nesting in May and leave off about August ... I go out bird-nesting three times a week ... I start between one and two in the morning and walk all night.[12]

At his rural destination, an entire day was spent searching for nests, something the farmers were generally happy about, thinking that it stopped birds plundering their crops: 'I climb the trees, often I go up a dozen in the day, and many a time there's nothing in the nest when I get up. I only fell once; I got on the end of the bough and slipped off ... I wasn't much hurt, nothing to speak of.' Some nests were awkward to obtain: 'The house-sparrow is the worst nest of all to take; it's no value either when it *is* got, and is the most difficult of all to get at. You has to get up a sparapet (a parapet) of a house, and either to get permission, or run the risk of going after it without ... The owl is a very difficult nest to get, they builds so high in the trees.'[13] He took about three hundred different nests each season, and like most birdcatchers, he had a deep knowledge of wildlife, but no opportunity to use it constructively.

Boys in London's streets bought most of his nests and eggs, though some were for particular customers:

> I sell the birds'-nesties in the streets; the three-penny ones has six eggs, a half-penny a egg. The linnets has mostly four eggs, they're 4*d*. the nest; they're for putting under canaries, and being hatched by them. The thrushes has from four to five – five is the most; they're 2*d*.; they're merely for cur'osity – glass cases or anything like that ... Chaffinches has five eggs; they're 3*d*., and is for cur'osity. Hedge-sparrows, five eggs; they're the same price ... and is for cur'osity ... The most of my customers is stray ones in the streets. They're generally boys. I sells a nest now and then to a lady with a child; but the boys of twelve to fifteen years of age is my best friends. They buy 'em only for cur'osity.[14]

The birdcatcher recited to Mayhew the names of another two dozen birds whose nests he raided, including the long-tailed tit or bottletit: 'the nest and the bough are always put in glass cases; it's a long hanging nest, like a bottle, with a hole about as big as a sixpence, and there's mostly as many as eighteen eggs; they've been known to lay thirty-three'.[15]

Eggs could also be purchased from dealers or direct from country people, and nests formed part of the stock-in-trade of the popular Victorian 'curiosity shops' that were made famous by Charles Dickens in his novel *The Old Curiosity Shop*. When the sportsman and naturalist Charles St John stayed at Inchnadamph, at the start of a tour of northern Scotland in 1848, he observed that all the shepherds and gamekeepers were keen on searching for eggs of wild birds, which they sold at a very high price. As the collectors were unable to see the actual nests, they were often duped into buying eggs that were falsely labelled: 'Indeed I am very sure that many of the eggs sold by London dealers are acquired in this way, and are not to be in the least depended on as to their identity.'[16] Collectors could be ruthless in their pursuit of eggs, even if it meant making rare birds extinct, and after St John died suddenly in 1856, he gained the reputation of having wiped out the last ospreys in Scotland, even though they continued breeding there for several decades.[17]

Collectors also acquired eggs through exchange and trade. Writing from Iona to Robert Gray in Glasgow in November 1852, Henry Graham apologised: 'I am very sorry to hear of the fate of the Cormorant's egg. I put it into too frail a box, and the stamping in the post office must have broken it. I have forwarded another by a private hand, but it is not such a fine specimen.'[18] Montagu Browne, who in the late nineteenth century was a museum curator at Leicester, gave practical advice on entrusting parcels to the post: 'Eggs, when being sent any distance, should be separately wrapped in cotton wool, and packed in a strong box, and interstices being lightly filled with wool also. Sawdust or bran should never be used as a packing medium, as the eggs shake together and break each other in travelling.'[19]

In order to prevent eggs becoming rotten, collectors pierced a hole at each end and blew out the viscous contents, the procedure for which James Gardner explained in his 1865 manual on taxidermy:

To prepare the eggs of birds for cabinets is not a difficult matter, and if the eggs be properly labelled and catalogued, a very instructive sort of amusement is always at hand ... If the eggs are to be strung on a string, make a little hole at the small end of the egg with a pointed stick or a sharp knife, and another, but much smaller, hole at the other end of the egg. This done, apply your lips to the blunt or large end, and blow the contents out. If, however, it is intended to preserve the eggs for exhibition in a case or cabinet, the proper plan is, not to pierce the ends, but the side on which they are intended to lie. For this purpose, make two small holes in the side of the egg, and, with a blowpipe at one of the holes, blow the contents out of the other. To make the yolk run freely, run a needle or bodkin through it and stir it well about.[20]

Later on, small steel egg-drills became standard, which Browne described: 'Drills are to be procured from the various dealers, but can be made from steel wire softened in the fire and filed to a sharp three-cornered point . . . or filed up for the larger eggs to the pattern of a "countersink" used for wood.'[21]

After blowing, Gardner recommended the thorough cleaning of the interior of the egg:

get a glass of warm water and with your lips suck up some of it into the egg through the larger hole, then closing both holes with your fingers, shake the egg well about. By this means, repeated till the shell is quite clean, the contents of the egg may be perfectly got rid of. Should the outer shell be dirty, it may be cleaned with a little soap and water and a soft brush. In order to get rid of the thin white membrane which still remains inside, the egg must be rinsed with a solution of corrosive sublimate and alcohol. Without this precaution the film will corrupt, and your egg-shell will be useless in a short time. The solution is to be sucked up into the egg in the same way as the water. You need not be afraid of the solution entering your mouth, for as soon as it rises into the shell the cold will strike your finger and thumb, when you must cease sucking. Shake the shell as before, and afterwards blow the solution back into the glass. Your shell is now perfectly clean and beyond the reach of corruption.[22]

Browne recommended a solution of six grains of corrosive sublimate to an ounce of rectified [distilled] spirits of wine: 'This may be sucked up into the bulb of the "egg-blower" [blowpipe] and thence ejected into the egg, which is to be rotated, and what solution is left may then be sucked back and thrown away, or returned to the bottle. Great care must be taken, however, that the mixture shall not pass the bulb and be drawn up into the mouth, as it is, of course, a deadly poison.'[23] He was quite right to advise caution, as corrosive sublimate is today better known as mercuric chloride, which is highly toxic.

The delicate egg shells were displayed in drawers, in their nests or even strung together, as the naturalist John Knapp mentioned in the 1820s: 'the blue eggs of the hedge sparrow are always found in such numbers on his [a boy's] string, that it is surprising how any of the race are remaining'.[24] Nests with their eggs could be arranged in glass cases, and Mayhew's 'gypsy-looking lad' supplied nests to a wholesaler in window glass: 'He puts 'em into glass cases, and makes presents of 'em to his friends. He has been one of my best customers. I've sold him a hundred nesties, I'm sure. There's a doctor at Dalston I sell a great number to – he's taking one of every kind of me now.'[25]

Egg and nest collecting developed into an extremely popular hobby for children, especially boys who had greater freedom, but with concerns being raised about the harm caused by birdnesting, the writer William Howitt made an impassioned plea: 'It has been denounced as cruel and savage … but the fact is, that while there are boys and birds'-nests, there always will be birds'-nesting … It is an instinct, a second nature, a part and parcel of the very constitution of a lad. There is nothing in all country life that is so fascinating, that so absorbs and swallows up in its charms the whole boy, as birds'-nesting … What is spring and what is the country without birds'-nesting?'[26] When Captain Horatio Nelson was ashore and on half-pay in Norfolk from 1788, his first biographer Robert Southey said that he amused himself with various occupations: 'Sometimes he went birds'-nesting, like a boy: and in these expeditions Mrs. Nelson always, by his express desire, accompanied him.'[27]

The Wild Birds Protection Acts of 1880 and 1881 did give some safeguards to wild birds, from March to the end of July each year, but not to the eggs. As a keen supporter of education for children, William Warde Fowler thought that this was the right approach, 'for as the offenders are usually of tender age, they must be appealed to rather by education and moral suasion than by the terrors of the law. It lies with the clergyman and the schoolmaster to see that gross cruelty meets its proper punishment.'[28] Birds' nests were still destroyed by boys out of sheer devilry or when paid to do so by farmers, and in Oxfordshire in the 1940s, tearing down the nests of magpies and crows proved daring adventures for the village boys. If a landowner declared a tree to be impossible to climb, they always found a way to do so, and the local blacksmith made special climbing irons for them, so that no bird was safe. Montagu Browne even published instructions on how to make a pair of climbing irons 'for those individuals who do not possess the agility of a cat or of a schoolboy'.[29]

Because it was acceptable behaviour for boys to raid nests, they were rarely stopped. When Cornwall Simeon was waiting for a train one July afternoon at Weybridge station in Surrey, he watched a multitude of sand martins on both sides of the deep railway cutting and said to a porter that he presumed the boys robbed the nests: 'Oh! Sir, they would if they was allowed, but the birds are such good friends to us, that we won't let anybody meddle with them.'[30] He explained to Simeon that the sand martins constantly flew up and down the station, keeping the flies at bay, which would be otherwise intolerable, just as they had been in the few hot spring days before the sand martins arrived. 'Now,' he said, 'we may now and then see one, but that is all.'

In 1876 William Adams, who was editor of the *Newcastle Weekly Chronicle*, introduced a Children's Corner under the pen-name 'Uncle Toby', based on the character of 'My Uncle Toby' in the novel *Tristram Shandy* by Laurence Sterne – because Sterne's character had shown kindness to a fly. As Uncle Toby, Adams persuaded children that cruelty towards birds, including robbing nests, was wrong, and he set up the Dicky Bird Society (D.B.S.), creating a Big Book with the names and locations of every member.

Ten years later, with a hundred thousand members, huge celebrations took place in Newcastle, culminating at the Tyne Theatre.
By 1914, the membership reached 366,000, with some honorary
celebrity members such as Lord Tennyson and Robert Louis
Stevenson.[31]

Adams described the society's phenomenal growth: 'Although
the Dicky Bird Society was initiated in the North of England,
it very soon extended to all parts of the civilized world. The
name of Uncle Toby, as the founder and president of a great
organization of children intended to promote the principles of
kindness and humanity, is almost as well known in Cumberland,
Lancashire, Cheshire, and Yorkshire, as it is in Northumberland
and Durham.'[32] Even more surprising was its spread to the British
colonies and beyond:

> The first branch of the Dicky Bird Society established outside
> of the British Isles was commenced in Norway on the 3rd of
> February, 1877. A few weeks afterwards, a branch was estab-
> lished in Victoria, Australia. Then the cause was taken up in
> Nova Scotia, in New Zealand, in Tasmania, in South Africa, and
> in other of our distant colonies. Besides all these widespread
> localities, as the pages of the Big Book show, the D.B.S. can
> boast of members in France, Germany, Italy, Sweden, Gibraltar,
> Constantinople, Hong Kong, Ceylon, South America, various
> parts of the Indian Empire, and almost all parts of Canada and
> the United States. Indeed, it may be said that there is scarcely
> a district in any quarter of the globe in which English people
> have settled that does not contain members of the Dicky Bird
> Society.[33]

On joining, each member was required to sign a pledge: 'I
hereby promise to be kind to all living things, to protect them
to the utmost of my power, to feed the birds in the winter time,
and never to take or destroy a nest. I also promise to get as many
boys and girls as possible to join the Dicky Bird Society.'[34] 'Uncle
Toby' received all kinds of letters from members and published

excerpts in his newspaper. 'I have a very nice letter from WILLIE and LEILA STAFFORD,' he wrote. 'I am sorry their poor pussy slipped into the cistern and was drowned, but I don't think it was because she ate the robin [that] they tell me she killed.'[35] Willie Stafford and his younger sister Leila were the society's 100th and 101st members and were at that time living in Newcastle, though as their father was a railway agent, the family constantly moved around.[36]

It was generally considered unlucky to harm a robin or touch its nest, though Mayhew's London bird-catching informant was happy to take their nests: 'The robin-redbreast has five eggs ... and is 3*d*.' He did, though, feel anguish about much of his plunder: 'After I takes a bird's nest, the old bird comes dancing over it, chirupping, and crying, and flying all about. When they lose their nest they wander about, and don't know where to go. Oftentimes I wouldn't take them if it wasn't for the want of the victuals, it seems such a pity to disturb 'em after they've made their little bits of places.'[37]

In his poem 'The Redbreast and Butterfly', Wordsworth summarised the reverence in which the robin was held:

> *Art thou the Bird whom Man loves best,*
> *The pious bird with the scarlet breast,*
> *Our little English robin.*

According to *The Book of Days* for 1863, a saying in Suffolk was 'You must not take robin's eggs; if you do, you will get your legs broken', and because the boys there never touched the nests, 'you will never find their eggs on the long strings of which boys are so proud'.[38]

The nests of rooks, far larger birds than the robins, were also linked to superstitions, as it was considered lucky to have them nesting nearby and equally unlucky to destroy their nests. This led to a scene witnessed by John Nicholson at the premises of William Cass, a stationer, printer and bookseller in Hull: 'In the spring of 1886, a rook [probably a crow] built its nest in the solitary tree standing in the yard of Mr. Cass, Prospect Street, Hull, and when any mischievous lads came to molest it, an old lady in Portland Street used to drive them away with her sweeping brush; for had the rook been compelled to forsake its nest, the whole neighbourhood would have participated in the consequent ill-luck.'[39] Swallow and martin nests were also a good sign, and according to Thomas Wilkie, 'It is a very fortunate omen when swallows take possession and build their nest in any person's premises; and unfortunate when they take their departure and never return.'[40]

Many superstitions and traditions were associated with the eggs of wild birds and poultry, and one widely held notion was that it was unlucky to keep the eggs of wild birds indoors. In 1869 a question was asked in *Notes and Queries*: 'A native of Kent lately gave me a collection of the eggs of British wild birds, but with a strict injunction not to retain the possession of them, as the keeping of them would be very unlucky. Is this superstition general?'[41] It had actually been mentioned forty years previously by the naturalist John Knapp:

> Bird-nesting boys, I suppose, are yet to be met with in many a rural village, being a habit from immemorial antiquity, pursued with eagerness in contention with their fellows for numbers and rarity ... regarding these birds' eggs we have a very foolish superstition here; the boys may take them unrestrained, but their mothers so dislike their being kept in the house, that they usually break them; their presence may be tolerated for a few days, but by the ensuing Sunday are frequently destroyed, under the idea that they bring bad luck, or prevent the coming of good fortune, as if in some way offensive to the domestic deity of the hearth.[42]

Two decades later in Northamptonshire, Thomas Sternberg said that keeping any wild bird eggs in the home was deemed unlucky, something that Mrs Latham knew from her village of Fittleworth: 'There is a superstitious feeling that will not allow birds' eggs to be brought into a house, though long strings of them may generally be seen in Spring hanging up in out-houses. It is thought they bring bad luck.[43] This idea persisted, and at the end of the nineteenth century in rural Yorkshire the boys competed on May Day to produce the longest string of birds' eggs, but it was unacceptable to take them into the house, though in East Anglia the ill-luck was caused only if the eggs were of small birds.[44]

Reginald Scot tried to expose superstitions in his book *The Discoverie of Witchcraft*, which was published in 1584. In some countries, he said, it was believed that 'to hang an egg laid on ascension daic in the roofe of the house, preserveth the same from all hurts [by witches]'.[45] The practice obviously took hold in Britain, because in a letter to *The Times* in 1934, the question was asked: 'Can explanation be offered why in Nottinghamshire it is believed that an egg laid on Ascension Day, if placed in the roof of a house, will ward off fire, lightning, and other calamities?'[46] Eggs laid on other devotional days were also valued – ones laid on Maundy Thursday (Holy Thursday, the day before Good Friday) were thought to protect a house against thunder and lightning and those laid on Whitsun Day would help to extinguish fires. Good Friday eggs also performed that role and could be kept all year, as they were said never to go stale, and keeping one would ensure the health of the poultry.[47]

In Northamptonshire, when a baby was taken for the first time to friends of the parents, Sternberg said that 'it is customary for them to give it an egg; this, if preserved, is held to be a source of good fortune to the future man'. This tradition was also known in Lincolnshire: 'At the birth of a child . . .when it is first taken to a neighbour's house, it is presented with *eggs*, the emblem of abundance, and *salt*, the symbol of friendship.'[48] In the West Riding, it was said that at the first house into which a baby is taken, the mistress must give it its blessing, which included salt, bread, a match and a coin, as well as an egg 'that it may never want meat', while at the village of Coxwold, also in Yorkshire, 'People are sometimes very angry if this is omitted.'[49]

The number of eggs that a hen sat on at any one time also

mattered, as Sternberg related: 'In what is technically termed "setting a hen," care is taken that the nest be composed of an odd number of eggs. If even, the chickens would not prosper.' This superstition was general, and in Cornwall Margaret Courtney noted: 'Hens must never be put to sit on an even number of eggs, eleven or thirteen are lucky numbers.'[50] In addition, each egg had to be marked with a black cross, 'ostensibly for the purpose of distinguishing them from the others, but also supposed to be instrumental in preserving them from the attack of the weasel and other farm-yard marauders'.[51] Thomas Wilkie in Scotland said that if hens laid eggs with double yolks or if they hatched out all hen birds, then somebody was likely to die, though it was lucky if they were all cock birds.[52]

Eggs could be misused by witches to cause harm. The parish records for Wells-next-the-Sea in Norfolk include the names of fourteen unmarried local men who were buried in St Nicholas churchyard on 26 December 1583. After sailing all the way from Spain, they lost their lives close to this port, and the parish clerk added a note to say that the shipwreck was caused by a witch from nearby King's Lynn: '[their] deathes were brought to pass by the detestable woorking of an execrable witch of Kings Lynn, whose name was Mother Gabley; by the boyling, or rather labouring of certeyne eggs in a payle full of cold water; afterwards approved sufficiently at the arraignment of the said witch.'[53] Tales of storm-raising and ship-sinking by witches have occurred since antiquity, but this is the earliest recorded one in northern Europe, and she may have been hanged for the offence.[54]

More than three hundred years later, another incident of boiling eggs in cold water by a witch occurred at Scarborough in Yorkshire. In September 1904, an inquest was held in the town into the death of an emaciated seventeen-month-old child, George Cooper, the son of Francis Cooper, a corporation carter who lived with his wife Sarah and several other children in a terraced house in Ewart Street. When Mrs Cooper was cross-examined by the coroner, she claimed that her son had been bewitched by the woman who had lived next door:

How did she bewitch the child when she had never been in the house?

—I suppose she did it with witchcraft; I don't know. She said she would bewitch the child by boiling eggs and mashing them.

—And she gave the child the eggs?

—No; she was never in the house.

—How could she bewitch it, then?

—She has done so. I can always hear her when there is nobody in the house, and there are terrible shadows from behind the doors ...

—You don't think death was due to improper feeding?

—No, sir: I think it is her. The landlord turned her out.

—Because she was bewitching people?

—Because I said something about her.

—And then she revenged herself by bewitching the child?

—Yes.

—You should be careful not to say anything to annoy your neighbours.[55]

The jury returned a verdict that death was caused by convulsions due to rickets resulting from improper feeding, but the coroner could not help saying that 'they had heard an extraordinary story about witchcraft, and perhaps it appeared rather foolish on his part to take notice, but he could not help it'. The parents were subsequently charged with neglecting all their children and found guilty, with Mrs Cooper sent to prison for three months and Francis for one month.[56]

Because it was widely believed that witches could appropriate empty shells from boiled eggs, they were commonly crushed after a meal, and in his *Vulgar Errors*, first published in the mid-seventeenth century, Thomas Browne commented:

To break the egg shell after the meat [cooked egg] is out, we are taught in our childhood, and practise it all our lives; which nevertheless is but a superstitious relict, according to the judgment of Pliny, *Huc pertinet ovorum, ut exrobuerit quisque, calices*

protinus frangi, aut eosdem cochlearibus perforari; and the intent
hereof was to prevent witchcraft; for lest witches should draw
or prick their names therein, and veneficiously mischief their
persons, they broke the shell, as *Dalecampius* hath observed.[57]

Jacobus Dalecampius was a French botanist who, in 1587, had
published a commentary on *Naturalis Historia* (*Natural History*)
by the ancient Roman writer Pliny the Elder, who died during the
volcanic eruption of Vesuvius in AD 79. When discussing divination,
Pliny said: 'There is indeed nobody who does not fear to be spellbound by imprecations. A similar feeling makes everybody break
the shells of eggs or snails immediately after eating them, or else
pierce them with the spoon that they have used.'[58]

In 1832 George Oliver, an antiquarian and Freemason, as well
as rector of Scopwick in Lincolnshire, pointed out the connection
between eggs and witches: 'Some people, after eating boiled eggs,
will break the shells to prevent the witches from converting them
into boats, because an ancient superstition gave to these unhappy
beings the power of crossing the sea in egg-shells.'[59] Mrs Latham
enthusiastically broke eggshells as a child in Sussex: 'the bottom
of the shell should be always broken through by you after you have
eaten the contents, and I remember with what energy at our nurse's
bidding we used to burst the bottom of egg-shells with spoons to
disappoint the witches, who, we were told, would otherwise put out
to sea in them'.[60] Witches were easy to blame for disasters at sea,
and with their connections to eggs, mariners considered it unlucky
to take any eggs on board their vessels.

With their apparent links with the heavens and the supernatural,
birds were reckoned to be endowed with a range of prophetic gifts
and even left messages on their eggs. Thomas Coward pointed
out that some eggs have irregular lines or 'scribbles', those of the
yellowhammer being particularly noticeable, causing them to be
called writing or scribbling larks. 'An old Lancashireman assured
me,' he said, 'a cryptic message was inscribed by the bird which,
interpreted, read, "Don't take my eggs." '[61]

According to Alfred Williams, country people in the Upper

Thames Valley considered it an evil omen to dream of eggs, though they used eggs to foretell marriage prospects and other details. Wilkie described one such method that was carried out on the last night of the year: 'take a new-laid egg and perforate the small end of it with a pin, and let fall three drops, into a bason filled with water, of the white (albumen) which diffuses itself upon the surface in beautiful and fantastic shapes of trees, etc., from which they augur what the fortune of the egg-dropper will be in the ensuing season, and what kind of temper his wife will have, and how many children she will bring to him'.[62]

For over sixty years until his death in 1870, Jonathan Couch was a doctor at Polperro in Cornwall, where, according to him, about two-thirds of the inhabitants were engaged in fishing and other maritime pursuits and the other third in agriculture, trade or commerce. He said that Midsummer's Day was when future husbands were revealed: 'Get a glass of water, and having broken an egg, and separated the white from the yolk, throw the former into it, and place it in the sunshine. You will soon see, with a little aid from your fancy [imagination], the ropes and yards of a vessel, if your husband is to be a sailor, or plough and team, if he is to be farmer.'[63]

Wirt Sikes was an American journalist who in 1867 was appointed the United States consul for Wales, based at Cardiff. Being fascinated by Welsh folklore, he published a book in 1880, in which he mentioned that Robert Ellis, a minister and poet whose Bardic name was Cynddelw, revealed how Welsh farm-women discovered their future husbands: 'The maiden would get hold of a pullet's first egg, cut it through the middle, fill one half-shell with wheaten flour and the other with salt, and make a cake out of the egg, the flour and the salt. One half of this she would eat; the other half was put in the foot of her left stocking under her pillow that night; and after offering up a suitable prayer, she would go to sleep.' Sikes added that with romantic thoughts and a raging thirst after eating such a salty cake, she would have a vision in the night of her future husband coming to her bedside with a vessel of water or other drink.[64]

A similar ceremony, told by Sternberg for Northamptonshire,

took place on 24 April, St Mark's Eve, and involved 'eating the yolk of an egg in silence, and then filling the shell with salt, when the sweetheart is sure to make his visit in some way or other before morning'.[65] St Agnes Eve, 20 January, was the time to carry out a similar rite at Stamfordham in Northumberland, as related by John Bigge, who was the vicar:

> Eat nothing all day till going to bed, boil as many hard eggs as there are fasters, extract the yoke, fill the cavity with salt, eat the egg, shell and all, then walk backwards to bed, uttering this invocation to the Saint –

> *Sweet St Agnes work thy fast;*
> *If ever I be to marry man,*
> *Or man be to marry me,*
> *I hope him this night to see.*[66]

Easter is another occasion when eggs have significance, and the earliest record in Britain is in the household expenses of King Edward I for 1290, which mentioned (in Latin) the purchase of four hundred and fifty eggs costing eighteen pence for the Sunday feast of Easter. When extracts from these accounts were presented to a meeting of the Society of Antiquaries in London in February 1805, John Brand commented that some of the eggs 'might have been purchased for the purpose of being stained with various colours, and given as Easter presents to the royal household, a custom which generally prevailed in Catholic times, in token of the resurrection ... In some parts of the north of England such eggs are still also presented to children at Easter, and called *paste* (*pasque*) eggs.'[67]

Because it was forbidden to eat eggs during Lent, those not needed for hatching chicks were preserved by boiling in water containing vegetable dye, which stained and marked them. They were also known as pasch or pace eggs, derived from the Greek and Latin words 'paska' and 'pascha', originally meaning the Passover and later on Easter. These Easter eggs became customary gifts, marking the end of Lent, and some were further decorated with painted designs and even the names of the recipients. Many were retained for years as keepsakes and even survive in museums.[68]

Various customs of 'pace-egging' developed, especially in northern England, the most common being groups of children going from house to house begging for eggs, which might then be rolled down a slope until they cracked open and were eaten, or used in a game similar to conkers, holding the egg in one hand and striking an opponent's until one was smashed. It was said that in Blackburn, Lancashire, in the 1840s, 'children, both male and female, with little baskets in their hands, dressed in all the tinsel-coloured paper, ribbons, and "doll rags" which they can command, go up and down from house to house; at some receiving pence, at others eggs, in others gingerbread . . . Houses are literally besieged by these juvenile troops from morning till night. "God's sake! a pace-egg," is the continual cry.'[69]

In East Yorkshire, John Nicholson said that one reason why boys raided birds' nests was to have enough eggs for another day of festivity – Royal Oak Day or Mobbing Day on 29 May: 'These eggs are expended by being thrown at other boys, but all boys who carry a sprig of Royal oak, not dog-oak, either in their cap or coat, are free from molestation. Not only wild birds' eggs, but the eggs of hens and ducks are used to "mob" (pelt) with, and the older and more unsavoury the eggs are, the better are they liked – by the thrower.'[70] On this day in 1660, his thirtieth birthday, Charles II rode into London in triumph at the restoration of the monarchy. A sprig of oak leaves was worn by royalists to mark the occasion when, as Prince Charles, he hid in an oak tree near Boscobel House in Shropshire. Although this public holiday was abolished in 1859, the tradition persisted in some areas.

Eggs made effective ammunition, not least because they turn rotten and stinking. Incredibly, in an archaeological exavation near Aylesbury in Buckinghamshire, four chicken eggs were found in a waterlogged Roman pit, possibly deposited as a votive offering. One was retrieved intact, a unique discovery, but being so fragile, the other three broke and emitted a sulphurous odour, even after nearly two thousand years.[71] Rotten eggs were commonly thrown at convicted criminals, and such attacks were especially vicious in London, as William Richards found, a licensed Methodist preacher of about sixty years of age, who was convicted in October 1807 of assaulting a boy: 'Thursday – Richards, a schoolmaster, stood in the pillory on Clerkenwell-Green, for unnatural propensities. The populace, particularly the female part, were so incensed against him, that, notwithstanding the efforts of the constables, they saluted him with rotten eggs, and filth of other descriptions, in such a manner, that scarcely any of the human features were discernable.'[72]

When Henry Graham was collecting eggs on the small islands around Iona, where the lesser black-backed gulls bred on the flat, marshy summits, they were checked by putting them in water, as fresh eggs sink and rotten ones float:

> [We] put them to the ordeal at the first convenient *dub* [pool] of rain-water. All that sink are *spolia opima* [booty]; those that float, when we cannot return them to their nests, are flung over the edge of the cliff, if possible in the direction of the patient boat-keeper some giddy fathoms down ... Meanwhile, the clamour overhead is frightful; you cannot make your comrade hear though you shout in his ear. The enraged birds dash at your head, and your dogs slink at your heels with lowered crest and tail, as if ashamed of being cowed by mere vociferous birds.[73]

Accustomed to scaling rigging high up on a man-of-war, Graham also fearlessly participated in gathering the eggs of herring gulls on the same small islands, though their nests were in clefts and on inaccessible ledges of the cliffs and precipices. Each year, the inhabitants of countless coastal communities around the British Isles risked their lives by climbing down sheer cliffs to gather hundreds of thousands of eggs. They were taken from nesting seabirds such as gulls and guillemots, mostly for food or industrial processes, though also with collectors in mind. During the summer of 1805, when Charles Fothergill was journeying around his native Yorkshire, he stopped at Flamborough:

I proceeded directly to the house of Bryan Spike facing the
church. This is the most famous man throughout the country
for getting the eggs of sea-fowl, having been a 'climmer' as it is
called, all his life; and he has a son [George] who assists him in
his labours ... He said that he could use or dispose of twice the
number he got and said it was scarcely possible for him to say
how many he got during a season, they were so numerous ...
many people in the vicinity get them besides him – even a
woman at one farm house lets down her son ... The birds begin
laying about May day, soon after which is the best time for
getting the eggs.[74]

Fothergill accompanied Bryan and George on that year's final
egg collection:

the men having equipped themselves with two set of ropes 80
yards long each, an iron pin of large size to stick in the ground,
a bag and a basket, we proceeded to the cliffs ... The most
dangerous part of the employment of climbing for sea birds
eggs does not appear to consist in a fear of the ropes giving way
but in stones loosened by the friction of the rope falling upon
the climber, which not unfrequently happens, the adventurer
getting many a broken head; the only mode they have of avoiding them, is when they hear them tumbling down (for they dare
not look up) to cling as close as possible to the side of the cliff
or under some fortunate projecting ledge.[75]

He next witnessed how the ropes were secured:

The pin is driven into the ground at some little distance from
the edge of the precipice: to this pin one of the ropes is tied and
thrown over the steep to be held by the climber for the purpose
of pulling himself up or letting himself down, the other rope is
tied round the loins of the climber at one end, while the other
is held by the assistant. The rope is chiefly for security and for
assistance in being helped up and down, a signal for which he

gives by a shake or pull of this rope . . . These men are fatalists or rather predestinarians and consequently have no particular fear of death from their employment.[76]

In the summer of 1841, the zoologist James Wilson accompanied Sir Thomas Dick Lauder, secretary of the Board of Fisheries, on an extensive tour of Scottish coasts and islands. By August they were at St Kilda. 'The natives,' Wilson said, 'collect a considerable quantity of sea-fowl eggs in spring and the earlier part of summer. They prefer them when *sour*, that is . . . when about ten or twelve days old, and just as the incipient bird when boiled forms in the centre "into a thickish flaky matter like milk." '[77] The collected eggs were stored in stone and wooden buildings, known as cleitean.

Although it was too late in the season to gather eggs, the island's minister, the Reverend Neil Mackenzie, arranged for them to sail round to one of the cliffs where, at an agreed signal, a few local men demonstrated their skills:

> Suddenly . . . three or four men, from different parts of the cliff, threw themselves into the air, and darted some distance downwards, just as spiders drop from the top of a wall. They then swung and capered along the face of the precipice, bounding off at intervals by striking their feet against it, and springing from side to side with as much fearless ease and agility as if they were so many school-boys exercising in a swing a few feet over a soft and balmy clover field. Now they were probably not less than seven hundred feet above the sea . . . shouting and dancing, they descended a long way towards us, though still suspended at a vast height in air.[78]

The men worked in pairs, with the upper one supporting the weight of the man below:

> the cragsmen, having each a rope securely looped beneath his arms, rested occasionally upon his toes, or even crawled with a spider-like motion along projecting ledges, and ever and anon

we could see them waving a small white fluttering object, which we might have taken for a pocket-handkerchief, had we not been told it was a feathery fulmar ... But to see them dangling in the air, like spiders from webs of gossamer, the ropes being scarcely visible, owing to the great height from which they were suspended, was in truth a surprising sight.[79]

St Kilda is now a World Heritage Site, having been uninhabited since 1930, and archaeological excavations have yielded fragments of egg membrane and shell.[80]

Inevitably, accidents occasionally happened, as in Pembrokeshire in June 1826: 'A poor man [a servant] lost his life last week, in consequence of having been suspended by a rope over the cliff, near Linney Head, to the southward of Milford Harbour, for the purpose of collecting gulls' eggs. When hauled up, his body presented a very appalling spectacle, being cut and galled dreadfully by the pressure of the rope, and his swinging to and from about the rocks. There have been several instances of the loss of life in a similar way near that place.'[81] Children were at risk as well, as occurred in the Orkneys in May 1854: 'On Friday week, a young lad named James Linklater, supposed to have been looking for wild birds' eggs, fell over the crags at Yesnaby, in Sandwick, Orkney, and was either drowned, or killed amongst the rocks.'[82]

When growing up at Woodbury in Devon during the Great Depression of the 1930s, Eric Ware said that he and other children roamed the countryside in the spring looking for birds' nests, and they also collected dipchick (moorhen) eggs and took them home to be cooked.[83] Nowadays, the eggs most people eat are laid by hens, but eggs of many types of wild birds used to be eaten, and those of lapwings, a type of plover also known as peewits, were an expensive delicacy. In the first half of the nineteenth century, William Yarrell, the son of a London newsagent and bookseller, acquired an incredible knowledge of fishes and birds, and in 1843 he published *The History of Birds* in three volumes. In a description of lapwing eggs, he included information given to him by Francis Plomley:

The marshes of Lincolnshire, Norfolk, Cambridgeshire, Essex, and Kent, afford a large proportion of the quantity with which the London market is supplied ... The trade of collecting them continues for about two months ... all the most likely ground is carefully searched for eggs, once every day, by women and children ... Mr. Plomley sends me word that two hundred dozens of Plovers' eggs were sent from Romney Marsh to Dover in the season of 1839; and that dogs are trained for the purpose of finding the eggs.[84]

Yarrell died in 1856, and some years later an expanded, revised edition of his work appeared, in which the editor added a note: 'The earliest [plover] eggs fetch such fancy prices as fifteen shillings *apiece*: and a leading West-End poulterer recently informed the Editor that if he were assured of having the first ten eggs he would not hesitate to give £5 for them. As the supply increases, the value falls rapidly, until it reaches 4s. 6d. per dozen, which is the average London price in the season.'[85]

Richard Lubbock detailed the trade in Norfolk:

THE LAPWING is ... still common in the fens in the breeding season, but greatly reduced in numbers. Its nest is too often plundered entirely, without leaving a single egg to encourage it to lay again ... In 1821 a single egger, resident at Potter Heigham, took an hundred and sixty dozen in the adjacent marshes. In those days nearly a bushel of eggs have been gathered by two men in a morning, principally from this bird ... The egg is here worth more than the bird. Many of the very earliest are sold by those who take them at eightpence each, and they seldom get lower than three shillings a dozen.[86]

Methods of serving plovers' eggs were featured in *The Magazine for Domestic Economy* for 1838:

in London and other populous cities, they are supplied by the poulterers in the state in which they most generally appear for sale, namely, boiled hard, and left till they are cold, like the coloured eggs of Lent in Roman Catholic countries. Sometimes, however, they are obtained fresh, when the usual mode of dressing them for fashionable tables is, to boil them during ten minutes, and serve them up, either hot or cold, in a napkin. It is thought that any other mode of cookery would destroy their flavour; though we ourselves have found them very good with white sauce.[87]

A few years earlier, John Knapp highlighted food snobbery: 'the glareous liquor or white of the egg of this bird, upon being boiled, becomes gelatinous and translucent, not a thick opake substance like that of the hen; a circumstance that is likewise observable in the eggs of the rook, and of many of our smaller birds ... the eggs of the poor rook, though bearing little resemblance to those of this plover, are in some places not uncommonly taken and sold conjointly with them in the London market'. In his view, rook eggs tasted much the same and 'probably the habitual eater of them only can distinguish a sensible difference'. Lubbock also knew of other eggs being passed off as plovers' eggs: 'In those days, indeed, various were the eggs which joined to fill a basket.'[88]

With so many lapwing eggs being gathered, the numbers of these birds plummeted, and action had to be taken. In mid-February 1928, when introducing the second reading of the Protection of Lapwings Bill, an exasperated Lord Buckmaster asked why the government was so unsupportive, particularly as plovers were extremely valuable to agriculture in eradicating wireworms and leatherjackets. He had even received letters from farmers wanting this bill to pass, because they were suffering so much from pests now that lapwings were disappearing. He was sure he knew why the bill was being blocked: 'I think the answer is to be found in what is, to me, the abominable spectacle that every poulterer's shop will exhibit in the next few weeks if this Bill is not passed. From end to end of the country, not merely in London but in every little seaside place, poulterers' shops will be smothered with plovers' eggs. It seems to me that the necessary result to the plovers must be disastrous.'[89] The bill was passed and greatly restricted the taking of those eggs, although lapwings are yet again endangered, due largely to changing agricultural practices.

Another inland stretch of water in Norfolk is Scoulton Mere, which once had a breeding colony of thousands of black-headed gulls. Their eggs were also considered to be delicacies, eaten cold like those of plovers. Lubbock wrote about how egg collecting was a commercial operation here:

> I went lately to visit the gullery at Scoulton. The swampy island upon which they breed occupies a great portion of the mere, and the Gulls are indeed in myriads upon it. The worthy proprietor does not suffer them to be unfairly molested. A portion of the eggs is always taken; and their numbers may be judged of, from the fact that an average season produces more than 30,000 eggs – five years back, the keeper said, they took 44,000. Parts of their abode are so swampy, that no one can walk there to gather eggs, which of course tends to the maintenance of their numbers. Now and then a year of jubilee is given, and no eggs taken; this was done lately at the instance of the neighbouring farmers, who justly value the services of these birds in the destruction of grubs, &c.[90]

Some years later John Gurney described two visits he made:

> In 1860, on the day of the great gale – the memorable 28th of May, I visited Scoulton Mere, in Norfolk, with my father, the largest Gullery of its kind in England. About 16,000 eggs had been gathered. In 1872, when I went again, only 4,000 eggs were taken. This sad falling off was due to dry seasons. Brown [James Brown], the keeper, told me that once the farmers spread the fields in the neighbourhood with manure sown with salt, which poisoned the worms, etc., upon which the Gulls fed, and that a great number died in consequence.[91]

The application of salt (sodium chloride) as a fertiliser increased through the eighteenth century, though the tax on salt made it expensive until 1819, after which farmers were permitted to use it tax free if mixed with one-quarter ashes or soot. The introduction of pike and the incursion of rats proved damaging, and by about 1950, the gulls had gone from Scoulton Mere, and it is now a fishing lake.[92]

Many eggs of poultry and wild birds, especially seabirds, were required for industrial processes and for the production of other foodstuffs. The leather industry was one customer. In June 1805, Charles Fothergill watched eggs being collected at Flamborough: 'The price at which they are sold is 3 for a penny, and besides their excellence as food, they are bought and used by curriers and leather breeches makers.'[93] In preparing animal skins for leather, the fat and hair were removed before various further treatments, and skins such as goat and sheep were tanned with alum, salt, egg yolk and flour, known as tawing, to produce a soft, flexible white 'kid leather' for gloves, breeches and bookbinding.

George Dodd, a prolific writer of books and magazine articles in the mid-nineteenth century, visited the Bevington factory in Neckinger Road, Bermondsey, where he described skins being dressed with an egg mixture: 'the eggs are broken, in the proportion of one to each skin, and the yolks only are mixed with water and a little meal in a tub: the skins are then introduced, and are trampled by the naked feet of a man until the egg has been thoroughly imbibed.' He learned that the factory relied on eggs imported from France: 'sixty or seventy thousand are purchased for the Neckinger factory every spring, and are preserved in lime-water till wanted, a mode by which they may be kept perfectly sweet for two years'.[94]

Adding egg whites to wine helped to refine it, producing a clear liquid free of particles, and in 1758, in his memoranda books relating to the management of the Eccleston estate in Lancashire, Basil Thomas Eccleston wrote down a recipe:

Take the whites of eight eggs and beat them into a froth and then put a pint of red wine to it and work all together in the cask with a stick, first takeing out about two or three gallons of the wine, and work it for ten minutes and then fill the cask up, and make it close, and in a weeks time will be bright. Put a peg at the end of the cask and you will see at any time when it is fine [refined], and then bottle it in clean dry bottles. NB the above receipt is for half a hogshead [27 gallons].[95]

The same process was done with cider, and the agricultural writer William Marshall, when in Gloucestershire in the 1780s, learned that eggs or isinglass, or a combination, were used there to make cider look brighter.[96]

Eggs were also needed by sugar refineries to clarify or clean the raw brown sugar. When heated with egg white or bullock's blood, a thick scum formed on the surface, trapping most of the impurities. This scum was skimmed off, resulting in sugar that was much lighter in colour and able to form conical loaves in moulds.[97] An 1823 directory of Yorkshire said that at Flamborough, 'boys are let down the rocks by ropes fastened to stakes, and bring away bushels of eggs for the use of the sugar house in Hull, without seeming to diminish their countless number'.[98]

Ramsey Island is less than a mile off St David's Head in Pembrokeshire, and many of its eggs also went to sugar refineries. When Murray Mathew lived at nearby Wolf's Castle, he found that guillemots were known there as 'Eligoogs': 'The farmers around St. David's are said to feed their calves in the summer with a custard made from the "Eligoogs"' eggs obtained on Ramsey.'[99] Robert Rising, a magistrate and farmer at Horsey Hall in Norfolk, near the coast, mentioned that a fair was held in the Horsey area every spring in mid-Victorian times. It was known as 'coot custard fair', because all the sweets were made from eggs of coots and black-headed gulls obtained from what was once a vast gullery, until the marshland was drained and enclosed.[100]

As well as eggs being used for food functions, a particular type of nest was prized for making soup. It was one of the delicacies from around the world that were on offer at the first dinner of the Acclimatisation Society of Great Britain in July 1862, which was held at Willis's Rooms, known formerly as Almack's, the famous gentleman's club in St James's, London. The dining room was decorated with stuffed animals,

and a case containing the little swallow [swiftlet] which builds the edible birds' nests, with a specimen of the nest and an egg. This was handed round to the guests when the soup produced from the nests was under discussion, and afforded an interesting corollary to the culinary lesson which the society were illustrating with plates . . . Of the bird's-nest soup . . . only about a quart could be made, and it was obvious that although everybody would wish to taste this unaccustomed delicacy, a hundred people could not be served with a large plateful each.[101]

Instead, everyone was given a small sample of the soup: 'By the most part pronounced to be excellent. The birds'-nest gives a strong gelatinous quality, and the flavour is – what you please. In this case (thanks to Messrs Willis's *chef*) it was excellent.'

In the years after World War One, Thomas Coward said, this edible bird's-nest soup continued to intrigue the western world:

The bird-nest soup which is occasionally imported from the east, and which the Chinese consume with avidity, is usually described as being manufactured from the nests of Swallows, but this is not so, the bird is a Swift. Our own Swift makes a nest which, if cleaner, would be edible, for a few straws and other rubbish are glued together in a mass of salivary secretion, which rapidly hardens, and it is this hardened jelly which the Chinese boil into soup.[102]

CHAPTER TEN

———— •◆• ————

FOOD

In the midst of the bay is an island inhabited by
puffins, which, at the proper season are caught,
and pickled for exportation.

Puffin Island, off Anglesey, in 1784[1]

Henry Graham, on the island of Iona, constantly studied, shot and
ate birds: 'I make it a rule in general to eat what I shoot, unless
killing for the paramount duty of obtaining a skin or specimen.'[2]
Most ornithologists eagerly tasted different birds to see if they
were edible, and when Murray Mathew was shooting bar-tailed
godwits – a long-billed winter visitor – on the river mudflats of
north Devon, he said: 'as these birds are excellent for the table, we
always found them to be greatly appreciated by the friends among
whom we distributed our spoils'.[3]

The poor in society have always relied on a few cheap, staple
foods, with meat and dairy produce being expensive luxuries.
In 1841 an assistant Poor Law Commissioner, Edward Carleton
Tufnell, presented his findings about households in Kent and
Sussex. One family in Sussex, he said, had a quarter of an acre for
a vegetable plot, but 'have no meat except on a Sunday, when a
meat pudding is made'.[4] This was most likely cooked with scraps
from butchers, as it was illegal for most of the population to trap or
shoot gamebirds, but smaller wild birds also provided a welcome
source of meat. Nowadays, to eat birds such as skylarks, blackbirds,

thrushes, rooks and sparrows is widely considered abhorrent (and illegal), but those suffering hardship would try anything.

Before being enveloped by Oxford's suburbs, Headington Quarry was a remote village to the east of the city, with stone quarries, claypits, brickfields, wasteland and scattered cottages and hovels. In the late nineteenth century, Waggle Ward would go out with his uncle Snuffer Webb, who was an expert at felling blackbirds with a catapult:

> Used to go out on the Clayhills and ketch blackbirds an' thrush-ers and Granny Webb 'ud do 'em. Blackbirds and nothing else . . . They used to joke – Granny Webb – our Mam did – she used to say 'This pie'll start whistling when you bring him the oven' . . . And I can remember eatin' 'em. It was all right – waren't a big bite on 'em but them old blackbirds when Granny Webb used to do 'em, you could see all the fat, all round the bodies like, yeller fat, you 'ud . . . But you never eat starlings, never eat starlings, Granny Webb 'ud'nt . . . you get a dozen thrushers in a pie, and blackbirds, well that was a bit of a feed, waren't it.[5]

Like most recipes for poultry and game, pies were cooked without removing the bones. Granny Webb probably had a small cast-iron range with an oven for cooking pies with pastry, but until the Victorian era most cooking was done by boiling in pots suspended over open fires, while meat was roasted on spits in front of a fire or taken to the local baker's oven.

In the Cotswold villages around Fairford in the early years of the twentieth century, Alfred Williams recorded that the poor struggled to survive. He knew that they ate badgers, squirrels, hedgehogs and mice, though never rats, except on one occasion when two young ploughmen roasted a rat in the brew-house fire, after removing its head and tail. On seeing Henry, a fellow farm labourer, approach, they feigned a quarrel about who should eat the bird, 'and while they were so engaged Henry quietly took the rat from the stick and ate it with a good relish, grinning at his supposed smartness. He felt sure it was a blackbird, and when his mates laughed at him and asked him how he liked "the rat", he still grinned and said it was "very good mate." [6]

Blackbirds were not the sole preserve of the poor. For many, they were a treat, and in mid-January 1812 William Holland in Somerset recorded that his son and two servants 'are gone out a bird baiting. They caught a good many and some blackbirds, and we had them afterwards made into a dumpling and very good they were.' [7] In his depiction of late nineteenth-century rural life in southern England, Richard Jefferies described a convivial gathering of neighbouring farmers:

About Christmas-time, half in joke and half in earnest, a small party often agree to shoot as many blackbirds as they can, if possible to make up the traditional twenty-four for a pie. The blackbird pie is, of course, really an occasion for a social gathering, at which cards and music are forthcoming. Though blackbirds abound in every hedge, it is by no means an easy task to get the required number just when wanted. [8]

A pie with twenty-four blackbirds was the theme of the old nursery rhyme 'Sing a song of sixpence', which includes the lines:

Four and twenty blackbirds,
Baked in a pie.
When the pie was opened,
The birds began to sing;
Was not that a dainty dish,
To set before the king? [9]

Possibly of sixteenth-century origin, the rhyme refers to the exotic practice of baking pies with surprises for lavish feasts. An Italian recipe for a blackbird pie, which was translated into English in 1598, advised filling a hollow uncooked 'coffin' (a pastry case) with flour through a hole in the base. After baking, the flour was emptied out and replaced with numerous live birds: 'this is to be done at such time as you send the pie to the table, and set before the guests: where, uncovering or cutting up the great lid of the pie, all the birds will fly out, which is to delight and pleasure show to the company'.[10]

All manner of small birds have been regarded as a delicacy since at least ancient Greek times, generally cooked by spit-roasting. A cookery book of 1768 included a recipe for skylarks:

To dress Larks.

Spit them on a little bird-spit, and roast them. When enough, have a good many crumbs of bread fried, and throw all over them; and lay them thick round the dish.

Or they make a very pretty ragoo with fowls livers: first fry the larks and livers very nicely, then put them into some good gravy to stew, just enough for sauce, with a little red wine. Garnish with lemon.[11]

Such recipes, aimed largely at the middle classes, give a glimpse of the many species of birds that were regularly eaten, though not by the poor who were generally unable to read or afford to purchase cookbooks. Instead, they relied on traditional recipes passed down through their families.

The winter of 1645–6 brought a windfall of skylarks to Exeter in Devon. Early in the Civil War, the city supported the Parliamentarians, before falling to the Royalists in September 1643. In June 1644, Henrietta, the daughter of Charles I, was born there, and a few weeks later the clergyman Thomas Fuller was appointed as her chaplain. The following year, with the New Model Army defeating Royalist positions from Hampshire to Somerset, Exeter improved its defensive walls and remained in a state of siege until surrendering in April 1646. Fuller remained in Exeter during the siege and related how, in the bitterly cold winter months, huge flocks of skylarks prevented starvation:

When the City of Exeter was besieged by the Parliament's Forces, so that only the South side thereof towards the Sea was open unto it, incredible number of Larks were found in that open quarter ... hereof I was an *eye* and *mouth* witnesse. I will save my credit in not conjecturing any number ... They were as fat as plentifull, so that, being sold for two pence a dozen, and under, the Poor (who could have no *cheaper*, as the Rich no *better meat*) used to make pottage of them, boyling them down therein.[12]

Many reasons were suggested for the appearance of these sky-larks, but Fuller was convinced that 'the *Cause of causes was Divine Providence*, thereby providing a *Feast* for many poor people, who otherwise had been *pinched* for *provision*'.

In January 1814, during a harsh winter, Peter Hawker could not resist a similar windfall while he was lodging at Poole Harbour: 'The weather became so intensely severe, that the people of the house were busily employed in preparing puddings of the larks and other birds, which flocked into the house and sheds, and were not only there, but even in the furze and on the shore, easily taken with the hand ... Such was the intensity of the cold that I picked up pocketfuls of larks that had perished and fallen in the water.'[13]

In the mid-nineteenth century Henry Mayhew explained how skylarks for the London market were obtained largely from rural regions to the north: 'The larks for the supply of fashion-able tables are never provided by the London bird-catchers, who catch only "singing larks," for the shop and street-traffic ... They are principally the produce of Cambridgeshire, with some from Bedfordshire, and are sent direct (killed) to Leadenhall-market, where about 215,000 are sold yearly, being nearly two-thirds of the gross London consumption.'[14]

Around this time, Arthur Knox saw skylarks being lured to their death on the Downs near Brighton from late September to the end of October. Although many skylarks were resident all year, large flocks migrated from the Continent to spend the winter in England. A decoy device or lark-glass was used, constructed from wood and fragments of reflective looking-glass fastened on a red cloth. By pulling on a length of string, the contraption rotated, reflecting the sun's rays, and unable to resist, the larks would instantly descend and be shot: 'The fields in the neighbourhood of the coast on both sides of the town, are haunted by various parties of shooters from the hour of sunrise until ten or eleven o'clock, about which time the great flights of larks cease or diminish ... The birds thus killed are comparatively lean and worthless, not fetching, in the market, within fourpence a dozen of the price usually demanded for those which are taken by lark-nets during the winter months.' He explained that skylarks of better quality were caught at night from March to October with immense drag-nets, when hunters swept the fields where they roosted: 'these birds form a considerable article of traffic, and hang in numerous bunches at all the poulterers' stalls in the town and market'.[15]

Another writer lamented how skylarks, adored for their beautiful song, were cruelly served up as food: 'Some people are so fond of larks as to eat them ... Around Dunstable [Bedfordshire] the birds are caught in great numbers, for the hideous purpose of mastic-ation, and sent to the metropolitan market, where they sell at from three to four shillings the dozen. There is certainly no accounting for taste.'[16] In a description of the countryside in 1890, Charles Dixon was more poetic in his damnation:

Who can view the Lark's tiny body, all stripped of its beauti-ful plumage, without sending a thought to the breezy fields, and the wild uplands where this charming songster lives, and where he gladdens the rural scene with his matchless melody. No songster is more thoroughly English than the Sky Lark. Its voice is the language of the fields, the music that is in keeping with the cattle grazing peacefully in the meadows, with the distant chimes of village bells, and the blue sky and white drifting clouds, that are such a beautiful feature of an English landscape ... As we look at the festoons and bunches of Larks in all the game-dealers' shops, we cannot help feeling regret at the sacrifice of so much musical life. Passing sad it is to think that so much that is beautiful and musical should come to such an untimely end, to feed the pampered tastes of nineteenth century civilisation![17]

London's Cheshire Cheese inn in Wine Office Court was famed for its winter season of steak and kidney puddings, cooked with mushrooms, skylarks and oysters in the filling. Early in 1891, Sarah Morton, an American journalist, decided to try the illustrious dish that had been enjoyed by so many celebrated writers and journalists, including Samuel Johnson and Oliver Goldsmith. She was the first woman to do so in this male preserve, and her account was published in several American newspapers:

> 'Twas the night of the beefsteak pudding, a delicacy served only twice a week, and in precisely the same way that it has been served in this very place for two hundred years.
> Every seat is occupied.
> 'Tis just six.
> The door swings slowly open.[18]

She watched the enormous pie being carried ceremoniously into the room and the diners served: 'There is my big dinner plate piled high with – what on earth! Birds – yes, tiny bits of birds – skylarks, kidneys, strips of beef just smothered in pastry, like sea-foam, and dark brown gravy, steaming with fragrance, as seasoning.'[19]

Only the year before, in March 1890, Scottish newspapers reported the opening of the Forth Bridge: 'An Edinburgh poulterer (says the *Evening Dispatch*) is said to have received an order for 400 larks, to make lark pie for the luncheon at the Forth Bridge in connection with the Prince of Wales' visit.'[20] That event led Sarah Morton to comment: 'There were only four hundred skylarks put into the pudding made for the Prince of Wales at the banquet of the Forth Bridge opening in Edinburgh. How many thousands of the "blithe spirits" have been put into the Cheshire Cheese pudding for 200 years! Shades of Shelley and Keats!'[21] The long tradition at the Cheshire Cheese ended only on the eve of World War Two.

Wheatears were also considered highly desirable, and the preferred method of cooking them was the same as for most skylarks and other small birds – not in pies, but by roasting on a spit in front of the fire. In August 1833 Peter Hawker sailed across the Solent from Keyhaven to the Isle of Wight, noting in his diary: 'In an excursion this month to that paradise, Alum Bay, where I took a gun to get a few dozen of those delicious birds the wheatears.'[22] These birds were caught in vast numbers on the South Downs during the start of their autumn migration to central Africa from late July to late September. Many were captured by shepherds using traps, and in 1788, when John Byng and a friend were exploring the downland around Hastings, they took a few wheatears for their dinner:

> we ascended the steep hill ... searching many turf-traps set for wheat-ears, when the custom is to leave a penny for every caught bird you take away. Within the castle, we seated ourselves for some time ... till the shepherd came to survey his traps, when we paid him sevenpence for his capture of seven birds, whom we sat instantly to pluck in preparation of our dinner spit; and it wou'd have made others laugh to have seen us at our poulterers work; which, being finish'd, we hasten'd back to the inn [to have them cooked].[23]

That afternoon, at the Swan Inn in Hastings, they dined on 'a roasted duck, cold beef, and plumb pye, with 3 wheatears and a half each'.[24]

Arthur Savory grew up at Brighton in Sussex, where in the 1860s he also saw how wheatears were taken: 'They were once very plentiful on the Sussex Downs and seaside cliffs, and as a boy walking from my first school at Rottingdean to visit my people at Brighton, from Saturday to Sunday night, I have passed hundreds of traps consisting of rectangular holes cut in the turf, having horsehair nooses inside, set by the shepherds who took thousands of wheatears to the poulterers' shops in the town. They were then considered a great delicacy.'[25]

Mrs Beeton's mid-nineteenth-century cookery book included a wheatear recipe: 'After the birds are picked, gutted, and cleaned, truss them like larks, put them down to a quick fire, and baste them well with fresh butter. When done, which will be in about 20 minutes, dish them on fried bread crumbs, and garnish the dish with slices of lemon.' She added that the larger wheatears were sent to London to be potted, 'where they are by many as much esteemed as the ortolans of the continent'.[26] Until recently, the French famously – or infamously – would devour almost all of the ortolan, including its skull and bones, so larks and wheatears were most likely eaten in a similar manner.

Ortolan buntings pay rare visits to Britain if they fly off course when migrating to and from Africa. In Europe, they were frequently caught with nets and then fattened for the table as a niche luxury. In 1850 John Baily, a respectable poulterer of Mount Street near Grosvenor Square in London, wrote to Edmund Dixon in Norfolk about these birds: 'I was the first importer of Ortolans into this country, as a matter of trade; some years since I used to get them one or two at a time, and then sold them easily at a guinea each ... We now have them by hundreds, and fat them for the table. They are Buntings, breaking their seed against a small projection from the roof of their mouth ... they live happily in confinement, neither loving nor fearing man.'[27]

Baily explained to Dixon that ortolans became extremely fat when kept in a cage and were best eaten in July and August. Until two years previously, he said, they were enjoyed by visitors to Paris, but that market vanished with the 1848 revolution in France. Instead, many live birds were brought to Britain, which had resulted in prices collapsing, so that he now charged just 1s. 6d. each.[28] For Robert Gray in Glasgow, this cost was still expensive: 'To those who cannot afford even sixpence an ounce for this costly viand, I would recommend Sanderlings [a small wading bird] ... if properly cooked, will not disappoint any one choosing to make the experiment ... The birds, on their arrival in autumn, are loaded with fat, and are then at their best.'[29]

He was not alone in thinking that some birds were overrated. Although it was fashionable for the wealthy to shoot red grouse on the Scottish moors, William MacGillivray was not enthusiastic about eating them: 'As an article of food, the Brown Ptarmigan [red grouse] is highly esteemed. The flesh is very dark-coloured, and has a peculiar somewhat bitter taste, which by some is considered as extremely pleasant, while others affect to relish it because it is fashionable to do so.'[30] Woodcocks were also once highly prized and a great delicacy. A largely nocturnal bird with a long bill, they spend the winter in most areas of Britain, and their arrival from northern Europe each autumn was a real event, as related by *The Sporting Magazine* in October 1794:

Woodcocks are reported already to have made their emigration, by their early appearance in the north, particularly on the mountains in the vicinity of Ambleside, Westmoreland [Lake District]; and the last moon has also brought a flight into the eastern canton of Dorsetshire. The first of these birds was said to have been shot a few days ago at Craneburne [Cranborne, Dorset], by one of the game-keepers of the Marquis of Salisbury. A hard frost and clear night are the seasons these rarities are caught in great abundance upon the northern fells, where they run into snares called springs, whence they supply the markets.[31]

Three months later, the magazine elaborated: 'Tis very remarkable, that when the woodcock first arrives here, the taste of its flesh is quite different from what it is afterwards, it is very white, short, and tender, and seems to have no blood in it, but after it has been in this country a considerable time, it becomes more tough, stringy and fibrous.'[32] A contemporary London cookery book gave practical advice:

They are best about a fortnight or three weeks after their first appearance, when they have rested after their long passage over the ocean. If they be fat, they will feel firm and thick, which is a proof of their good condition. Their vent will be also thick and hard, and a vein of fat will run by the side of the breast; but a lean one will feel thin in the vent. If it be newly killed, its feet will be limber, and the head and throat clean; but the contrary, if stale.[33]

Elizabeth Raffald suggested a simple way of cooking woodcocks. She had once been housekeeper to Lady Elizabeth Warburton at Arley Hall in Cheshire before marrying the estate gardener and moving to Manchester, where she published a popular recipe book in 1769. Woodcock and snipe, she advised, should be plucked of feathers, but not gutted: 'put them on a small spit, dust and baste them well with butter, toast a few slices of a penny loaf, put them on a clean plate, and set it under the birds while they are roasting, if the fire be good, they will take about ten minutes roasting.'[34]

Many other species of wild birds were also considered to be suitable for the table, and in 1802 William Daniel gave his opinion of how the golden plover tasted: 'Their flesh is sweet and tender, they are dressed like the woodcock with their trail [not gutted], and are excellent eating.'[35] Similarly, Peter Hawker was full of praise for the flavour of the fieldfares that he easily shot when thousands took refuge at Keyhaven during the very heavy snowfall of February 1831: 'It was somewhat singular that, although these fieldfares were tamer than sparrows, yet they were as fat as butter, and I never ate any more delicious birds in my life. I had about four dozen in all, and might have had a sack full if I had made a business of it.'[36]

For the poorer classes, sparrows provided a cheap source of meat, and they could even be passed off as ortolans, which Charles Dixon described: 'It should here be remarked ... that too often the birds mentioned in the menu as "Ortolans" or "Larks" are not those birds at all, but Sparrows; yes, in many cases pert little London Sparrows, that are netted at night amongst ivy, and in hayricks where they roost.'[37] Just after World War Two, Norman Wymer said that Oxfordshire folk would trap sparrows at night by lantern light using clap-nets, often along the eaves of a thatched roof: 'there was nothing very difficult about the operation, but sparrow pie – especially when the sparrows were skinned rather than plucked – tasted uncommonly good to the Oxfordshire countryman!' Over a century and a half earlier, Mrs Raffald gave a recipe for 'sparrow dumplin': 'Mix half a pint of good milk, with three eggs, a little salt, and as much flour as will make it a thick batter, put a lump of butter rolled in pepper and salt in every sparrow, mix them in the batter, and tie them in a cloth, boil them one hour and a half, pour melted butter over them, and serve it up.'[38]

From at least Tudor times, hemispherical earthenware pots with a projecting spout were fixed to the exterior walls of buildings to encourage sparrows and starlings to nest, which stopped the birds nesting in the thatched roofs, causing damage. They were especially common in Kent and Sussex, in rows under the eaves of barns and other farm outbuildings, and may have been introduced from the Netherlands. Each pot was fixed to a wall by a nick or perforation in the flat base slotting over a nail, and a large opening enabled sparrow eggs and fledglings to be removed for food.[39] The poet Thomas Randolph, who died in 1635, wrote 'To his well timbred mistresse', comparing parts of her body to elements of a thatched wooden building. The shape of these commonplace pots inspired the lines:

> *Another story make from wast to chin*
> *With breasts like Pots to nest young sparrows in;*
> *Then place the Garret of her head above,*
> *Thatcht with a yellow hair to keep in Love.*[40]

These pots were also known throughout Europe, and Giovanni Pietro Olina, an Italian naturalist who died about 1645, published a picture of one with a spout that was intended for starlings. In Britain, starlings or stares tended not to be eaten, and having spent several months in Italy, from September 1663, Francis Willughby noted: '*Stares* are not eaten in *England* by reason of the bitterness of their flesh: The *Italians*, and other Outlandish people are not so squeamish, but they can away with them, and make a dish of them for all that.'[41] MacGillivray took the opposite view and approved of eating starlings:

> The flesh of the starling is not much inferior to that of a thrush, although tougher ... The Hebridean shooters always twist off its head the moment they get hold of it, alleging that in the blood of that part there is something of a poisonous nature.

Others maintain that the bitter taste which they attribute to the starling's flesh resides in the blood, and that this affords the true reason for the decapitation practised. For my own part I never could perceive any difference between the flesh of a starling merely shot, and that of another both shot and beheaded; and all that I can say on the subject is that both are very good, and not at all inferior to the flesh of the wild pigeon.[42]

Lord Lilford knew about this practice, and in his view 'they are extremely nasty as food ... repulsive on the table in any circumstances'.[43]

The rook was another bird that repulsed some and was eaten with relish by others. According to Daniel, 'The young birds are very good when skinned, steeped in milk, and afterwards put in a pie.' Writing decades later, William D'Urban and Murray Mathew said: 'Young rooks are extremely delicate eating and make delicious pies; Devonshire country people will go any distance to fetch a bunch of them when a great shoot comes off at some large rookery.'[44] Mrs Raffald had a recipe for them: 'A ROOK PYE. SKIN and draw six young rooks, and cut out the back bones, season them well with pepper and salt, put them in a deep dish.' The dish was then filled with water and half a pound of butter before being covered in puff pastry and paper to prevent it burning, 'for it requires a good deal of baking'. The historian Dorothy Hartley advised against making a rook pie, although in the spring, if a large rookery needed clearing, she suggested that the breast and thighs of the young rooks might be used, as long as the rest, being bitter and black, was discarded.[45]

Some birds were truly disgusting to eat, especially carrion or fish eaters. Herons eat not just fish but also creatures like frogs, mice, rats and eels, and in 1854 Grantley Berkeley published a book in which he described how he went fowl-shooting some years previously in Christchurch Harbour along the south coast of Dorset. At times he caught herons by baiting hooks with eels on a fishing line: 'In the maw of one heron so caught were seven small trout, a thrush and a mouse.' Another incident put him off serving heron at dinner: 'I once killed a heron with a huge water-rat in his belly, which effectually prevented my dressing them again at table.' According to D'Urban and Mathew, 'Young Herons are by no means despised for the table, their brown meat closely resembling that of a hare in flavour.'[46]

Another bird that tasted like hare, Henry Graham said, was the green-crested cormorant (shag), which he called a scart. It bred in the sea caves and cliffs, such as Fingal's Cave on the Isle of Staffa, where he would venture on excursions from Iona. Although many people thought that they were offensive and rank, he disagreed:

> Now, from many years' experience, corroborated by my most fastidious friends and *bon-vivants*, who have tried the experiment, I aver that a couple of scarts are equal to a good plump hare. As a proof of this, I have heard a gentleman of property in the islands say to his gamekeeper, 'Donald, the mistress expects a friend to dinner to-day, so you must bring home a hare or a couple of scarts.'[47]

He preferred not to eat them straightaway: 'The scarts should ... be hung for a week or more, according to the weather, then skinned, and treated exactly like a hare, for making that pride of the Scottish *cuisine* – hare soup. Any good recipe for making this should be exactly followed.'

For people living round the coast, the rank taste of sea-fowl might be acceptable, even enjoyable, especially if they knew nothing else, and there were ways of reducing the unpleasantness. D'Urban and Mathew said that they once shot three cormorants on the Taw estuary at Barnstaple in Devon:

> Not knowing what to do with the three heavy birds, when we had them on the sand, we went up to a labourer, who was working in a field at no great distance, and offered to present them to him. He at once accepted them with great delight, assuring us that after they had been buried a little time in the earth they would be as good as turkeys! Like most sea-fowl, if skinned

and soaked for a night in milk-and-water, a Cormorant is by no means bad eating, and we have known an old lady to appreciate a 'black goose,' as she called it. A younger lady, unaccustomed to game, who partook of one, declared it tasted so like grouse that she would not have known the difference but for its size. When any of the sailors on the North Devon coast shoot cormorants or shags, they carry them from house to house endeavouring to sell them as 'Muscovy Ducks'![48]

Robert Gray had never tried eating cormorants: 'From living exclusively upon fish, its flesh, as I have been informed by those who have had the courage to taste it, is peculiarly rank and unpleasant. An old friend of mine told me lately that he had cooked one and eaten part of it about forty years ago, and that the terribly fishy flavour was in his mouth still. Perhaps Milton had some such gastronomic recollections of the bird.'[49] Gray was referring to the lines in Milton's 1667 epic poem 'Paradise Lost', where he compares Satan to a cormorant:

> *Thence up he flew, and on the tree of life,*
> *The middle tree, and highest there that grew,*
> *Sat like a cormorant, yet not true life.*

Gray also wryly commented on the velvet scoter, which is a black seaduck, a winter visitor to Britain that feeds mainly on shellfish: 'Living on such a diet, it is not to be expected that the flesh of this bird can be much esteemed. Some persons seem to think that fresh earth will effectually remove the strongest flavour natural to a fish-eating bird. I would therefore recommend any one desirous of trying the edible qualities of a Velvet Scoter to bury it in a convenient place, *and leave it there.*'[50]

His friend Henry Graham was equally dismissive of the gannet, a large, white seabird also known as the solan goose: 'Though unquestionably the noblest of our water-fowl, not even excepting the swan, yet he is of no value for the pot. It is the only water bird we could not eat. It has a strong, rank smell peculiar to itself, and to the fulmar, storm petrel, and shearwater, which they never lose, even after the skin has been stuffed some years, and which clings to any pocket, bag, or box which has contained them.'[51] In 1769 Thomas Pennant, in a tour of Scotland, mentioned kittiwakes being eaten: 'The young are a favourite dish in North Britain, being served up a little before dinner, as a whet for the appetite; but, from the rank smell and taste, seem as if they were more likely to have a contrary effect.'[52]

Considerable numbers of wild geese were shot or captured each year when they migrated to Britain for the winter, where they fed mainly on grasses in saltmarshes and estuaries. In the West Country, brent geese were once the most common species. 'As we write its name,' D'Urban and Mathew said, 'we seem once more to hear the crunch, crunch of our boat as we are slowly beating against wind and tide towards a flock of these little black Geese on the mudbanks on the Barnstaple river, where it was numerous enough in the old days before the railway ran on either side, when the salt-marshes were still undrained and unenclosed, and when shore-shooters were few.' They used to shoot about six or eight, and sometimes 'out of these obtained, we would find one which was fairly palatable when cooked, but, in general, the flavour of the flesh was strong and fishy'.[53]

A decade or two earlier, during the severe winter of December 1838 and January 1839, Arthur Knox shot several brent geese at Pagham Harbour, south of Chichester, where they were unusually abundant. His opinion of eating them was quite different: 'This is the best bird I ever tasted: the flesh is as tender and juicy as that of a teal, and there is a total absence of the fishy flavour which renders so many of our water-fowl unfit for the table.'[54] Some seabirds were actually classified as fish when it came to religious fasting. The common scoter is a winter visitor, feeding on molluscs and formerly known in eastern England as the black duck. In 1862 Charles Johns wrote: 'The flesh of the Black Duck is said to be oily and fishy; on this account it is in some Roman Catholic countries classed with fish and allowed to be eaten during Lent.'[55]

Almost three centuries earlier, in 1570, John Caius wrote about various animals, including the puffin, and because of its rank, fishy taste, he said that the Church authorities allowed it at times of fasting, when meat was not eaten: 'it is driven out from a rabbit's burrow by a ferret turned in by any hunter in a place situated not far from the sea. It is used as fish among us during the solemn fast of Lent: being in substance and taste not unlike a seal.'[56] In the rabbit burrows of Priestholm or Puffin Island, off Anglesey, thousands of puffins once nested, and during his visits there in 1798 and 1801, William Bingley saw them being captured and preserved: 'Their usual food is sprats and sea-weeds, which render the flesh of the old birds excessively rank. The young, however, are pickled for sale by the renters of the island, and form an article of traffic peculiar to this neighbourhood. The oil is extracted ... and the bones are taken out, after which the skin is closed round the flesh, and they are immersed in vinegar impregnated with spices.'[57]

When James Wilson sailed to St Kilda with Sir Thomas Dick Lauder in the summer of 1841, on approaching the coast, 'an unexpected feature of the island became apparent in thousands of puffins ... flying overhead ... It is a fat little bird.'[58] After being shown round the village, they rested in a house inhabited by a widow and her daughter. The woman had been out that morning to collect some food:

> Their chief sustenance at this time consisted of the small sea-fowl before mentioned under the name of *puffin*. The widow had snared about a score, and having already eaten *a few* for breakfast, was now employed in boiling a corresponding number for dinner. We saw their little fat bodies turning round and round in the pot ... These birds are caught by stretching a piece of cord along the stony places where they chiefly congregate. To this cord are fastened, at intervals of a few inches, numerous hair nooses, and from time to time, when the countless puffins are paddling upon the surface, in go their little web feet, they get noosed round the ancle [ankle], and no sooner begin to flap and flutter than down rushes a ruthless widow woman and twists their neck.[59]

Wilson said that in the summer the natives ate young puffins, fulmars and eggs, and towards the autumn they collected more birds, just before they flew off, for salting and drying.[60]

A few years later, in 1860, Isabella Bird toured the Outer Hebrides. The first woman to be admitted to the Royal Geographical Society, she anonymously published five reports about her findings in *The Leisure Hour*. On Bernera, she said, the islanders relied mainly on seabirds for their living, descending the rocks at the risk of their lives: 'The flesh they salt and eat ... The birds which tempt to those perilous deeds are chiefly guillemots, kittewakes, and puffins. Puffins are procured here, as elsewhere, by dragging them from the holes on the summits of the cliffs on which they breed; but they are obtained also, in large numbers, by a singular method practised nowhere but in Bernera.'[61]

She explained that when fierce south-westerly winds were blowing, puffins would frequently be carried inland, forcing them to wheel back round to their nests, which was a good opportunity to catch them: 'A man lies on his back close to the verge of the cliff, with his head to the sea, and holds a light pole in his hand, which is projected over the edge of the rock. In this position he remains patiently until a bird, driven by the wind, comes within his reach, when one dexterous stroke of the pole brings it stunned to the ground.' During incessant gales, she said, many puffins were caught:

> How true it is that one half of the world does not know how the other half lives, and, perhaps, does not care to know! ... of what interest to the loungers of our clubs and drawing-rooms are the perils of their Bernera kinsmen? ... It would be hard to imagine anything more desperate than the position of these birdcatchers, with the hurricane raging round them, the wrack drifting over them, the sea-birds shrieking as they circle round their heads, and, seven hundred feet below, the Atlantic hurling itself in thunder against the base of the cliffs with violence enough to send jets of spray over their summits.[62]

In winter months, the inhabitants of the Outer Hebrides and other island communities survived on dried and salted seabirds. Elsewhere, the lack of fodder in winter meant that many animals were slaughtered in November and December, and the meat was preserved. Alternative sources of fresh meat included wild game and rabbits, and it was once assumed that domesticated pigeons also fulfilled this role. The exploitation of pigeons extends far into prehistory, as shown by evidence dating back 67,000 years from Neanderthal deposits in Gorham's Cave, Gibraltar. The domestication of pigeons probably began about 10,000 years ago in the Middle East, and some 5,000 years ago pigeons were being eaten in Egypt, where dovecotes developed. Although the Romans used dovecotes, they may only have been introduced to Britain after the Norman invasion.[63]

From the Middle Ages pigeons were widely reared for the nobility and gentry in dovecotes, which were also referred to as pigeon houses, columbaria, doocots and culverhouses. The words 'pigeon', 'doo', 'dove' and 'culver' are interchangeable. Domesticated pigeons were descended from rock doves (also known as blue rock pigeons) that nest in rocky cliffs – they were not wood pigeons. Over time, pigeons became mundane food, but for centuries they were regarded as a luxury and a sign of status, not intended for the poor. From the twelfth century, harsh penalties were imposed for stealing or killing pigeons, and those tenant farmers who were not allowed to have their own dovecotes were particularly resentful if they saw their crops being plundered by the landowner's pigeons.

To prevent dovecotes being raided at night, they were sited in secure positions, often in view of manor houses, farms or priories. Thousands of dovecotes filled with pigeons could once be found across Britain, with many more all over Europe. They were often freestanding structures, built in an array of shapes, primarily circular, square, rectangular, hexagonal or octagonal.[64] Writing from Walton Hall in West Yorkshire in 1836, Charles Waterton commented:

No farm-yard can be considered complete without a well-stocked dovecot, the contents of which make the owner a most ample return, and repay him abundantly for the depredations which the pigeons are wont to make upon his ripening corn. He commands a supply of delicious young birds for his table; and he has the tillage [dung] from the dovecot, which is of vast advantage ... A dovecot ought to be well lighted; and it should be whitewashed once a year.[65]

The interior of dovecotes had hundreds of nesting boxes, in tiers, which have given rise to the word 'pigeon-hole' for any open-fronted recesses or compartments in shelving or desks for filing documents, while the term 'pigeon-holed' means literally 'placed or filed in a particular category'. In the seventeenth and eighteenth centuries, 'pigeon-holes' was a popular table-top game in which contestants had to bowl small balls on a board into a number of arches that resembled pigeon-holes in dovecotes.[66]

Pigeons might be given supplementary nourishment, but by and large they foraged for food, mainly seed, in the neighbouring countryside and, with increased cultivation, pillaged growing crops. A pair of dovecote pigeons was likely to produce two chicks (squabs) six times each year, usually March to September, which they fed with their own regurgitated milk.[67] After four weeks, before they were ready to fly and when their flesh was still tender, the sizeable squabs were collected for eating by positioning ladders against the walls so as to reach the nesting boxes. In some dovecotes, a contraption known as a potence was installed (from the French for 'gallows' or 'gibbet'), comprising revolving beams and one or more ladders.

Squabs could be killed and cooked for a quick meal. In August 1735, four Cambridge gentlemen were touring Kent, among them the Reverend John Whaley, a poet and private tutor to Horace Walpole. From Canterbury they accompanied the Reverend Peter Vallavine to Monkton, where Whaley wrote: 'On Monday Aug : 4. we took a coach & 4, and went with a very honest clergyman to his little Vicarage House in the Isle of Thanet, where we went into the Dove House, killed a dozen pigeons, pluckt them, spitted 'em, roasted 'em and eat them ourselves.' This dovecote was actually mentioned some four hundred years earlier, in a description of the vicarage: 'the vicar shall have one hall with two chambers, a kitchen, a dovecote, a court lodge, or curtilagium, with fit garden or enclosure'.[68]

In his 1768 cookery book, James Jenks said: '*Dove-house Pigeons* are old when they grow red legged; if stale their vents are flabby and green; but if they be new and fat, they will be limber-footed, and feel fat in the vent. *N.B.* By the same rule you choose all kinds of doves, plovers, the fieldfare, thrush, lark, blackbird, *&c.*' Pies made with pigeon meat and other game proved popular, and Jenks gave a recipe:

> Pick and clean the pigeons very nicely, both within and without: cut off the pinions [the wings] and neck. Season each pigeon with pepper and salt, and put a good piece of fresh butter, with pepper and salt, in their bellies. Lay them regular in the dish covered all over with paste [pastry]; and in the center of them all leave a space for the necks, pinions, gizzards, livers and hearts. Then put in as much water as will almost fill the dish. Lay on the top crust, and bake it well. – Some add a beef steak, or a piece of beef skirt at the bottom, to help the gravy. This is the best way to make a pigeon pie.[69]

The rising price of wheat in the Napoleonic Wars led to a reduction in the number of active dovecotes, and abandoned domesticated pigeons from dovecotes moved into towns and cities, where they survive as feral pigeons.[70]

Chickens, turkeys, ducks and geese – poultry – were all domestically reared, though turkeys were only introduced to Britain from America in the late fifteenth or sixteenth century.[71] From the Middle Ages, geese were kept for eggs, feathers and meat, and most domestic geese were descended from the wild greylag geese, with their wings clipped to prevent them flying. Green geese were those killed at twelve to fourteen weeks – 'green' meant young and tender. In late June 1791, James Woodforde, rector of Weston Longville near Norwich, entertained guests: 'We had for dinner two dishes of fryed soals [soles], ham & 2 boiled chicken, a leg of mutton rosted, a green goose rosted, peas and beans, plain batter pudding & gooseberry tarts.'[72] The diaries that Woodforde kept over many years show that he ate more green geese earlier in the year than he did 'stubble geese', which were fattened by being turned out into the stubble fields to eat anything left behind after harvesting. Stubble geese, also known as Michaelmas geese, were eaten from late September.[73]

Geese were of particular significance on 29 September, which was Michaelmas, the Feast of St Michael and All Angels. On that day in 1798, Woodforde wrote: 'Dinner to day, beef-steaks, and a Michaelmas Goose &c.'[74] This was an important date in the calendar, being the first day of the new farming year and one of the quarter days when rents were due and servants were hired. It could also be celebrated on 10 or 11 October, reflecting the calendar change in 1752. A common superstition was that anyone eating goose at Michaelmas would be prosperous in the months ahead, which was reflected in an East Anglian saying: 'If you do not baste the goose on Michaelmas-day, you will want money all the year.'[75] Eating goose was therefore a sign of forthcoming success, and Cassandra Austen would have understood what her sister Jane meant in the letter she wrote to her on 12 October 1813: 'I dined upon goose yesterday, which, I hope, will secure a good sale of my second edition.'[76] Jane Austen was longing for success for her novel *Sense and Sensibility*.

Annual goose fairs were held in or close to market towns and cities, some of which were green goose fairs and others held around Michaelmas, the most famous being at Nottingham and Tavistock. Nicholas Blundell often attended 'goos feasts' during the annual October fair at Great Crosby in Lancashire, as on 18 October 1708: 'It being Crosby Goosfeast I dined at Mr Wairing with him, his two Sons, one Williamson the Schoolmaster of Formby, Thomas Syer of the Ford, &c; My Children dined at Richard Newhouses. The Maids stayed late and were locked out.'[77]

From the eighteenth century, flocks of around a thousand geese were reared in Lincolnshire and Norfolk and then driven to cities like Nottingham and London to be sold, while the geese on Salisbury Plain went to Winchester. Once the geese reached their destination, they were often slaughtered and plucked in the streets, creating a nuisance that city officials periodically tried to control. In London in 1416, it was agreed by the mayor and aldermen that 'geese should in future be sold in the Poultry, and elsewhere within the City of London, whole, together with the heads, feet, and intestines'.[78] When Thomas Pennant was travelling through Lincolnshire in 1769, he learned that all the tough old geese, no longer wanted for feathers, were sent to London with the vast flocks, so that 'all the super-annuated geese and ganders (called here *Cagmags*) ... serve to fatigue the jaws of the good Citizens, who are so unfortunate as to meet with them'.[79]

A century later, *The Leisure Hour* related that many geese were still being driven to market: 'In the eastern approaches to London, we have before now met large flocks of them waddling in their own funeral procession, under the charge of the goose-herd – forlorn hopes, we may call them, coming up to meet the sage and onions which are to consummate their career. A wretched figure they cut, their sleek plumage matted and clotted with mire, and their hungry throats agape, after a march without rations from the neighbourhood of Epping Forest. If, however, they are not in the best condition on arrival, they can be fattened before killing.'[80]

The geese moved at about one mile an hour, grazing along the way, and their feet were shod beforehand by walking them through pools of tar, pitch or clay, followed by fine sand or sawdust, which gave some protection. When cattle were driven immense distances to market, they were shod with iron shoes, and so shoeing the goose with iron horseshoes – an impossible task – was a popular motif in late medieval art in Britain and on the Continent, one of many subjects showing role reversal and absurdity. It developed into a phrase 'Go shoe the goose', indicating derision or incredulity.[81]

In a letter to *Country Life* in 1946, the author Margaret S. Smith from Soberton in Hampshire recalled how geese were imported from Ireland: 'I remember buying young geese fresh from Ireland, as recently as 1900, still shod for the road journey to the port of embarkation. The process was to drive the birds through a pool of Stockholm tar, and then, I think, over loose sand. This gave each bird a pair of stout, flexible plimsolls, that saved its feet from being hurt by the rough road. The feet and legs gradually peeled off, leaving a nice, clean skin underneath and the birds were no worse.'[82]

Geese became popular at Christmas, and in December 1863 *The Leisure Hour* magazine published an account of the different Christmas clubs: 'The poulterer has his goose-club ... the poor-man's butcher has his roast-beef club ... All these clubs are on the simplest plan: you pay sixpence or a shilling a week for thirteen weeks before Christmas, and when Christmas comes you get your poultry or plums, your beef or your pudding, at a good bargain price, without feeling that it has cost you much.'[83] At Christmas, although countless turkeys, mainly from Norfolk, were brought to London, geese remained more desirable: 'They are borne by steam and rail from France and Ireland; they come by truck-loads from near and distant counties; one sees them unpacking from boxes and hampers, ready plucked; while at the same time and place they are pitching by hundreds, with their feathers on, out of waggons and carts, into underground cellars, where fifty women and girls are plucking away at them day and night.'[84]

The displays of poultry being sold in London at this time were impressive:

The consumption of poultry in London always reaches its climax at Christmas time ... the shop of the poulterer – and not only his shop but his entire housefront – undergoes a striking trans-formation; by degrees it envelopes itself in plumage up to the fourth story, it happen to be so high, as though it were prepar-ing to fly away. The geese, turkeys, and barn-door fowls may be reckoned the staple of his store; but besides these, there is every species of British game, from grouse to larks, with no small collection of foreign birds from France, Belgium and Holland. More than this – the poulterer, in the pride of his profession, will exhibit anything rare or curious that has wings to fly, independent of its adaptation to English appetites; a plump seagull, a sprawling stork, a brilliant peacock, a bustard, a huge jack raven – any or all of them he will hang out to view.[85]

One poulterer even managed to obtain a rare specimen: 'Some years back, a fortunate tradesman actually displayed an albatross, or rather the skin and plumage of one, measuring over ten feet between the tips of the extended wings. Such displays gratify the tradesman, who thus hints to the public that the whole domain of earth and air is his warren.' Elsewhere, the geese were sold at all manner of locations: 'they hang, heads downwards, in dense battalions, on bulks and window-boards, not only in poulterers' shops, but at grocers', at milkshops, and dairies, at the pork-butcher's, at the greengrocer's, at the fishmonger's, and even occasionally at the publican's. In no inconsiderable proportion they flank the thronged thoroughfares.'[86]

Inevitably, all the meat on offer in these shops was available only to those who could afford it. When the Swing Riots against agricultural mechanisation led to uprisings in the county of Dorset in November and December 1830, there were arson attacks and fears of more serious destruction. Moreton House near Dorchester, the home of James Frampton, was partly boarded up as a precaution, and some of the Dorset militia was sent there, as he feared revolution. Yet although agricultural workers were suffering, those within Moreton House enjoyed the exotic Christmas splendour, including peacock, as James's sister, Mary Frampton, recorded in her diary:

Our Christmas was passed with a large family party at Moreton. The house was unbarred and unblockaded with the exception of one large window on the staircase . . . The peacock in full plummage with its fiery mouth was placed on the dinner table, with of course the boar's head; the immense candles were well covered with laurel. The hare appeared with the red herring astride on its back, and the wassail bowl and lamb's wool [a wassail drink] were not inferior to former years.[87]

In ignorance, poor people at times ate unusual rare birds that could have been profitably sold to collectors, or else they were eaten in error. Edward Rodd in Cornwall thought it incredible that a red-crested pochard was bought for food: 'A single specimen of this rare Duck was shot at the Swanpool in February 1845, and sold in Falmouth market for sixpence! Unfortunately instead of being preserved it was eaten.' Near Lydd in Kent in May 1856, over a period of three weeks, two labourers kept seeing a large heron: 'Heron makes an uncommon pudding, let us try to shoot it.'[88] Having done so, they found out that it was actually a black stork, rarely seen in Britain, and it ended up being preserved by a bird stuffer there, rather than eaten.

When wandering round the reservoir at Kingbury in Middlesex on 30 August 1864, James Edmund Harting shot a young little ringed plover, which he was able to establish was a different species to the common ringed plover. Subsequently, Robert Mitford, a keen ornithologist and collector from Hampstead, told him that ten days earlier he too had shot a little ringed plover at the same reservoir: 'This specimen, which was also immature, he intended to have preserved, but being called from home before he had time to skin it, it was unfortunately cooked and eaten.'[89]

During the winter of 1878, Murray Mathew learned from his friend Robert Browne, who was vicar at North Curry in the Somerset Levels, that a flock of about sixty Bewick's swans had been observed in the locality for about a month, but nobody could get close:

and although many gunners were on the watch for them, only one succeeded in obtaining a shot. This was a labourer who with a single-barrelled duck-gun knocked over four; two were obtained, two were only slightly wounded, and escaped. One of the two was sold to Mr. Foster, of North Curry; the other my friend the Vicar was anxious to secure, but arrived at the labourer's cottage just as he and his family were sitting down to a dinner of roast swan. A slice off the breast, although tasting both juicy and tender, seemed but a poor equivalent for the loss of what would have been valued as an interesting local specimen.[90]

Far more common than Bewick's swans were mute swans, which were resident all year, because for centuries most had been unable to fly, as their wings were deliberately clipped. On the River Thames and elsewhere, these swans were protected, with penalties for killing them, taking eggs or destroying nests. When visiting London at the end of the fifteenth century, a Venetian nobleman wrote: 'It is truly a beautiful thing to behold one or two thousand tame swans upon the river Thames, as I, and also your Magnificence have seen, which are eaten by the English like ducks and geese.'[91] In 1543, in Henry VIII's reign, the Spanish Duke of Nájera was in England, and his secretary kept a journal: 'A river runs through London, one of the largest I have ever seen ... Never did I see a river so thickly covered with swans as this.'[92]

From medieval times, swans adorned any reputable feast as a status symbol, but to be palatable they needed to be young swans. The monarch had a swan master and deputies throughout Britain to oversee the annual swan-upping (the taking up of swans), usually in July or August, when all the swans in each locality were captured with a shepherd's crook and counted. The cygnets were taken from the parents to have their wings clipped, so that they could not fly away. They also had their upper bills marked with a knife to denote ownership, and sometimes the webbed feet as well. Hundreds of marks were used, such as crosses, circles, notches and chevrons, and the designs were recorded on parchment swan-rolls. The cygnets were fattened up in ponds or stews, and those not needed for fattening were released back into the rivers. Swan-keeping declined in the eighteenth century once turkeys became easier to rear.[93]

Most of the population never ate swan, as it was reserved for special banquets, and that was even the case with the Reverend James Woodforde. He appreciated his food, taking an interest in his meals and recording many more details in his diary than his contemporaries. In January 1780, he noted that he had dined with the squire at Weston Longville, where they had swan: 'We had for dinner a calfs head, boiled fowl and tongue, a saddle of mutton rosted on the side table, and a fine swan rosted with currant jelly sauce for the first course ... I never eat a bit of swan before, and I think it good eating with sweet sauce. The swan was killed 2 weeks before it was eat and yet not the least bad taste in it.'[94]

Woodforde was lucky not to fall ill, but residents of Windsor had a less happy culinary outcome, which was described to Alexander Clark Kennedy, who lived in nearby Eton, by a correspondent who described himself as 'a Windsor lad of 1831':

> I remember about twenty-five years back a Mr. Hughes, connected with the Royal household and well known in Windsor as 'old Buffy Hughes', shooting in the neighbourhood of Clewer Point, Windsor, a couple of Hoopers or Wild Swans [whooper swans]; and I well remember that one was converted into soup, and most of the people who partook of it were very much disturbed in their internal economy.[95]

CHAPTER ELEVEN

———— ·◆· ————

SICKNESS AND DEATH

The Cheviot shepherds say they hear the raven
laugh when some one is about to die.

Newcastle Weekly Chronicle February 1899[1]

The prevention and treatment of ailments and injuries in the past
was based largely on religious and superstitious beliefs, as well
as long-held tradition, and the remedies to us today can appear
bizarre, ineffective, dangerous and repugnant. Some of them
made use of birds in various shapes and forms, from ointments
containing eggs or bird blood to the sacrifice of living hens and
cockerels. Medicine that was based on science developed from the
late eighteenth century, but superstition and quackery continued
to play a major role.

From the time of the ancient Greeks, health problems were
widely imagined to be caused by an imbalance of the four humours
of the body – blood, phlegm, black bile and yellow bile. The stock
treatment was purging, mainly by bloodletting, an idea that fell out
of favour only from the late nineteenth century, and it was done by
applying leeches to the skin or by incising or scarifying part of the
skin with a lancet. When the Welsh naturalist Thomas Pennant
was in Scotland in 1769, he questioned his friend Lachlan Shaw, a
Church of Scotland minister and local historian, about his parish of
Elgin. In response to medical matters, Shaw explained that their
physicians used feather quills, not lancets, for bloodletting: 'they

scarify the flesh about the ancle [ankle], and they take blood from the nasal vein by cleaving the quill of a hen and binding it into four branches, and scarifying the nostrils thereby'.[2]

In 1615 the prolific writer Gervase Markham published *Countrey Contentments*, containing *The English Hus-Wife*, which went through many editions. It claimed to contain all necessary knowledge for a woman to run a household, including medical matters. For toothache, he recommended preparing crushed daisy roots and sea salt, then 'put it into a quill, and snuffe it up into your nose, and you shall find ease'.[3] To look after teeth, Francis Willughby, also in the seventeenth century, mentioned the claw of a bittern: 'The back-claw, which is remarkably thick and long above the rest, is wont to be set in silver for a pick-tooth, and is thought to have a singular property of preserving the teeth.' According to the *Oswestry Advertizer* in 1872, 'There is a small bone in the head of a goose, which is commonly called the "goose's tooth." This is carried in the pocket or about the person for luck, and a sure preventative against toothache. I have seen one of these popular charms so polished and worn that it must have been in the possession of the credulous person for years!'[4]

For treating eyes, Willughby stated that swallows were a remedy for 'dimness of sight' and 'blear eyes', prepared by burning their bodies and 'their ashes mingled with honey and so applied'. In addition, blood from swallows was used for the eyes, 'and they prefer that which is drawn from under the left wing'. Eyes were even covered with their nests in order to heal the redness.[5] The medicinal herb known as greater celandine or swallow-wort was likewise favoured, and its scientific name of *Chelidonium maius* derived from the ancient Greek word for a swallow, since the plant started flowering when swallows reappeared in the spring and then died off on their autumn departure. It was believed that swallows applied the sap to the eyes of their fledglings to give them sight and that they had revealed to humans the plant's alleged curative properties for eye troubles.[6]

The involvement of swallow stones in folk medicine for various disorders including eye conditions dates back hundreds of years.

They were found in the crop or stomach of the swallow, though tiny stones including fossils were also collected as so-called swallow stones. The geologist George Lebour came across them during a visit to Brittany in 1865:

> there exists a wide-spread belief among the peasantry that certain stones found in swallows' nests are sovereign cures for certain diseases of the eye. I think the same notion holds in many other parts of France, and also in some of our English counties. These stones are held in high estimation, and the happy possessor usually lets them on hire at a sous or so a day. Now I had the good fortune to see some of these 'swallow-stones,' and to examine them ... The largest I have seen was three-eighths of an inch long and one-fourth of an inch broad; one side is flat, or nearly so, and the other is convex, more or less so in different specimens. Their peculiar shape enables one to push them under the eye-lid across the eye-ball, and thus they remove any eye-lash or other foreign substance which may have got in one's eye; further than this they have no curing power: the peasants, however, believe they are omnipotent.[7]

Swallow stones were often carried or worn as amulets to cure or prevent other conditions, and Robert Burton, an Oxford scholar and writer (best known for his popular book *The Anatomy of Melancholy*, first published in 1621) said of swallow stones: 'In the belly of the Swallow there is a stone called the *Chelidonius*, which if it be lapped in a fair cloth, and tyed to the right arm, will cure lunaticks, mad men, make them amiable and merry.'[8] In the first century AD, the Roman writer Pliny the Elder described swallow stones in his *Natural History*, and he advised wearing one, especially for epilepsy, which was often referred to as the 'falling sickness'. Willughby also said that swallow stones were worn as amulets 'to help the falling sickness in Children (*bound to the arm, or hung about the neck*)', before adding: 'They report this stone to be found especially in the increase of the Moon, and in the first hatch'd young one. Others take it out in *August* about the full of the moon.'[9]

Numerous other so-called cures existed for epilepsy, a condition that was greatly feared, with the person affected often thought to be possessed by the devil. Burying a live cock or hen was widely practised. While chaplain to the English factory at Algiers in the 1720s, Thomas Shaw travelled across the Middle East and observed everyday superstitious practices, including the sacrifice and burial of cocks and other animals for persistent ailments,[10] but in Britain similar rites were also being carried out. Arthur Mitchell specialised in mental illness and was a commissioner in lunacy for Scotland. As a keen antiquary, he was particularly interested in superstitions, many of which seemed to him pagan in origin. When he was in the Scottish Highlands in 1859, he recorded what he saw being done to cure epilepsy:

> there is still practised, in the North of Scotland a formal sacrifice – not an oblique, but a literal and downright sacrifice to a nameless but secretly acknowledged power, whose propitiation is desired. On the spot where the epileptic first falls a black cock is buried alive, along with a lock of the patient's hair and some parings of his nails. I have seen at least three epileptic idiots for whom this is said to have been done. A woman who assisted at such a sacrifice minutely described to me the order of procedure. In this instance ... three coins were also buried, and a '*curn*' of red onions, pounded small, were applied to the patient's navel.[11]

He was convinced that he had seen more examples of cock burial rituals than most people: 'It would be a great mistake to suppose that the persons referred to are the grossly ignorant, and a still greater mistake to suppose that they are the irreligious. On the contrary, they are often church-attending, sacrament-observing, and tolerably well-educated people.'[12] He learned from Dr John Grigor of the seaside town of Nairn that, some years previously, he was requested to visit a poor man from the fishing community who was prone to epileptic seizures and had died suddenly:

His friends told the doctor that at least they had the comfort of knowing that everything had been done for him which could have been done. On asking what remedies they had tried, he was told that among other things a cock had been buried alive below his bed, and the spot was pointed out ... An old fisherman was asked by the Doctor if he knew of other cases in which this heathen ceremony had been performed, and he at once pointed out two spots on the public road or street where epileptics had fallen, and where living cocks had been cruelly buried, to appease the power which had struck them down.[13]

A healing well dedicated to St Tecla of Iconium (Konya in Turkey), close to the parish church at Llandegla in north Wales, was popular with those seeking a cure for epilepsy, as Pennant recorded in the early 1770s:

About two hundred yards from the church ... rises a small spring ... The water is under the tutelage of the saint; and to this day held to be extremely beneficial in the *Clwyf Tegla*, St. *Tecla*'s disease, or the falling-sickness. The patient washes his limbs in the well; makes an offering into it of four pence; walks round it three times; and thrice repeats the Lord's prayer. These ceremonies are never begun till after sun-set, in order to inspire the votaries with greater awe. If the afflicted be of the male-sex, like *Socrates*, he makes an offering of a cock to his *Esculapius*, or rather to *Tecla Hygeia*; if of the fair-sex, a hen.[14]

Pennant was alluding to Aesculapius, the ancient Greek and Roman god of healing and the son of the god Apollo, whose own attributes included healing (and for whom the cock was sacred). Hygeia was the daughter of Aesculapius and was worshipped as part of his cult.

This Welsh ritual began by carrying the cock or hen in a basket round the well and into the churchyard, before doing a circuit of the church: 'The votary then enters the church; gets under the communion-table; lies down with the Bible under his or her head; is covered with the carpet or cloth, and rests there till break of day; departing after offering six pence, and leaving the fowl in the church. If the bird dies, the cure is supposed to have been effected, and the disease transferred to the devoted victim.'[15] Wirt Sikes heard of an eyewitness to this ritual: 'The parish clerk of Llandegla in 1855 said that an old man of his acquaintance "remembered quite well seeing the birds staggering about from the effects of the fits" which had been transferred to them.'[16]

A similar rite was recorded in another Celtic area: 'in Cornwall a black cock is buried alive on the spot where a person is first attacked by epilepsy; or ... one is drowned in a sacred well'.[17] Transferring disease and pain also involved other birds. In 1883 the Folk-Lore Society published a volume on medicine, written by William Black, a Scottish antiquary and lawyer, who said that Pliny the Elder talked about transferring ailments to a puppy or duck. He then added: 'To inhale the cold breath of a duck is still recommended in England.'[18]

For anyone affected by the plague, Gervase Markham advised a series of measures, including a pigeon: 'with hot cloathes or bricks made extreme hot, and layd to the soales of your feet, after you have been wrapt in woollen cloathes, compel yourselfe to sweat, which if you do, keep your selfe moderately therein till the sore begin to rise; then to the same apply a live Pidgeon cut in two parts.'[19] Samuel Pepys was also familiar with the practice of transferring illnesses to pigeons. In his diary for 19 October 1663, he mentioned Catherine of Braganza, the wife of Charles II: 'It seems she was so ill as to be shaved and pidgeons put to her feet, and to have the extreme unction given her by the priests, who were so long about it that the doctors were angry.'[20]

Pepys subsequently noted another incident: 'Up, and while at the office comes news from Kate Joyce that if I would see her husband alive [Anthony Joyce, his cousin], I must come presently. So, after the office was up, I to him ... and find him in his sick bed ... and his breath rattled in his throate, and they did lay pigeons to his feet while I was in the house, and all despair of him.'[21] Jeremy Taylor was born in Cambridge and ended up as a bishop in Ireland, where he died in 1667. In one of his devotional books, he wrote about this practice as if it was routine: 'though we do not eat them, yet we cut living pigeons in halves and apply them to the feet of men in fevers'.[22]

The writer Samuel Pratt referred to a similar therapy over a century later: 'It is said that a pigeon applied warm to the stomach of children in some dangerous maladies, will attract the disease, and with the loss of its own, save the infant's life.' He cited a recent case, in about 1800, of a baby who had appeared to be dying. As a last resort the parents took a dove, stripped its breast of feathers, and placed it on the child's chest: 'By degrees, the infant recovered; while the bird, receiving the disease, expired in agonies.'[23] In late nineteenth-century East Yorkshire, John Nicholson heard of a related treatment for a painful knee, which had been tried out on animals by vets, but also proved effective on people: 'Kill a cat, split the body length-wise, and apply it while warm to the knee, bind it thereon, and there let it stay, until the cure is complete. A fowl answers quite as well.'[24]

In 1786, when Richard Polwhele was a curate at Kenton in Devon, a village 5 miles south of Exeter, he wrote 'The Cottage-Girl', a poem in which a young woman tries to transfer her 'ague-fits' (probably malaria) to her dog. He added a note to the poem:

> These customs have lately fallen under the Author's observation. It is also usual in this neighbourhood, with those who are affected by an ague, to visit at dead of night the nearest cross-road, five different times, and there bury a new-laid egg. The visit is paid about an hour before the cold-fit is expected; and they are persuaded that, with the egg, they shall bury their ague. If the experiment fail (the agitation it occasions may often render it successful), they attribute it to some unlucky accident that may have befallen them on the way. In the execution of this matter, they observe the strictest silence, taking care not to speak to any one whom they may happen to meet.[25]

Nearly four decades later and now living in Cornwall, Polwhele lamented the fact that traditional village herbalists and other healers had fallen into disrepute, while the fashionable medical fraternity, with nothing better to offer, had abandoned their patients. For an adder bite, he said, the cottage-doctors were successful in cutting open the bite and applying a mixture of plantain and salad oil. In addition, 'The anus of a young pigeon imbibing the poison has been known to effect a cure.'[26] The transference to the pigeon was much like a cure for plague sores mentioned a few years after the Great Plague of London by Joseph Blagrave from Reading, who studied astrological medicine: 'Take a living chick and apply the fundament of the chick into the plague sore, it will draw forth the venom, kill the chick, and cure the patient.'[27]

Donald Morrison, a tenant in the parish of Edderachillis on the north-west Scottish coast, was bitten by an adder, probably in the 1830s:

the effects gave rise to apprehensions of immediate death. When in the greatest agony, the captain of a strange vessel landed on the coast, who prescribed the following singular cure: a young chicken to be split or cut up alive, and instantly applied to the stung part. After the same treatment had been repeated by cutting up alive and applying nine chickens without intermission, the patient was relieved; each chicken which was applied indicating by its swelling that it had absorbed poison. The individual who underwent the treatment recovered, is still alive, and enjoys perfect health.[28]

Towards the end of the century, William Black described a similar practice: 'To cure snake bites, it is said in Worcestershire that the warm entrails of a fowl, newly killed, should be applied to the poisoned part.'[29]

In past times, one single remedy or medicine was often trusted to cure all manner of ills, so it is not surprising that, apart from treating epilepsy and eye complaints, other benefits of swallows were listed by Willughby: 'Swallows *entire* ... cure also the squinancy, and inflammation of the *uvula*, (being eaten, or their ashes taken inwardly) ... A swallows *heart* is also said to be good for the falling sickness, and to strengthen the memory. Some eat it against the quartan ague.' For good measure, he threw in the nest: 'The *nest*, outwardly applied gives relief to the squinancy ... and is good for the biting of an adder or viper.'[30]

Decades earlier, Markham gave details on how to make 'the oil of swallows', which he said was extremely good for broken bones, bones out of joint or any pain or injury in the bones or sinews. A mixture of vegetation was needed, such as walnut tree leaves, violet leaves and rosemary, as well as twenty live swallows:

> beate them altogether in a great morter, & put to them a quart of Neats-foote oyle, or May butter, and grind them all well together with two ounces of cloves well beaten, then put them all together in an earthen pot, and stop it very close that no ayer [air] come into it, and set it nine dayes in a seller [cellar] or cold place, then open your pot and put into it halfe a pound of white or yellow waxe cut very small, and a pint of oyle or butter, then set your pot close stopped into a panne of water, & let it boyle six or eight houres and then straine it.[31]

In 1692 Mistress Jane Hussey of Doddington Hall near Lincoln wrote down recipes for curing ailments in a still-room book, including one for 'oyle of swallows'. She also penned a recipe for 'My Aunt Markam's swallow-water': 'Take forty or fifty swallows when they are ready to fly, bruise them to pieces in a morter, feathers and all together: you should put them alive into the mortar. Add to them one ounce of castorum in pouder, put all these things in a still with three pints of white wine vinegar; distill it as any other water, there will be a pint of very good water, the other will be weaker: you may give two or three spoonfuls at a time with sugar.' This swallow remedy, she said, was 'very good for the passion of the mother, for the passion of the Heart, for the falling-sickness, for sudden sounding fitts, for the dead palsie, for apoplexies, lethargies, and any other distemper of the head, it comforteth the braine, it is good for those that are distracted, and in great extremity of weakness, one of the best things that can be administered: it's very good for convulsions.'[32]

The ingredients of potions often contained the dung of birds. Of swallows, Willughby said: 'The dung heats very much ... and is acrimonious. Its chief use is against the bitings of a made [mad] dog, taken outwardly and inwardly; in a colic and nephritic pains taken inwardly, put up it provokes excretion.'[33] Markham suggested making a plaister for a stitch in the side or elsewhere: 'take doves dung, red rose leaves and put them into a bagge, & quilt it: then thoroughly heate it upon a chaffing dish of coales with vinegar in a platter: Then lay it to the pained place as hot as may bee suffered, & when it cooleth heate it againe.'[34]

While the ritual of transference was a key treatment for epilepsy, other remedies were tried as well, and in a 1677 manual for maid-servants, the dung of peacocks was recommended: 'make it into a powder, and give so much of it to the Patient as will lye upon a shilling, in a little succory [a herb] water fasting'.[35] A shilling was then about one inch in diameter. Willughby mentioned using a mistle thrush for epilepsy: 'For convulsions or the falling sickness, kill this bird, dry him to a powder, and take the quantity of a penny weight every morning in six spoonfuls of black cherry water, or the distilled water of *Misselto*-berries [mistletoe]. The reason of this conceit is, because this bird feeds upon *Misselto*, which is an approved remedy for the *Epilepsie*.'[36]

Eggs have long been regarded as ideal nourishment for a sick person, especially if consumed raw. In 1812 William Holland was visiting a sick parishioner – 'young White' at Plainsfield in Somerset. The son of a farmer, he was about to be married, but was now dangerously ill: 'Miss North ... gave him with a tea spoon a boiled egg. As many as you can, said I, of those eggs but if they were raw it would be better. They are almost raw, said she.'[37] By the end of the month he was greatly improved, contrary to everyone's expectations. Raw eggs were also a frequent ingredient in concoctions, and many examples were given by Markham, including one for scrofula: 'Take *Frankinsence, Doves dung*, and *Wheate flower*, of each an ounce, and mixe them well with the white of an egge, then plasterwise apply it where the paine is.'[38]

Not everything about eggs was beneficial. There was a wide-spread conviction that warts resulted from washing hands in boiled egg water, as in Lancashire during the 1860s:

It is commonly held that washing the hands in water in which eggs have been boiled will produce a plentiful crop of warts. Not long ago two young and intelligent ladies stated that they had inadvertently washed their hands and arms in egg water, and in each case this had been followed by large numbers of warts. This sequence they affirmed to be a consequence, and the warts were shown as an ocular demonstration of the unpleasant results of such lavation.[39]

Parts of various birds were used in making medicines, and oil from seabirds had numerous medicinal applications. When Martin Martin visited St Kilda in 1697, he recorded that the inhabitants valued the oil from fulmars 'and use it as a *Catholicon* [universal remedy] for diseases, especially for pains in the bones, stitches, &c. Some in the adjacent isles use it as a purge, others as an emetic.'[40] Goose fat or oil was an ingredient in cures, and in northern Ireland the fat of a heron killed at full moon was supposedly an excellent remedy for rheumatism.[41] In a letter of December 1826, Matthew Culley described how eagles shot in Sutherland were normally dealt with: 'The head and claws are sent to the Sutherland Society ... The feathers are little worth, but the fat considered a sovereign remedy in all cases of sprains, bruises, &c.'[42] In the Welsh Borders, it was said that the eagle's flesh contained properties that would cure shingles: 'Persons having eaten of this flesh, transmit to their descendants for ever the power of curing the patient by simply breathing on the affected parts.'[43]

By the start of the twentieth century, when medicine was becoming more scientific, some old superstitions felt weird to the ornithologist Thomas Nelson in Redcar: 'Yorkshire folk-lore connected with the Owl family embraces some curious superstitions formerly prevalent in the Cleveland dales, though at the present day these ideas are almost forgotten and exist only in the memories of the oldest dalespeople. The concoction called "Owl-broth" was at one time used medicinally in cases of palsy, but with what effect it would be impossible now to say.' A few years earlier, it was said that for whooping cough, 'here and there in Yorkshire, "Owl-broth" is considered as a certain specific'.[44]

Especially because most treatments throughout history were ineffective, death has been an ever-present fear, and many birds have been linked to this fear. In order to find out how many years they had left to live, children morbidly chanted a version of the cuckoo rhyme:

> *Cuckoo, cherry tree,*
> *Come down, tell me,*
> *How many years I have to live.*[45]

An alternative middle line was 'Good bird, tell me', and the number of times the cuckoo called in response gave the number of years.

In what was once a largely rural world, people would have been able to recognise the many different birds that were present, along with their patterns of behaviour, giving plenty of scope for interpreting them as signs of good fortune or more likely of ill-luck, disaster and death for family, friends and livestock. A constant state of anxiety must have existed, and Wirt Sikes commented:

There are death portents in every country, and in endless variety; in Wales these portents assume distinct and striking individualities ... The Adern y Corph is a bird which chirps at the door of the person who is about to die, and makes a noise that sounds like the Welsh word for 'Come! come!' the summons to death. In ancient tradition, it had no feathers nor wings, soaring without support high in the heavens, and, when not engaged upon some earthly message, dwelling in the land of illusion and phantasy. This corpse-bird may properly be associated with the superstition regarding the screech-owl, whose cry near a sickbed inevitably portends death.[46]

The screech-owl was the name given to the barn owl, and in 1725 Henry Bourne, the curate of All Saints church in Newcastle upon Tyne, wrote that beliefs in such corpse birds were heathen: 'Omens and prognostications of things are still in the mouths of all, tho' only observed by the vulgar ... If an Owl, which they reckon a most abominable and unlucky bird, sends forth its hoarse and dismal voice, it is an omen of the approach of some terrible thing; that some dire calamity, and some great misfortune is near at hand.'[47] In about 1800, when growing up in the village of Painswick in Gloucestershire, Mary Roberts compiled notes about nature, including the owl: 'To some her cry is melancholy and predictive of ill: she is a bird of omen, and of reverential dread; and if the larger species chance to cry beside the cottage-door, they expect to see the master pine away with a slow consuming sickness, or that the favourite bairn, the little one, will be laid under the daisy sheet.'[48]

Nelson described how in the Cleveland dales, an old sign of ill-omen related to the barn owl, known there as an ullot: 'An Ullot's cry thrice heard after rush-light, soon followed by a "fire-fraught" (a hot cinder flying out of the fire), which dies before the one nearest the fire can cast their breath upon it, is a sure sign beyond all doubt the ill one shall die. If there be no ill person at that time, then surely shall one under that thack (thatch) fall suddenly sick, beyond all saving help.'[49] Charlotte Latham at Fittleworth in West Sussex detailed an encounter with a barn owl:

That the screech-owl should be regarded as a messenger of evil by the ignorant can excite no surprise in any one who is acquainted with its unearthly note. I was walking rather late one summer's evening close by the village church, when a strange, startling sound seemed suddenly to come from the low belfry, and continued so long that the inhabitants of the neighbouring cottages ran to the spot, to see what had occasioned it. At length an owl was seen sitting solemnly on the roof-ridge of the church. Satisfied with this discovery of the cause of the excitement, I was walking off, when some women stopped me to express their fears lest we should soon hear of a death in the parish. This, as we might expect, is one amongst the oldest and universal of all superstitions.[50]

Some three decades before, on 13 March 1834, in neighbour-ing Hampshire, Peter Hawker wrote in his diary: 'Poor Leech departed for "another and a better world" at about four o'clock this morning.'[51] As a cherished friend of the Hawker family, Richard Leech had lived with them at Longparish House in Hampshire for over fifty years. Hawker added a note: 'N.B. – I am not over-superstitious, but it was a singular occurrence that about half-past one, shortly before Leech died, the bells in our town house [in London] rang so as to alarm all the neighbours and the police, and not a soul was about, though our people got up to see; and the previous night Longparish House was literally assailed and attacked at the windows by a screech owl – a bird I had not heard of or seen there for many years.'[52] As if further proof, Hawker said that his daughter Mary received a note from her mother who was in London: 'towards four o'clock on the morning of the 13th she was awoke by a most tremendous rattling of her shutters, without anyone or any wind occasioning it; and that she was dreadfully frightened, and observed that "some one was going to die!" she knowing nothing of what had occurred here. These mysteries are facts, to be accounted for as we please.'[53]

Birds probably peck at window glass with the idea that their reflection is a rival, but in times past this had a more sinister interpretation. An undergraduate student at Aberystwyth in the 1920s, who was from the neighbourhood of Mathry in north Pembrokeshire, said: 'In our district birds often give omens; if one taps on the window it is a sign that a death in the house will soon follow.' Another student agreed: 'The people in our neighbourhood [north Cardiganshire] believe that if a robin attaches himself to a house and afterwards pecks upon the window, there will certainly be a death in the house. It happened in our house; a robin pecked upon the window and a death did follow.'[54]

Although swallows were associated with healing, in certain circumstances they were regarded as omens of death, especially when they came down a chimney. One poignant incident was communicated by a correspondent to *Notes and Queries* in 1867, involving a mother who was certain that her daughter would die:

> A lady was mentioning, the other day, a superstition relating to this bird which I do not remember to have heard before, and which is opposed to the general notion of good luck attending it. She was visiting the sick child of a poor woman – a girl about twelve years old – and the child had said something about a hope of soon being able to get out again, when the mother replied, 'You know you never will get well again;' and turning to my informant, said – 'A swallow lit upon her shoulder, ma'am, a short time since, as she was walking home from church, and that is a sure sign of death.'[55]

The Reverend Charles Swainson, in his 1885 work on the folk-lore of birds, mentioned a saying from Sheringham in Norfolk: 'When swallows gather before they leave, and sit in long rows on the church leads, they are settling who is to die before they come again.'[56]

Another unlikely bird to instill fear was the wheatear. In 1835 the writer and polymath Robert Mudie explained that in Scotland (where they were known as clacherans) they often perched on top of mounds beneath which were buried those who had perished along the coast or committed suicide. They were also seen on the top of walls surrounding graveyards, and so imaginations roamed freely due to their association with death, with the birds being persecuted.[57] Some years later, a Scottish zoologist, Edward Alston, referred to what Mudie had written and then quoted from a letter he had received from Robert Gray in Glasgow:

> Being a migratory bird, this species is also the subject of anxious consideration on its arrival. When seen for the first time perched on a rock or stone, it portends a bad year to the person who sees it; but when found perched on a grassy mound, it is looked upon as a happy omen. Two friends, for example, who have not seen

each other for some time happen to meet, and the one asks the other eagerly, 'Have you seen the clacharen yet?' 'Oh, yes!' is the reply, 'I have seen it on a sod – all right! But it sometimes happens that an ominous shake of the head tells its own story.[58]

After serving as a surgeon with the East India Company, Thomas Wilkie set up a medical practice in 1821 at Innerleithen in the Scottish Borders, but even before then he was collecting old ballads, songs and traditions and sharing them with Sir Walter Scott. Numerous tokens of death, he said, were associated with birds, including 'magpies flying before you on the road in going to sermon, or surrounding the house; ravens croaking on the house-top or neighbouring trees to your house',[59] while 'crows or rooks sitting down in the streets of a town or village betokeneth that there will be much death shortly after among the inhabitants'.[60] William Lee in 1887 was living on the outskirts of Newcastle: 'For a magpie to be seen near anyone's doorstep is an omen of death. I was told some time ago by an old but intelligent gentleman, a resident of Winlaton, that when he had seen a magpie on three different occasions fly close to the door of people's dwellings a death took place shortly after.'[61]

Legends were told of birds such as magpies and ravens trying to give warnings about impending danger, including a Cornish one told by Sabine Baring-Gould in which a raven twice dropped a pebble on a quarryman's head, but was ignored. The bird instead fetched some wood from a wreck on the coast and dropped that on his head, making him rush to the beach to salvage more, so avoiding death when a massive rock crashed down in his absence.[62] Lee related a comparable story that he had heard from a long-deceased friend. It happened around 1815 near Haydon Bridge, where they grew up, and highlighted the dangers of ignoring the birds:

a quarryman named T——, and who resided in the above village, proceeded to work early one morning, accompanied by other fellow-workmen. On walking up Cleatby Bank, opposite West Mill Hills, a magpie crossed and re-crossed their path several times. It then disappeared, but, when nearing East Brokenheugh, it suddenly flew in amongst them, nearly knocking the hat off Mr. T——'s head. One of the poor fellows was much alarmed, and advised T—— to return home. 'Not I,' said T——; 'I don't believe in such nonsense.' Arriving at the scene of their labour, they commenced work, and had only been working a short time when an alarm was raised, and each shouted to the other to run for his life. On T—— endeavouring to save himself, his hat fell off, when he stumbled over it and fell, and before he could rise a large stone came rolling down, crushing him to death.[63]

The Scottish ornithologist William MacGillivray was not inclined to believe such signs:

I have no faith in the faculty which ravens, crows, and magpies, are alleged by some to possess, of discovering by the sense of smell or otherwise the existence in a house of disease or death. It is certain that ravens can have no experience in this matter; and if their natural instinct or sagacity should enable them to discover approaching death in a human being, how does it happen that they never settle on the back or in the neighbourhood of a sickly animal, until it has presented visible indications of decay?[64]

In Sussex, Mrs Latham disagreed, as magpies there were generally taken as a bad sign for any animal they perched on: 'and it has perhaps some truth in it: for before the farmer or the shepherd is aware of it, the magpie often smells out a lurking disease, and is known to attack and tear out the eyes of weakly sheep and lambs'.[65]

Robert Hunt in his 1865 book on Cornish folklore mentioned the raven as being a bird of ill-omen:

There is a common feeling that the croaking of a raven over the house bodes evil to some members of the family. The following incident given to me by a really intelligent man illustrates the

feeling: – One day our family were much annoyed by the continued croaking of a raven over our house. Some of us believed
it to be a token [of death]; others derided the idea; but one good
lady, our next-door neighbour, said, 'Just mark the day, and see if
something does not come of it.' The day and hour were carefully
noted. Months passed away, and unbelievers were loud in their
boastings and inquiries after the token. The fifth month arrived,
and with it a black-edged letter from Australia, announcing the
death of one of the members of the family in that country. On
comparing the dates of the death and the raven's croak, they
were found to have occurred on the same day.[66]

In 1593, Thomas Nashe, a writer and friend of Ben Jonson, published a religious lament, in which he said that the plague was a
punishment from God, though the 'vulgar meniality conclude ... it
is like to encrease, because a Hearneshaw [heron] sate on top of S.
Peters Church in Cornehill [London]'.[67] The *Lincoln Herald* related
that when a cormorant settled on the tall tower of St Botolph's
church at Boston, it was watched for hours:

Though seen for the first time by the mass of people on the day
just named [Sunday 9 September 1860], we are informed that
it settled upon the tower on Saturday afternoon, and remained
an hour or two and then flew away, returning again some time
during the night. It left its position again for about two hours on
Sunday afternoon ... and returned in the evening. On Monday
morning Mr Hackford [the caretaker] rose between five and six
o'clock, and finding it still seated upon a corbel of the tower, he
loaded a gun and shot it ... It measured 4ft. 6in. from tip to tip
of the wings.[68]

Later reports suggested that everyone was agitated, fearing
a calamity as the cormorant was a bird of ill-omen. Shocking
news was indeed received about a fortnight later – in the early
hours of the 8th, the *Lady Elgin* steamer had sunk in a collision
on Lake Michigan, with the loss of over three hundred lives,

including Herbert Ingram and his son. Ingram was the founder of the *Illustrated London News* and the Member of Parliament for Boston.[69]

In Yorkshire, if white pigeons alighted on a house or close to somebody, a death was thought likely. In the 1870s and 1880s, Miss Mary Fowler collected various superstitions from the manufacturing districts and colliery villages of the West Riding of Yorkshire, where, she said, a strange pigeon entering a house was interpreted as a sign of death:

> A lady would not believe the doctor when he reported the sinking condition of her husband, for no pigeon had as yet come to warn her. She had caught one in her room before the death of her father, and her sister had been visited by another before the sudden and unexpected death of a daughter, and until one appeared she could not consider her husband a hopeless case. As soon as the doctor called on the following day, she greeted him with the words, 'I can believe you now, for a pigeon came into his dressing-room early this morning.'[70]

Such was the link between pigeons and dying that it was judged to be a bad sign for a sick person to request some pigeon meat to eat, as the Reverend Jonathan Eastwood, curate at Ecclesfield in Derbyshire, learned in 1851: 'On applying the other day to a highly respectable farmer's wife to know if she had any pigeons ready to eat, as a sick person had expressed a longing for one, she said "Ah! poor fellow! is he so far gone? A pigeon is generally almost the last thing they want. I have supplied many a one for the like purpose." '[71] Other morbid beliefs involved chickens, with Wilkie recording that 'Hens falling down dead suddenly and without having been previously ill, is a bad omen to the owner, as he will soon after die.' It was considered unnatural for a hen to crow, something that Thomas Sternberg observed in mid-Victorian Northamptonshire: 'The crowing of a hen . . . is frequently followed by the death of some member of the family.' The only way to avert the calamity, he said, was to cut off the hen's head.[72]

Cocks crowing between nightfall and midnight also meant that someone would die, and in a book on the dialect of the East Riding of Yorkshire, it was said that the cock had a foreknowledge of death: 'Within the last dozen years [about 1865] a Holderness farmer, conversing with a sceptic, exclaimed, "Then dis thoo meean ti say oor awd *cock* disn't knaw when there's boon ti be a deeth i famaly!" [Then do you mean to say that our old *cock* doesn't know when there's going to be a death in the family].'[73] Even in the late 1940s, one village schoolmaster in northern Shropshire said that for those who kept poultry, the crowing of cockerels had to be heeded – if it happened at midnight, then that always meant death.[74]

Robins also caused concern, and not just for tapping on window glass. According to William Henderson, in his 1866 book on folklore in northern England, 'at St. John's College, Hurstpierpoint [near Brighton], the boys maintain that when a death takes place a robin will enter the chapel, light upon the altar, and begin to sing'.[75] About a decade earlier, when the college was still under construction, Baring-Gould was a teacher there, but moved to Horbury in Yorkshire as a curate eight years later, close to Henderson who was preparing his research for publication. He therefore incorporated much material supplied by Baring-Gould, including an incident with a robin at St John's College: 'Singularly enough, I saw this happen myself on one occasion. I happened to be in the chapel one evening at six o'clock, when a robin entered at the open circular east window in the temporary apse, and lighting on the altar began to chirp. A few minutes later, the passing bell began to toll for a boy who had just died.'[76]

Edward Capern wrote a melancholy poem 'The Robin is Weeping' after his daughter Milly died in 1860. He prefaced it with an explanation: 'This Ballad is founded upon a very old superstition, which is still prevalent in North Devon. When a Robin perches on top of a cottage, or on a wall or gate belonging to it, and utters its plaintive monotone, which I have known it to do for a day together, the cottagers say it is "weeping," and that it is a certain token that the baby in the house will die.'[77] Capern was a postman in rural north Devon, often seen stopping to jot down ideas for poems and ballads based on everyday life, for which he achieved much acclaim. A few years later he added a more uplifting final verse and also noted that if a mother lost her infant after such signs from a robin, which was often the case, then all was not lost. Whenever a robin was heard singing in a thorn [bush] on the return of the funeral party from the grave, then 'it is taken as a comforting sign that the little lost one is safe in Heaven'.[78]

CHAPTER TWELVE

———— · ◆ · ————

MYTHS AND MAGIC

In the North Riding it is said that to cast your eye
upon the first Robin through glass, after the winter
quarter has set in, is unlucky.

THOMAS NELSON, Yorkshire
ornithologist, writing in 1907[1]

Throughout prehistory and history, different versions of deities
and spirits have influenced everyday lives, not just in sickness and
death, but in countless other curious ways. The official beliefs from
the religion of one civilisation would inevitably become the superstitions of the next, a process summarised as 'the tattered survival
of a discredited religion'.[2] From time immemorial, birds have been
revered or feared, treated as deities or as messengers of good and
evil from the gods, or simply sent by an all-powerful God for the
benefit of humankind. In Britain, pagan beliefs have co-existed
with Christianity, alongside a host of irrational ideas and fears.
Because most gods were imagined as living in or beyond the sky,
it is no wonder that birds were considered as a supernatural link
between the heavens and earth. In the twenty-first century, these
superstitions are fading from the collective memory, so that they
are now the forgotten history, heritage and folklore of the country,
no longer part of the pattern of everyday life in which birds once
played an intrinsic role, though here and there some birds retain a
mystical significance, and remnants of witchcraft cling on.

There are hints of beliefs in birds possessing mystical powers as far back as the Palaeolithic period, the Old Stone Age. The Lascaux caves of south-western France include paintings of birds that are estimated to be about seventeen thousand years old, and one of them appears to depict a man with the head of a bird, which has been interpreted as a shaman, a combination of priest and sorcerer. The Lascaux paintings have been eclipsed by much older finds elsewhere, including on the Indonesian island of Sulawesi, where hunting scenes that depict humans with bird and animal heads are more than forty thousand years old.

From prehistoric times, people have been accustomed to untold numbers of birds, with large flocks capable of flying very high, swooping about and changing direction as if controlled by some unseen presence. Their behaviour made no sense unless other forces were at play, a belief that was commonly held before human flight first became possible in balloons towards the end of the eighteenth century. Birds mysteriously disappeared for months on end, arriving and leaving at the same time each year, and were capable of singing, mimicry and, remarkably, of feigning injury so as to draw intruders away from their nests.

Farming folk especially were constantly in peril from the weather, diseases and pests, and so they relied on the accumulated experience of generations of ancestors to interpret the messages from nature and the gods. Only a few traditions were intended to invite good luck and fortune, as the overriding need was to avert death, disaster and demon forces, as well as glimpse the future. In 1725 Henry Bourne in Newcastle complained that the common people were far too superstitious and did not have the correct religious beliefs. Instead, they misguidedly imagined that evil spirits wandered about at night until banished at dawn by the cock crowing:

> It is a received tradition among the Vulgar, that at the time of cock-crowing, the Midnight Spirits forsake these lower regions, and go to their proper places. They wander, say they, about the world, from the dead hour of night, when all things are buried in

sleep and darkness, till the time of cock-crowing, and then they depart. Hence it is, that in country-places, where the way of life requires more early labour, they always go chearfully to work at that time; whereas if they are called abroad sooner, they are apt to imagine every thing they see or hear, to be a wandring Ghost.[3]

This belief is used by Shakespeare in the opening scene of *Hamlet* – the ghost of the late King Hamlet appears on the ramparts of Elsinore and is just about to speak when the cock crows, signifying dawn:

> *And then it started like a guilty thing*
> *Upon a fearful summons. I have heard*
> *The cock, that is the trumpet to the morn,*
> *Doth with his lofty and shrill-sounding throat*
> *Awake the god of day; and at his warning,*
> *Whether in sea or fire, in earth or air,*
> *Th'extravagant and erring spirit hies*
> *To his confine.*[4]

Countless prehistoric stone circles and standing stones, as well as strange geological formations, acquired a supernatural association, many of them with cocks. At Looe in Cornwall, it was said that a white-coloured rock on the edge of the harbour was once located above the town, on top of another rock. Each time the cock crowed in a neighbouring farmyard, it rotated three times. Elsewhere, other stones apparently possessed the same gift, including the topmost stone of the Cheesewring, an outcrop of granite slabs on Bodmin Moor.[5] Sidney Addy was a native of Derbyshire and a solicitor who, in 1895, published a book on his region's folklore. He said that a conspicuous pillar or standing stone, about 15 to 20 feet high, was located on a hill at Hollow Meadows near Sheffield and known as the 'Cock-crowing Stone'. When the cock crowed on a certain morning of the year, it turned round. He was unable to establish the particular day, though he did discover that other stones in the vicinity were called cock-crowing stones, while one at

Curbar, known as the Eagle Stone, was also believed to turn round when the cock crowed.[6]

Birds have long played a part in foretelling the future, and one strange incident was a combination of Christianity and pagan forces when a duck became key to the future of a university college. Henry Chichele, Archbishop of Canterbury, decided to found a college in memory of those who had lost their lives in the wars with France. In a dream, he was advised to choose a site in Oxford's High Street, next to the Church of our Blessed Lady the Virgin, and on digging the foundations he would find a huge mallard imprisoned in the cesspool or sewer, well fattened and almost bursting – 'a schwoppinge mallarde imprison'd in the sinke or sewere, wele yfattened and almost ybosten. Sure token of the thrivaunce of his future collidge'.[7]

In February 1438, Chichele came to Oxford, said mass and began digging the foundations of All Souls College: 'But long had they not digged ere they herde … within the wam of the Erthe horrid Strugglinges and Flutteringes, and anon violent Quaakinges of the distressyed mallarde … Now when they brought him forth behold the Size of his Bodie was as that of a Bustarde or an Ostridge.'[8] The duck flew away, never to be seen again. Another version of the story has the Fellows hunting and then feasting on this sewer-fattened creature. Benjamin Buckler, a Fellow of All Souls, was outraged when it was suggested that the mallard was actually a goose, and in 1750 he published *A Complete Vindication of the Mallard of All Souls College*.[9] Whatever the truth, the mallard became a symbol of the good fortune of the college, and a hunting of the mallard ceremony takes place once every century, including the singing of their old ballad:

Griffin, bustard, turkey, capon,
Let other hungry mortals gape on;
And on the bones their stomach fall hard,
But let All-Souls' men have their mallard.
Oh! by the blood of King Edward,
Oh! by the blood of King Edward,
It was a swapping, swapping mallard.[10]

During the Roman republic, augury was critical in foretelling the future and to ascertain if a proposed course of action had divine approval. The reading and interpretation of signs from the gods, known as *auguria*, relied on examining the flight pattern of wild birds or the feeding habits of sacred chickens held in captivity. The auspices were taken by magistrates or priests before any event such as a voyage, battle or Senate meeting, and sacred chickens were even carried by armies in the field, so that they could be observed eating before a battle.[11] In 249 BC, during the First Punic War, when on the point of engaging the Carthaginian fleet in the naval battle of Drepana, the Roman commander Publius Claudius Pulcher was told that the sacred chickens would not eat. Furious, he threw them overboard, with the retort 'let them drink instead'. There was little surprise when he was defeated, losing 93 of his 123 warships.

Cockerels and other birds were seemingly able to encourage or sense victory in warfare. In the Georgian era, ships of the Royal Navy carried chickens as a source of food, and in battle they were stowed in coops and put in small boats. One cock was free to roam during the naval Battle of the Saintes, where the British fleet, led by Admiral Sir George Rodney, heavily defeated the French: 'The little bantam cock which, in the action of the 12th of April [1782], perched himself upon the poop, and, at every broadside poured into the Ville de Paris [the French flagship], cheered the crew with his "shrill clarion," and clapped his wings, as if in approbation, was ordered by the Admiral to be pampered and protected during life.'[12]

At the Battle of the Nile on 1 August 1798, a small bird was observed on board Rear-Admiral Horatio Nelson's flagship, the *Vanguard*, and apparently brought good fortune, as the French were again defeated. A few weeks later, the vessel arrived at Naples, and Miss Cornelia Knight went on board: 'I remarked a little bird hopping about on the table,' she said. 'This bird had come on board the Vanguard the evening before the action and had remained in her ever since. The admiral's cabin was its chief residence, but it was fed and petted by all who came near it, for sailors regard the arrival of a bird as a promise of victory, or at least an excellent omen. It

flew away, I believe, soon after the ship reached Naples.'[13] During
the Battle of Trafalgar on 21 October 1805, the *Colossus* became
badly damaged, but Captain James Morris refused to give up: 'She
had a hen-coop on board, and during the battle the cock flew out
and perched on the captain's shoulder and crowed loudly, much to
the amusement of the crew, who cheered while they kept up the
fighting.'[14] The French and Spanish fleet was decisively defeated,
and the *Colossus* even captured two ships.

Sparrows allegedly foretold the last battle that was fought on
British soil, at Culloden near Inverness on 16 April 1746. This
battle was between the Jacobite army of Charles Edward Stuart
and the English army under the Duke of Cumberland, and it
brought an end to the Jacobite rebellion. When John Spooner was
sixteen years old, he spent some time with William Thornton
and the troop of volunteers he had raised in 1745 to help fight the
Scottish rebels. Later on, Spooner kept a hostelry at Brimham
Rocks in Yorkshire, and in 1805 he told a guest, Charles Fothergill,
what he had heard when the entire army was returning after the
Battle of Culloden:

> It was related by the natives of the country immediately about
> Culloden moor that on the very day 12 months preceding the
> decisive battle fought on that moor between the English and the
> Rebels, a vast number of the common house-sparrows assembled
> on the moor and fought so long and so desperately that the
> ground for a considerable space was strewed with the dead,
> insomuch that many people who witnessed the circumstance
> took up many skuttles full, amounting in the whole to several
> bushels: this was told to the British army after the victory in a
> superstitious manner.[15]

Whether or not this sparrow battle was an omen, it felt sinister
enough to Fothergill to note in his diary during his Yorkshire tour.

According to legend, sparrows themselves were known as a
weapon of war. Burning arrows and other incendiary missiles fired
at ships' sails or thatched roofs was a known tactic, and sparrows

and other birds allegedly played a similar role in early medieval times, when flammable materials were tied to them and ignited before they were released.[16] Alexander Neckam was abbot of Cirencester (the old Roman town of Corinium Dubonnorum) from 1213 until his death in 1217, and he related that the Saxons besieged the town in the sixth century in the wars against the Britons. Their leader Gormund noticed that sparrows nested in the thatched roofs, and so he had his men catch large numbers when they flew into the surrounding fields to feed. By tying burning straw to their tails, the sparrows returned with fire to the thatched buildings, and after the Britons abandoned the burnt-out town, Cirencester was supposedly known as Urbs (or Civitas) Passerum, or Sparrowcester.

A few years earlier, in the late twelfth century, Gerald of Wales had described this same Gormund coming from Africa to Ireland, and then being invited to England by the Saxons to besiege Cirencester, which he reduced to ashes by deploying sparrows.[17] When the antiquary Thomas Wright edited the writings of Gerald of Wales, he added a note that a similar legend existed 100 miles to the north, at Wroxeter:

> The people of Wroxeter in Shropshire still tell how, when the barbarians laid siege to the Roman city of Uriconium (of which Wroxeter is the site), and could make no impression on its walls, they collected all the sparrows from the surrounding country, and having tied burning matches to their legs, set them at liberty. The sparrows flew into the city, and settled on the roofs of the houses, which, being thatched with straw, took fire immediately, and during the confusion caused by the general conflagration, the besiegers forced their way into the city. The same story is told of Silchester, the Roman Calleva [in north Hampshire].[18]

Wright was involved in archaeological excavations at Wroxeter from 1859. 'When I was first at Wroxeter to watch the excavations,' he said, 'one of the inhabitants came to me and offered to conduct me to the field where the sparrows were let loose.'[19] There is actually no evidence of Saxon sieges ever taking place at these old Roman towns. Instead, these were local versions of an ancient legend that is found in Icelandic sagas and across Europe.

The eagle played a role during the Great Siege of Gibraltar when, in September 1782, the Spanish and French were about to launch an attack on the British-held territory. Suddenly, the British soldiers cheered, thinking the Royal Navy had arrived, especially as they could see a flag for a fleet hoisted on the signal-house flagpole. The flag turned out to be an eagle, and the fleet was a French-Spanish one. Eagles from Africa did occasionally fly into Gibraltar, and some bred in remote places at the top of the Rock, but those officers and soldiers who knew some classical history decided to interpret this eagle as a sign of victory. John Drinkwater of the 72nd Regiment commented: 'Though less superstitious than the ancient Romans, many could not help fancying it a favourable omen to the Garrison.'[20] The French and Spanish forces were decisively defeated and failed to capture Gibraltar.

Eagles were important to the Roman army some two thousand years earlier. As a rallying point during battle, army units had standards with images that included eagles, wolves and boars, but in 104 BC the consul Gaius Marius reformed the legionary standards, keeping only the eagle as a motif, since it represented strength and power. They became such significant cult symbols that a legion might be disbanded if their eagle was lost. The single-headed eagle of these standards was adopted as a symbol by the Holy Roman Emperors, starting with Charlemagne who was crowned by the Pope in 800. The eagle then came to be widely adopted as a symbol of empire and a heraldic device.

In heraldry, the eagle was depicted in different ways, but usually as an 'eagle displayed', with its legs, wings and tail outstretched. Those with a single head were ultimately derived from the Roman legionary eagles, but two-headed eagles are known from ancient Near Eastern civilisations. The heraldic 'eagle displayed' could have one or two heads and was informally known as a spread eagle, which became a common inn sign in England from at least the seventeenth century, often reflecting the coat of arms of the local lord. In everyday language, 'spread-eagled' meant outstretched for punishment by flogging.

When American independence was declared on 4 July 1776, it was resolved to devise a national coat of arms for a Great Seal, a task that took six years. The Great Siege of Gibraltar was part of the American Revolution, and on 16 September 1782, three days after the failed French and Spanish attack, the United States of America used its Great Seal for the first time, on a document giving George Washington the authority to negotiate with Britain over the exchange and better treatment of prisoners-of-war. The design featured the displayed eagle – or spread eagle – with its head looking to the right. A bundle of thirteen arrows is clasped by the left talon, representing the original states, and an olive branch of peace by the right talon. Held in its beak is a scroll with the Latin motto *E pluribus unum* ('out of many, one'). Rather than the imperial eagle, the native American bald eagle was chosen, which became the national symbol of the United States. It was once a very common bird, about 3 feet in length and 6 to 8 feet in wingspan, with a white head and tail – the word 'bald' has the old meaning of 'white'.

Twenty-one years after the United States adopted the eagle symbol, Napoleon Bonaparte was preparing to invade Britain. His troops were constructing a huge camp on the French coast at Boulogne, and a Roman legionary eagle standard was unearthed near the location earmarked for Napoleon's pavilion. This perhaps influenced him in his choice of an imperial symbol, because in July 1804 he wrote to Marshal Berthier from this camp announcing that each regiment would be issued with an individual eagle standard.[21] On 2 December Napoleon was crowned emperor, and three days later he distributed these standards to his regiments. They had cast gilded bronze eagles that were based on the Roman eagle, with its talons clasping the thunderbolt of Jupiter, who was the god of all gods – reflecting Napoleon's rise to power.

Eagles were favoured by political and military leaders, especially emperors, because they appeared to fly higher than any other bird and approach the sun, and as such they were significant in Christianity, including as a symbol of St John the Evangelist. The eagle could supposedly stare unflinching at the sun, and so it was likened to believers contemplating the holy word. Medieval bestiaries illustrated animals in order to show how each one fitted into Christian doctrine, and the eagle was identified as the king of all birds, looking up to the sun to rejuvenate itself, just as the faithful were supposed to look to God for renewal. These bestiaries also told the mythical story of old eagles flying up to the sun until their wings caught fire, so that they fell back into water, from which they would emerge renewed.[22]

The phoenix was another creature featured in bestiaries, a mythical, eagle-like bird that became a symbol of the resurrection. The main myth can be traced back to ancient Egypt, and it was encountered in the fifth century BC by the Greek historian Herodotus when he visited Heliopolis in the Nile Delta. This was the centre of worship of Re-Atum, god of the sun and creation, who was closely associated with the sacred mythological benu-bird that was usually portrayed as a grey heron. At Heliopolis, Herodotus learned about a legend involving a sacred bird called the phoenix, possibly the benu-bird, though it had red and golden plumage in images and supposedly visited Egypt from Arabia every five hundred years, when the old phoenix died.

Although Herodotus did not believe the legend, other Greeks embellished it to create the myth of the phoenix reproducing itself by dying in a fire, with a new phoenix rising from the ashes. Belief in a such a bird continued through the medieval era, when it was often portrayed with a sun-like halo, and in the Renaissance it was adopted as a symbol by poets and playwrights. During Elizabeth I's reign, she was compared to the phoenix, and in the early Stuart play *Henry VIII*, a collaboration between John Fletcher and William Shakespeare, Archbishop Thomas Cranmer baptises the infant Elizabeth and equates her with the phoenix:

> *The bird of wonder dies, the maiden phoenix*
> *Her ashes new create another heir*
> *As great in admiration as herself*
> *So shall she leave her blessedness to one –*

When heaven shall call her from this cloud of darkness –
Who from the sacred ashes of her honor
Shall starlike rise, as great in fame as she was.[23]

These lines probably referred to James I succeeding to the throne after the death of Elizabeth I, 'the maiden phoenix'. Today the phoenix is a general symbol of hope – of something rising up from disaster to be successful once more.

Within churches, sculptures of eagles were often incorporated into freestanding lecterns or bookstands. Early ones were of wood, but from the fifteenth century brass eagles were popular. They were placed in the choir, though after the Reformation the east end of the nave was preferred, where they supported the Bible, which was now printed in English.[24] Clergymen interpreted the eagle symbolism in different ways, and in 1922 William Smalley Law, the vicar of Oundle in Northamptonshire, wrote: 'Having regard to the date of printing, and the story of the English Bible, we must not imagine the first purpose of the lectern to have been that for which it is used to-day. It would probably stand in the choir to hold the books from which the chanters chanted the sung portions of "The Hours." The ancient custom of placing the book-stand on the wings of an eagle comes from the symbolisation of St. John's gospel by the eagle.'[25] His own church of St Peter's at Oundle still has a fifteenth-century brass lectern in the form of an eagle, which supposedly came from the church at nearby Fotheringay.

In the seventeenth century, particularly the period 1640 to 1660, the Puritans in England were a powerful iconoclastic force, ridding churches of popish ornaments such as paintings, candlesticks, stained glass windows and brass eagles, as happened at St Stephen's church in Norwich, where the brass eagle, probably a lectern, was sold off in 1656.[26] Even after the restoration of the monarchy in 1660, church furnishings were not safe from the clergy. Writing about his native Bristol, the Reverend John Evans, a schoolmaster and minister, said: 'In St. Mary Le Port Church is placed the Eagle which once stood in the Cathedral, and which in 1802 excited no inconsiderable attention in the city.'[27] The brass eagle had been given to Bristol Cathedral in 1683, but in 1802 the Dean and Chapter sold it as scrap metal, weighing 692 pounds. A horrified citizen discovered its whereabouts and purchased it, only to be forced to sell it at auction. The auctioneer Thomas Kift placed a lengthy advertisement in the newspapers:

> THE EAGLE from the BRISTOL CATHEDRAL. TO BE SOLD BY AUCTION, At the Exchange Coffee-Room, in this City, on Thursday, the 2d of September, 1802 ... A BEAUTIFUL BRAZEN SPREAD EAGLE, With a Ledge at the Tail, standing on a Brass Pedestal, supported by four Lions, one at each Corner ... Such a handsome Bird would be, as it has hitherto been, a very great ornament to the middle Aisle of a Church. It for many years stood in the Choir of the Bristol Cathedral, and upheld with its wings the Sacred Truths of the Blessed Gospel. The Minor Canons formerly read the Lessons on it, and in most Cathedrals the custom is kept up to this Day.[28]

Kift was nonetheless critical: 'N.B. – The Purchaser offered, previous to any Advertisement, to re-sell the Eagle, at the price he paid for it, provided it were replaced in the Choir; which offer was rejected.'[29] Evans was also not impressed, saying that the Dean and Chapter used the money from the scrap metal to purchase a piece of plate for the cathedral communion service without any public consultation. At the auction, the eagle lectern was bought for St Mary le Port, in the heart of Bristol, only to be destroyed when the church was bombed in World War Two.

Two of the smallest birds in Britain – robins and wrens – were also highly regarded and yet feared, a mixture of religious and superstitious sentiment. It was even argued that they were the male and female of the same species, probably because of their close association during winter. Both birds were credited with covering the face of anyone they found lying dead, and the opening lines of John Webster's 1612 poem 'A Land Dirge' mention them in that role:

Call for the robin-red-brest and the wren,
Since ore shadie groves they hover,
And with leaves and flowres doe cover
The friendlesse bodies of unburied men.[30]

This superstition is known from the late sixteenth century, when the population was vastly reduced after suffering so many deaths from the plague and other epidemics. It may have been common to stumble over corpses outdoors, strewn with leaves.

A contributor to *The Book of Days* in 1863 said it was unlucky to kill a robin and related a poignant tale. 'How badly you write,' I said one day to a boy in our parish school; 'your hand shakes so that you can't hold the pen steady. Have you been running hard, or anything of that sort?' 'No,' replied the lad, 'it always shakes; I once had a robin die in my hand; and they say that if a robin dies in your hand, it will always shake.'[31]

Charles Clough Robinson of Leeds in Yorkshire, an authority on the local folklore and dialect, told *Notes and Queries* in 1868 how country people believed that if a robin was killed, the cows belonging to the person or family would give 'bloody milk', as had occurred near Boroughbridge, some 25 miles north of Leeds:

A young woman, who had been living at a farm-house, one day told her relatives of the circumstance having occurred to a cow, belonging to her late master, giving bloody milk after one of the family had killed a robin. A male cousin of hers, disbelieving the tale, went out and shot a robin purposely. Next morning her uncle's best cow, a healthy one of thirteen years, that had borne nine calves without mishap, gave half a canful of this 'bloody' milk, and did so for three days in succession, morning and evening. The liquid was of a pink colour, which after standing in the can, became clearer, and when poured out, the 'blood' or the deep red something like it, was seen to have settled to the bottom. The young man who shot the robin milked the cow himself on the second morning, still incredulous. The farrier was sent for, and the matter furnished talk to the village.[32]

Robinson invited readers to contact him if they wished to verify the story, as he had the names and places of those involved. Another reader, C.A. Federer, informed him that the same superstition occurred in the Swiss Alps, where the herdsmen never killed robins for fear of their cows giving red milk.[33] The explanation was said to lay in the notion of the robin 'having attended our Lord upon the cross, when some of His blood was sprinkled on it', resulting in a red breast. A Newcastle newspaper of 1899 added: 'Robin got his red breast from being near the Cross, and some say he is a lucky visitor, but most people feel "creepy" at the sight of him.'[34]

Other superstitions involving birds were also linked to Easter, and Thomas Nelson mentioned that, in many districts of North Yorkshire, it was customary for country people to wear some new article of clothing on Easter Sunday for fear of offending the crows or rooks. This was summarised in a dialect poem by Elizabeth Tweddell:

> *An' if you've nowt te put on new*
> *There is a fine ta deea;*
> *Fer t' craws is seer te finnd it out,*
> *An' soil yer awd cleeas mair.*[35]

The flights of rooks were watched carefully at Easter, and if they stayed near their rookery in order to feed, rather than flying some distance, the farmer believed that his crops would be afflicted by grubs and pests that year. References to crows are difficult to interpret, because rooks may be meant, and Nelson pointed out the common belief in Yorkshire that a rook became a crow after its first moult: 'Of this idea I had oral proof from the wife of a country gentleman in the North Riding, who remarked, in driving past a Rookery late in the year, "I suppose they will soon be growing into Crows".' He also thought that the carrion crow was most likely meant in the children's rhyme:

> *Crow, Crow, get out of my sight,*
> *Or else I'll eat thy liver and lights.*[36]

In Suffolk it was fortunate to see two crows sitting in the road and even more so if six were encountered in the same place, while in Northamptonshire, Sternberg said, 'To see a crow flying alone is a token of bad luck. An odd one perched in the path of the observer is a sign of wrath.'[37] He referred to a few lines from 'Recollections after a Ramble', the 1821 poem of John Clare, who was from the same county:

> *Odd crows settled on the path,*
> *Dames from milking trotting home*
> *Said the sign foreboded wrath,*
> *And shook their heads at ills to come.*[38]

Some thought it lucky to see a flock of rooks fly over, and in Pembrokeshire in the 1920s, according to a student at Aberystwyth, 'A flock of birds flying over a house is an ill omen, unless indeed the birds are rooks, which are lucky.' Another student, known only as 'Miss P.', related that near Cardigan, 'Rooks are very much liked in our neighbourhood; they are supposed to keep crows away, and, as the crow steals chickens, the rook is regarded as a protector. We count omens by crows: "one for sorrow, etc." (The familiar magpie rime.)'[39]

In the Welsh Marches in 1872, a local newspaper, conflating crows and rooks, said that to see a certain number of rooks was understood to be full of significance:

> *One crow, bad luck*
> *Two crows, good luck;*
> *Three crows, a burying;*
> *Four crows, a wedding.*[40]

The variation in superstitions across the country about rooks and crows is a good example of how complex and confused different beliefs became over the centuries. By contrast, the range of beliefs about magpies is relatively narrow and consistent, though numerous versions exist of the fortune-telling rhyme about seeing magpies.

The best-known one has its roots in the 1970s, as the signature tune for the popular children's television programme *Magpie*:

> *One for sorrow,*
> *Two for joy,*
> *Three for a girl,*
> *And four for a boy,*
> *Five for silver,*
> *Six for gold,*
> *Seven for a secret,*
> *Never to be told.*

In 1838 Charles Waterton, of Walton Hall, published his thoughts on magpie beliefs:

> in general, the lower orders have an insurmountable prejudice against it, on the score of its supposed knowledge of their future destiny. They tell you that, when four of these ominous birds are seen together, it is a sure sign, that, ere long, there will be a funeral in the village; and that nine are quite a horrible sight. I have often heard countrymen say that they had rather see any bird than a magpie; but, upon my asking them the cause of their antipathy to the bird, all the answer that I could get was, that they knew it to be unlucky, and that it always contrived to know what was going to take place.[41]

Some three decades later, in West Sussex, Charlotte Latham recorded similar views: 'Whenever I questioned my poorer neighbours about their evident dislike of it, they always answered that it was a bad bird, and knew more than it should do, and was always looking about and prying into other people's affairs.'[42] The idea of birds knowing too much was also mentioned by Edward Armstrong, who was brought up in remote Irish countryside: 'In parts of Ireland birds which come into cottages are killed lest they carry away good luck or betray family secrets.'[43]

The servant girl of the Reverend Henry Humble in Perth once

told him why the magpie was unlucky: 'It was the only bird which did not go into the Ark with Noah; it liked better to sit outside, jabbering over the drowned world.'[44] When Arthur Savory was farming at Aldington in Worcestershire before World War One, he said that although some magpie rhymes made 'one' an unlucky number, the most complete rhyme he had heard was quite the opposite:

> *One's joy, two's grief,*
> *Three's marriage, four's death,*
> *Five's heaven, six is hell,*
> *Seven's the devil his own sel'.*[45]

A contemporary Yorkshire dialect rhyme was related by Thomas Nelson, also ending in the seventh magpie being the devil:

A rhyme in vogue in country districts runs: –

> *'One for sorrow, two for mirth,*
> *Three a wedding, four a birth,*
> *Five heaven, six hell,*
> *Seven the deil's own sel'.'*

The indications vary in different districts: 'Four for death, five for rain' being substituted in some places, though it appears to be a general custom to endeavour to avert the disaster thus liable to be brought by making as many crosses on the ground as there are birds seen.[46]

The mention of crosses links magpies to witches, because making crosses was one way of averting the evil of witchcraft. A single magpie was usually most feared, and a variety of actions might be taken in such circumstances, as Nelson explained:

If a single bird crosses the path of anyone setting out on a journey it is a sure sign of ill-luck for the day, and persons have been known to turn back from a contemplated journey for this reason; but to counteract the evil influence it is the practice in North Riding country districts to make a cross in the air, or to take off the hat and make a polite bow; and in the West Riding the custom is to cross the thumbs, in addition to crossing oneself, repeating the lines: –

> '*I cross the Magpie.*
> *The Magpie crosses me.*
> *Bad luck to the Magpie.*
> *And good luck to me.*'[47]

Peter Hawker was usually a pragmatic person, but on Friday 7 November 1845 he went into Andover for a day's business and his horse became lame: 'When within a mile of home [at Longparish] I turned round to bow to Counsellor Missing, and at that very instant down came the mare and broke her knees. Why was all this? Because we took Friday for a day of business.'[48] He had a seaman's aversion to setting out on a Friday, though that was not all, because he believed that 'a single magpie foretold our disaster'.

William Henderson related what happened to a more cautious gentleman in Yorkshire:

> a lady of that county, Mrs. L—, tells me a curious instance of the good effects of attending to the magpie's warning. It relates to a gentleman with whom she was well acquainted, a county magistrate, and a landowner. One day, in the year 1825, he was riding to York with the view of depositing his rents in Chaloner's Bank, when a magpie flew across his path. He drew up his horse, paused a moment, and turned homewards, resolving to defer his journey till the next day. That day, however, the bank failed, and it only remained for the gentleman to congratulate himself on his prudent attention to the magpie's warning.[49]

This occurred during the banking crisis of 1825–6, with *The Times* publishing a report from a Yorkshire newspaper on 13 December 1825: 'It is with the deepest and most unfeigned sorrow we announce that the bank of Messrs. Wentworth, Chaloner and Rishworths, has stopped payment.'[50] The bank had collapsed.

Savory said that in Worcestershire, 'there are many people … who never see a solitary magpie without touching their hats to avert the omen, and convert it to one of good-luck; as a man once said to me, "It is as well not to lose a chance." '[51] On another occasion, in about 1865, a gentleman was walking with a companion near Reading in Berkshire and spotted 'a country fellow' about 50 yards ahead: 'He suddenly pulled off his hat, and made a sort of bow, though there was no one in his sight, we being behind him. On asking the reason of this strange proceeding, my companion pointed out a magpie which had just quitted a wood, and was flying across the road, and told me it was a general practice there to pull off the hat to the magpie "for luck".'[52] This story was sent to *Notes and Queries*, with the question 'Does this superstition prevail anywhere else?', to which Charles Garth replied that many superstitions may be perceived as local, but in reality belonged to a widespread tradition:

> It would be shorter, I think, to say where this did not prevail; but I add a few counties where it has come within my own knowledge: Cumberland, Westmoreland, the whole of the north of Lancashire, some parts of Yorkshire, all Cheshire, Derbyshire, Staffordshire, Shropshire, Warwick, Hertfordshire, Oxford, Devonshire, and Somerset. I would add that in the 'High Peak' in Derbyshire, a crucial flourish on the breast is often substituted for a bow.[53]

A Liverpool correspondent discussed the same superstition in Lancashire and Yorkshire:

> Its appearance singly is still regarded in both these counties by many even of the educated representatives of the last generation, as an evil omen, and some of the customs supposed to break the charm are curious: one is simply to raise the hat as in salutation; another to sign the cross on the breast; another to

make the same sign by crossing the thumbs. This last custom is confined to Yorkshire, and I know one elderly gentleman who not only crosses his thumbs, but spits over them when in that position; a practice which was, he says, common in his youth.[54]

According to D'Urban and Mathew, this also occurred in Devon: 'we have known superstitious people spit on the appearance of a single Magpie to avert the evil omen!'[55] Spitting was a universal practice, done to seal deals or when money changed hands, and also to ward off evil spirits.

Along with fears of ill-omen from specific birds, a general fear of witchcraft prevailed, which is one reason why counter-witchcraft deposits were incorporated in houses. They were not deposited by witches to cause harm, but were intended to bring good luck and prevent witches and other malevolent spirits from gaining access. Concealed in walls and buried under thresholds and floors, with many more round chimneys and hearths, they included cocks and chickens, though more common were cats, horse skulls and artefacts such as shoes. Quite often, they appeared to be a hotch-potch of objects, deliberately concealed in a single event. Many of the objects resemble rubbish, but their ritual significance was in the mind of the person responsible, and the true reason may never be known.

Throughout prehistory and up to the medieval period, smoke from open domestic hearths found its own way out of the roof of a building. The first chimneys were constructed from the twelfth century, enabling smoke to be channelled up a flue, but they became commonplace in domestic buildings only from the sixteenth century, when bricks for building and coal for fuel were more readily available. It then became possible to subdivide the large open rooms vertically and horizontally with internal walls and ceilings, and many more hearths and fireplaces were installed.[56] From the sixteenth to eighteenth centuries, when this refurbishment was taking place, ritual deposits were often deliberately placed within the hearths or chimneys, as well as in voids that were created. Coincidentally, this refurbishment era was also obsessed by witchcraft, and much effort was made to prevent evil intruders. Doors and windows could be closed off, but the open flues of chimneys were deemed especially dangerous.

At Highgate Hill in London in 1963, a bricked-up recess was discovered close to a first-floor fireplace at Lauderdale House. It contained four desiccated chickens and an egg, as well as other objects, including a basket, a broken glass goblet, a candlestick and two odd shoes, one of the 1590s and one of 1610–15.[57] A desiccated chicken was also found in the great chimney of the late medieval Porch House at Potterne in Wiltshire, during restoration work in 1874, which may have been a ritual deposit.[58]

Cats, sometimes posed with rats, are frequent finds, and at the once moated medieval house of Hay Hall at Tyseley in Birmingham, a cat 'with gaping jaws and extended claws' was discovered that had been deliberately set up facing a bird. It was in a square cavity about 9 inches deep, between the outside wall and the lath-and-plaster interior wall (possibly a later addition).[59] A clue to the purpose of some of the deposits occurred in 1890 at Falmouth in Cornwall, when builders refused to continue work on a cottage until a hare was sacrificed to the 'outside gods'. Years later, during repairs to the roof, a well-crafted coffin containing the remains of a rabbit was found near the top of a wall of this cottage.[60]

Some remarkable finds have been discovered in the voids by the side of chimneys, which have been termed spiritual middens. At the bottom of one at Hestley Hall in Thorndon, Suffolk, was a complete desiccated chicken, possibly a cock, which was mounted by dung on a charred log and had been carefully lowered into position. Other finds at this property include shoes dating to the first half of the eighteenth century, eggs and numerous wings of geese. It has been suggested that cocks and geese may have been intended as alarms in order to ward off fire or witches. Several such middens in Suffolk have goose bones and long wooden sticks for driving geese.[61]

At the hamlet of Saveock Water in Cornwall, on the site of a spring, recent archaeological excavations have revealed an extraordinary sequence of ritual pits dating from the medieval period up to recent times. They were lined with different materials, including cat fur and swans' skins with feathers dating to about 1640. Among the contents of the pits were numerous eggs and claws from a variety of birds, and the remains of magpies. The pits seem to have functioned into the twentieth century, possibly as part of a pagan fertility or healing cult or associated with witchcraft.[62]

Black hens and cocks were linked with witches and were thought to have magical powers. In the late nineteenth century, a black hen was considered to be a suitable offering to the devil: 'If you have called up the devil by repeating the Lord's Prayer backwards, the only way to appease him, is to present him with a black hen.'[63] This was familiar to Dr Richard Wood, the medical officer for Driffield in the East Riding of Yorkshire, who recorded that a couple from the neighbouring village of Kirkburn had suffered various ailments and calamities, including the death of an old mare. One particular woman was suspected, and so a wise man known as J.S. from the village of Haisthorpe, who carried out much healing, was brought to Kirkburn in a carrier's cart: 'Towards midnight the ceremonies began. First, some chapters were read out of the Bible; the Lord's Prayer was read backwards, together with a series of other similar solemnities. A black hen was then brought upon the scene, and the heart taken out, whilst the poor bird was partly alive. This was stuck full of pins. The whole party then went into the garden, where a hole was dug in the ground, and the fowl put in.'[64] The carrier witnessed everything and told Dr Wood that the wise man subsequently put on a terrifying performance, addressing the evil one, which sent the couple back into their house in great alarm.

In a letter to a local newspaper in 1887, William Ashby of Portishead near Bristol mentioned several superstitions, including a related ritual that he thought barbarous and cruel: 'A young girl has a recreant sweetheart, so she takes a pigeon, and at midnight tears out its heart, sticks it full of pins, and roasts it, and the lover returns to his ladylove, and is faithful ever after, as he should be.'[65]

While some measures were intended as counter-witchcraft, other practices might be employed to cause harm. In about the summer of 1868, a gentleman set sail in a vessel that was due to return to Newcastle upon Tyne. His wife lived at nearby North Shields, and in mid-January 1869 newspapers across the country related the scandal of her accusing a friend of causing the loss of the vessel and his death. She said that she herself had been bewitched, because a fortune-teller had told her that 'a bad woman had roasted the heart of a black hen on me and my husband ... I asked him if he could tell me the woman's name, so he wrote down in full; but she would have been the last on earth that I ever would have thought of doing harm to me or mine.'[66]

In Ireland in 1939, a survey looked into the rituals observed at Martinmas. The shedding of blood, especially of chickens, was commonly carried out on St Martin's Eve, 10 November, and it was sometimes applied to the four corners of houses. In addition, millers did not work on the 11th, St Martin's Day. Associated beliefs involving cocks and hens were known across Europe, and in 1931 a resident of Newton Abbot in Devon told the *Morning Post* that in order to prevent mill accidents, a cock of unspecified colour was sacrificed at Martinmas each year:

My grandfather, who was born in 1787 and a miller by descent, told me of a curious custom observed by old-time millers at Martinmas. There was a rooted belief that the mill, in the course of the year, would demand blood, and to satisfy this sanguineous craving and thus ensure the safety of the miller from accident during the ensuing twelve months, a cock was killed on St. Martin's Eve (Martin being the patron saint of millers) at midnight, and the machinery *sprinkled* with its blood. This was known as 'blooding the mill'. The miller's friends and neighbours were invited to be present, and at the conclusion of the ritual the remainder of the night was spent in festivity.[67]

CHAPTER THIRTEEN

———•◆•———

A NATIONAL SPORT

COCK-FIGHTING. A LARGE MAIN of COCKS, at the Royal Cock-pit, St. James's Park, city of Westminster, will be fought three double days play, on Monday, Tuesday and Wednesday the 29th and 30th instant, and 1st of May, between the Gentlemen of Kent and Surrey, for TEN GUINEAS a battle and ONE HUNDRED the Main. Feeders, W. GLADDISH, for Kent. – M. FISHER, for Surrey.

Kentish Gazette April 1799[1]

Cockerels not only played a key part in superstitions and healing rituals, but they were also used in various games that, like most physical sports in the past, were violent in nature. The main one was cockfighting, which was for centuries the national sport of Britain, but other pastimes involving cocks, hens and occasionally other birds were staged at fairs and other festivals, very often associated with religious rituals.

During Elizabethan and early Jacobean times, a chronicle of events was written down by an unnamed inhabitant of the town of Shrewsbury in Shropshire, highlighting aspects of life and death at a time of profound religious change. Several entries were copied from other sources, while some record his own experiences or news that he had heard about or read. There are mentions of devastating fires, dreadful weather, floods, poor harvests, earth tremors, uncontrolled sickness, sinister crimes and freak events that the author obviously found unsettling. His manuscript takes universal piety for granted, along with a total belief in God Almighty and the imminent approach of the end of time and the day of reckoning.[2]

These were common experiences and attitudes, against which the welfare of animals was a trivial and incomprehensible idea. Throughout history, pastimes as a distraction from the day-to-day hardships frequently involved animals and cruelty, reflecting a deep-rooted blood lust and callousness. It was also taken for granted that God created the world for the benefit of humans and that animals and birds were provided for any appropriate use, especially for food and clothing. The interpretation of the Bible dictated human behaviour and beliefs, and although a view was gradually adopted that animals should not be subjected to unnecessary suffering, their well-being was barely a factor. Considering how badly ordinary men, women and children were treated, a great deal of room was left for theological excuses to justify continued barbarity towards animals. By the mid-eighteenth century attitudes were shifting, but it would take decades for anything substantial to change for the better.[3]

Children copied adults and grew up assuming that birds were expendable and could be ill-treated with impunity. When the Reverend Gilbert White wrote to Thomas Pennant from his home at Selborne in Hampshire in September 1767, he obviously felt powerless to confront such behaviour: 'The most unusual birds I ever observed in these parts were a pair of hoopoes ... which came several years ago in the summer, and frequented an ornamented piece of ground, which joins to my garden, for some weeks. They ... were frighted and persecuted by idle boys, who would never let them be at rest.'[4] On another occasion, in September 1849, Robert Cumming, who was a respected ornithologist and taxidermist in Exeter, recorded a large flight of black terns coming up the River Exe as far as the city, 'and numbers were knocked down by boys with their caps and sticks'.[5] Because boys were actively encouraged in schools to participate in annual cockfighting contests, such casual cruelty was to be expected.

Chickens had long been domesticated from the jungle fowl of south-east Asia, being ideal for rearing for eggs and meat because they can barely fly. Cockfighting used the male chickens, which are known as cocks or roosters and are naturally aggressive towards each other. Their bony leg spurs (or heels) just above their feet also provided a natural weapon. For hundreds of years cockfighting (or 'cocking') was the most popular national sport, even dubbed the 'royal sport'. It may well have been known before Julius Caesar invaded Britain in 55 BC, because he wrote: 'They do not think it right to eat hare, hens and and geese, but nevertheless raise them for amusement and pleasure'.

Shrove Tuesday was also referred to as Fastern's or Fasten's e'en (the eve or day before the fasting of Lent) or Mardi Gras (fat Tuesday), the day when everything was eaten up, especially meat and eggs. This was the last chance for enjoyment until Easter, with revelry, carnivals and violent games, in which cockfighting and other brutal contests were popular.[6] The link between cocks and Shrove Tuesday in the early nineteenth century was described by one man, W.G. Smith: 'My mother ... tells me that when she was a little girl it was the custom in the Ollerton district of Nottinghamshire to always give the first piece of the first pancake to the cock. If the country folk who were dining were not in possession of fowls of their own, a piece of the first pancake was always cut off as soon as cooked and carried by a child or some other person to a cock belonging to some near neighbour.'[7]

It was claimed that boys would learn about the heroic values of courage and fighting for their country from participating in the sport of cockfighting at school. The earliest mention of cockfighting in Britain was by William Fitzstephen in the twelfth century, when describing activities in London: 'we may begin with the pastimes of the boys, (as we have all been boys) annually, on the day which is called Shrove-Tuesday, the boys of the respective schools bring to the masters each one his fighting-cock, and they are indulged all the morning with seeing their cocks fight in the school-room. After dinner, all the youth of the city go into the field of the suburbs, and address themselves to the famous game of foot-ball.'[8] It would be many centuries before football surpassed cockfighting in popularity.

Such Shrove Tuesday customs endured for hundreds of years, and in the early nineteenth century the surgeon Thomas Wilkie recorded similar practices in the schools of the Scottish Borders: 'All the boys of each parish school provide themselves with a game cock, and pit them one against another in the school-house. The Domenie [schoolmaster] being judge, for which service he is entitled to all the cocks which are killed and the fugies. These cocks that run away, or shun the fight, are called fugies; and are all drawn . . . and laid aside till all the fighting is over – which his wife makes into the old Scotch dish cockie-leekie.'[9] The cocks that fled were also called hamies. Around that time, the Scottish geologist Hugh Miller was a pupil at Cromarty Grammar School:

The school, like almost all the other grammar-schools of the period in Scotland, had its yearly cock-fight, preceded by two holidays and a half, during which the boys occupied themselves in collecting and bringing up their cocks. And such always was the array of fighting birds mustered on the occasion, that the day of the festival, from morning till night, used to be spent in fighting out the battle. For weeks after it had passed, the school-floor would continue to retain its deeply-stained blotches of blood, and the boys would be full of exciting narratives regarding the glories of gallant birds, who had continued to fight until both their eyes had been picked out, or who, in the moment of victory, had dropped dead in the middle of the cock-pit. The yearly fight was the relic of a barbarous age.[10]

Without exception, Miller said, every schoolboy was listed as a cock-fighter and had to pay the master twopence each, though he himself was able to avoid active participation: 'I could not bear to look at the bleeding birds. And so I continued to pay my yearly sixpence, as a holder of three cocks, – the lowest sum deemed in any degree genteel, – but remained simply a fictitious or paper cock-fighter.'[11] His schoolmaster received the cockfighting fees, as well as the slain and fugitive birds. Elsewhere, the fee was called cock-pence or cock-penny. Henry Hingeston was a merchant and Quaker from Kingsbridge in south Devon and a rare early voice against cockfighting. In 1700 he wrote a letter to the master of the local grammar school, expressing concern that cockfighting was to take place there: 'It makes children unmerciful, and sows a stony spirit of hardheartedness in them, plainly destructive to the Nature of true Compassion.'[12]

Cockfighting was not restricted to schools or to Shrove Tuesday, but became a national obsession, even leading to the once common expression of 'That beats cockfighting', meaning that it surpasses everything. The sport was enjoyed by all classes, from paupers to royalty, as seen in 1597 when the Shrewsbury chronicler described a cockfight in his town: 'This year in April and in the Easter week was a great cockfight and other pastimes ... unto the which came Lords, knights and gentlemen.'[13] The Swiss traveller César de Saussure was in London in February 1728 and remarked on the frenzied atmosphere amongst all levels of society at cockfights: 'The spectators are ordinarily composed of common people, and the noise is terrible, and it is impossible to hear yourself speak unless you shout ... on the contrary, where the spectators are mostly persons of a certain rank, the noise is much less.'[14]

De Saussure also noted that 'Ladies never assist at these sports', which was not entirely true, as there are occasional mentions of women, such as a diary entry from 1704 that refers to a cockpit at Durham under the control of a woman: 'The first time that any cocks fought in Madam Softleye's new erected pitt was Easter Monday the 17th of Aprill.'[15] Lord Redesdale, who died in 1830, was reported to have said that in his younger days 'both ladies and gentlemen went full dressed to the cock-pit – the ladies being in hoops'.[16]

At its most basic, the only requirements were two or more fighting cocks and a cockpit, which might simply have been a ring or arena marked out on the ground, perhaps on the village green or in a field close to a village. Some were located within a natural hollow or inside the earthwork of a prehistoric monument, while indoor cockpits varied from rings or raised platforms in a room or barn to specially constructed outbuildings attached to inns.[17] In rural areas, the outdoor fighting rings generally had grass surfaces, but sods of turf were laid in indoor ones, though in urban areas where this might not be feasible, matting and carpets were used instead. Just as horse racing is called 'the turf', cockfighting was referred to as 'the sod'.[18]

Thousands of cockpits staged fights all across the British Isles, and fights were also an added attraction of events such as fairs or horse races. Some even took place inside churches. The parish register for St Mary's church at Hemingbrough, Yorkshire, recorded on 11 February 1661: 'Upon Fastens eaven last came with their cocks to the church and fought them in the church, namely, Tho. Midleton of Clife, John Coates, Ed. Widhous, and John Batley.'[19] The Shrewsbury event in 1597 was held 'at Richard Horton's house, being gaoler of the town upon whose backside a house and the pit was made for the people to stand and see, safe from the weather'.[20] The audience would assemble on the surrounding embankment or tiered benches, or else jostle around the arena for a decent view, and the London cockpits were described by de Saussure:

> The stage on which they fight is round and small. One of the cocks is released, and struts about proudly for a few seconds. He is then caught up, and his enemy appears. When the bets are made, one of the cocks is placed on either end of the stage; they are armed with silver spurs, and immediately rush at each other and fight furiously. It is surprising to see the ardour, the strength, and courage of these little animals, for they rarely give up till one of them is dead.[21]

He also revealed some of the excitement of the fighting:

> Cocks will sometimes fight a whole hour before one or the other is victorious; at other times one may get killed at once.

You sometimes see a cock ready to fall and apparently die, seeming to have no more strength, and suddenly it will regain all its vigour, fight with renewed courage, and kill his enemy. Sometimes a cock will be seen vanquishing his opponent, and, thinking he is dead (if cocks can think), jump on the body of the bird and crow noisily with triumph, when the fallen bird will unexpectedly revive and slay the victor. Of course, such cases are very rare, but their possibility makes the fight very exciting.[22]

Fighting cocks, referred to as 'game cocks', were bred for good shape, strength and stamina, and were somewhat different to today's breeds. To improve their prowess, the feathers of the neck, tail and wings were clipped, and the comb and wattles trimmed, as de Saussure described: 'The animals used are of a particular breed; they are large but short-legged birds, their feathers are scarce, they have no crests to speak of, and are very ugly to look at. Some of these fighting-cocks are celebrated, and have pedigrees like gentlemen of good family, some of them being worth five or six guineas.'[23]

Successful cocks were constant topics of conversation, such as that overheard by the poet Edwin Waugh in the Rochdale area, having stopped at the door of an ale house after crossing the canal at Firgrove:

a lot of raw-boned young fellows were talking with rude emphasis about the exploits of a fighting-cock of great local renown, known by the bland soubriquet of 'Crash-Bwons.' The theme was exciting, and in the course of it they gesticulated with great vehemence, and, in their own phrase, 'swore like horse-swappers.' Some were colliers, and sat on the ground, in that peculiar squat, with the knees up to a level with the chin, which is a favourite resting-attitude with them. At slack times they like to sit thus by the road side, and exchange cracks over a quart of ale.[24]

Men known as 'feeders' looked after and brought to fitness the fighting cocks, and they were critical to the success of the sport.

Cocks were taken to the fights in bags, usually linen, though the wealthiest owners might use silk. In the 1870s, a farmer at Chirbury in Shropshire, whose stock was in good condition, though his buildings were dilapidated, told the folklorist Georgina Jackson: 'There'll come a good cock out of a ragged bag', meaning that the quality of his stock was not dependent on fancy buildings.[25]

During a visit to a dying coal miner at his Northumberland home in the late eighteenth century, the Reverend John Brand found that a beloved fighting cock was suspended in a bag, so that it could be seen from the sick bed: 'In performing … the Service appropriated to the Visitation of the Sick … to my great astonishment I was interrupted by the crowing of a Game Cock, hung in a bag over his head. To this exultation an immediate answer was given by another Cock concealed in a closet, to which the first replied, and instantly the last rejoined. I never remember to have met with an incident so truly of the tragi-comical cast as this.'[26]

While being trained to spar against each other, the spurs of the cocks were muffled or bound with leather to prevent injury, but were sharpened for actual battles. From the late seventeenth century, artificial metal spurs of steel or silver became popular, which were bound to a cock's leg. Being longer and sharper than natural spurs, they were capable of inflicting terrible injuries or death. In 1703 Henry Hingeston published a tirade against cockfighting, in which he expressed his disgust that his fellow mortals fitted metal spurs, 'because God in his *Creating Wisdom*, hath not furnish'd those Creatures with Weapons to answer their *Devilish Expectation*, they will make up that Deficiency (*viz.*) by Metal Spurs, in order to make short work, and will kill them faster, that the Sport may go on with speed'.[27]

A 'main of cocks' was a large-scale event held over several days, comprising a series of battles between birds owned by two groups of gentlemen, often from different villages or counties. The term 'gentleman' did not necessarily refer to their status, as emphasised by Thomas Turner, a general shopkeeper from East Hoathly in Sussex. On 10 June 1761, he wrote in his diary: 'Was fought this day, at Jones's, a main of cocks, between the gentlemen of Hothly and Pevensey.' He then added a caustic note: 'Is there a gentleman in either of the places concerned?'[28] Each rival group had an uneven number of birds, and they were paired off against each other according to their weight. The number of fights was also uneven to ensure that one side would always win the main. Advertisements publicised the prizes for the individual fights, as well as the prize for winning the overall main.

A rougher version was the Welsh (or Welch) main, a knock-out contest that usually started with sixteen pairs of cocks of about the same weight. The winners of the first round were formed into eight pairs, then four, two and one, until only the winner was left, though it was not necessarily straightforward, as birds from the same side did not fight each other. A battle royal was a spectacular free-for-all between several cocks simultaneously, resulting in just one winner.[29]

Cocks that lost a fight were likely to be severely wounded or dead, and in northern England 'blenkards' or 'blinkards' (probably deriving from 'blink' or 'blinker') were fighting cocks that had lost an eye. Such birds were still used in cockfights, pitted against each other, along with stags, which were young cocks.[30] An issue of the *Newcastle Chronicle* in 1765 had three adjoining notices for fights involving blenkards, one to be held at Chester-le-Street, one at New Elvet in Durham and this one at the Flesh Market in Newcastle:

> To be FOUGHT for, *At* Mr LOFTUS's *New Pit, in the Flesh-Market, Newcastle, on Monday the 13th of May next,* FIFTY POUNDS, by Stags, 4lb. 1oz. the highest.
>
> Tuesday the 14th, FIFTY POUNDS, by Cocks, 4lb. 4oz. the highest.
>
> Wednesday the 15th, FIFTY POUNDS, by Cocks, Stags and Blenkards, 4 lb. 2oz. the highest.[31]

Winning was all-important, both for the owners of the cocks and the gamblers, who were mostly men, though in 1757 in Yorkshire, on the death of John Croft, a horse breeder, it was noted: 'His wife was an admirer of the diversion of cock-fighting and would bet her money freely.'[32] The sport was underpinned by prodigious amounts of gambling, with bets constantly placed during the action as the odds changed. Even at the Shrewsbury cockpit in 1597, it was reported that 'great sums of money [were] won and lost'.[33] Nicholas Blundell, the landowner from Little Crosby in Lancashire, kept a diary for many years from 1702, with several entries referring to gambling at cockfights, as on 29 June 1709 at Wigan: 'Cozen Scarisbrook and I were at Wigan cocking ... Lost at Wigan cocking £1 od. 6d. Going twice into the cockpit 2s. od.'[34]

In 1890, the village of Sancton in East Yorkshire still had a cockpit: 'Sancton was a place famous for cock-fighting, the sport being under the special patronage of the clergyman, of whom it is related that ... he fell asleep during the singing of a long psalm, and, on being awakened by the clerk, cried out "All right, a guinea on black cock! Black cock a guinea." Hence the Sancton people, especially the worse lot, are known as Sancton Cockins.'[35] This was probably more folklore than precise fact, as a similar story was told in Somerset. James Lackington (the future London bookseller) was born at Wellington in 1746, and his father related stories about his own childhood at the nearby village of Langford Budville, where the old parish clerk became known as Red Cock 'for having one Sunday slept in church, and dreaming that he was at a cock-fighting, he bawled out, "A shilling upon the red cock." And behold the family are called Red Cock unto this day.'[36]

Huge sums were gambled in London, and the Royal Cockpit was no exception, as Friedrich von Kielmansegge from Hanover found when he was visiting the city in 1762:

heavy bets, made by the Duke of Ancaster and others for more than 100 guineas, were at stake ... No one who has not seen such a sight can conceive the uproar by which it is accompanied, as everybody at the same time offers and accepts bets. You cannot hear yourself speak, and it is impossible for those who are betting to understand one another, therefore the men who take the bets, which are seldom even, but odds, such as 5 to 4, or 21 to 20, make themselves understood to the layers of the bets by signs. There is not the slightest doubt that the bets are duly paid, although frequently the parties do not know one another, or have never seen one another.[37]

The situation was repeated across the country, and one writer in Newcastle referred to the gambling there in about 1830: 'most of the cock-pits in Newcastle were in Forth Street, Newgate Street, and Gallowgate ... there was always a deal of betting over the various fights. The most celebrated of the bookmakers at one time was named Sinclair, and he was noted for his marvellous memory. He would take or give the odds thirty or forty times, and never use pen or pencil, without ever making a mistake as to the wagers.[38]

Apart from training the cocks to win, anything else possible was done to influence the outcome, with a firm belief in superstition and charms, including hatching eggs in a magpie's nest, because any cocks would be 'devil's birds' and therefore unbeatable.[39] At the National Eisteddfod of 1887, the winning essay was by the Reverend Elias Owen on Welsh folklore. According to him, 'the most successful cock-fighters fought the bird that resembled the colour of the day when the conflict took place; thus, the blue game-cock was brought out on cloudy days, black when the atmosphere was inky in colour, black-red on sunny days, and so on.'[40] He had also heard about the practice of writing a verse from the Bible on a slip of paper that was wrapped round the cock's leg, before placing the steel spur over it, and that an infallible charm was a crumb from the communion table of the parish church taken at midnight following the sacrament. At the village of Llansantffraid-ym-Mechain in mid-Wales, 'A cock-fighter did once upon a time enter Llansantffraid church at the proper time, and having found the precious crumb, placed it, the next cock fight, in the socket of the steel spur, which was then adjusted to the natural spur according to the rule laid down for the proper working of the spur.'[41]

In December 1806 an obituary was published of a conjuror and fortune-teller by the name of John Roberts, from Ruabon in

Wales, who claimed to have learned his skills in Egypt. To those seeking love, he sold charms 'in dark and hieroglyphic characters, which were also in much request to ensure success in any enterprise – a hat race, or a cock fight'.[42] John Randall from Broseley in Shropshire, a writer, geologist and painter of birds for the local china industry, reckoned that Joe Crow, from the village of Upper Arley along the River Severn, had some mystical tricks: 'He knew the great "breeders," assisted at "burning the feathers," and casting spells at mains of cock, when that gentlemanly sport was at a premium ... He knew all the great "cockers," feeders, and fighters of the time; and could cast spells, and "raise the devil," with any of them.'[43] Fighting cocks were also used in magic rituals, and George Lawley came across a strange one. He had been a traveller for Butler's Springfield Brewery near Wolverhampton and was ninety when he died in 1935. At Bilston in Staffordshire, he heard that 'to release one from the influence of a witch's spell, it was only necessary to break the neck of a fighting-cock and suck the victim's blood while it was warm'.[44]

Cockfighting was such an important strand of life over the centuries that innumerable phrases and words have left an indelible mark on the English language, including 'battle royal', 'cocky', 'cocksure', 'pit against', 'get your spurs on' and, for anyone acting in a cowardly way, 'show a clean pair of heels' (unstained by blood). 'Cock-eyed' referred to a fighting cock with one eye, so therefore anything askew, while in the Midlands the expression 'No cock's eyes out yet!' became common, signifying that a particular calamity was not too bad.[45] The word 'cockpit' was simply a pit for fights, but came to be used by Shakespeare and his contemporaries as a metaphor for a cramped, round space, and from the seventeenth century it was even applied to the confined dark quarters of the midshipmen on board Royal Navy warships, down below the waterline. Cockpit was also applied to areas where contests were fought, with Belgium once being described as the 'Cockpit of Europe'.[46]

When Edwin Waugh stopped at the village of Smallbridge, north-east of Rochdale, he reminisced, in cockfighting jargon,

about the old way of life, when the men, mostly weavers, occasionally ventured to neighbouring villages 'to have a drinking-bout, and
challenge "th' cocks o' the clod" in some neighbouring hamlet'.
Individual fights between the men ensued, he said, but sometimes
'a general *melée*, or "Welsh main," took place; often ending in painful journies, with broken bones, over the moors, to the "Whitworth
Doctors." [47] James and John Taylor from Whitworth were famous
surgeons and bonesetters in the late eighteenth century.

Apart from cockfighting, other contests frequently involved
cocks. When, in 1801, the antiquary Joseph Strutt published his
research into sports and pastimes in England, he mentioned a
street game where a cock was the prize:

> In some places it was a common practice to put the cock in an
> earthen [pottery] vessel made for the purpose, and to place him
> in such a position that his head and tail might be exposed to
> view; the vessel, with the bird in it, was then suspended across
> the street, about twelve or fourteen feet from the ground, to
> be thrown at by such as chose to make trial of their skill: two-
> pence was paid for four throws, and he who broke the pot, and
> delivered the cock from his confinement, had him for a reward. [48]

He mentioned the same game in an incident that had occurred
some forty years earlier at North Walsham in Norfolk:

> some wags put an owl into one of these vessels; and having
> procured the head and tail of a dead cock, they placed them in
> the same position as if they had appertained to a living one: the
> deception was successful, and at last, a labouring man belonging
> to the town, after several fruitless attempts, broke the pot, but
> missed his prize; for the owl being set at liberty, instantly flew
> away to his great astonishment, and left him nothing more than
> the head and tail of the dead bird, with the potsherds. [49]

A more vicious game known as 'throwing at cocks' took place
throughout Britain on Shrove Tuesday, and the winner would have

taken home the remnants of the bird for a final feast before Lent. The revelry involved hurling sticks at a cock or hen buried up to its neck or tethered to a stake until it was killed. Owners even trained their cocks to dodge these missiles by throwing smaller sticks at them for a week or so beforehand. In his history of London, published in 1810, Daniel Lysons included details dating from 1622 in the parish registers for Pinner in Middlesex: 'The cruel custom of throwing at cocks was formerly made a matter of public celebrity at this place, as appears by an ancient account of receipts and expenditure in the hamlet of Pinner. The money collected at this sport was applied in aid of the poor's-rates.'[50] At Little Crosby near Liverpool, Nicholas Blundell found the spectacle perfectly normal, noting on Shrove Tuesday in 1708: 'My wife & I saw throw at the Cock in the Townfield', while in 1720 he said: 'This being Shrove Tuesday I saw Richard Syer, John Ainsworth, &c: throwing a Cock before my gaites [gates] in the foulds [fields].'[51]

Whereas cockfighting involved evenly matched birds, throwing at cocks was mismatched cruelty, human against poultry. It therefore tended to attract more criticism, especially as it was a pastime of the poorer classes. Charles Manning, a friend of the Methodist John Wesley, was vicar of St Mary the Virgin at Hayes in Middlesex from 1738 to 1756, and he recorded his displeasure in the parish register for February 1754: 'Being Shrove Tuesday, divine service was performed in the afternoon, and no care was taken to prevent the throwing at cocks, rioting, and swearing in the churchyard at the same time; though I gave notice to the churchwardens and the magistrate, and desired that it might be prevented for the honour of God and Public good; but his answer was this – I know no law against throwing at cocks, even in the churchyard.'[52]

John Skelly was the vicar at Stockton-on-Tees for three decades from 1742, and he did succeed in suppressing this sport: 'As he was walking down the high-street, a little below the Church, he observed a party of men engaged in this diversion. He endeavoured to reason the matter with them, and asked them – "If a being larger and stronger than themselves were to tie them to a stake, and use them in the same manner, how would they like it?"

It is said, they were struck with the force of his argument, untied the poor animal, and departed quietly.'[53] Cock throwing lasted for several more decades, though various towns did begin to prohibit it, including Bath in 1763, when a notice was placed in the local newspaper by the mayor and justices of the peace, threatening to punish anyone involved in this sport.[54]

In a dictionary of common language that he published in 1785, Francis Grose, an antiquary and retired army officer, gave a definition of an associated game called 'To whip the cock', explaining that it was 'a piece of sport practised at wakes, horse races, and fairs in Leicestershire, a cock being tied or fastened into a hat, bag or basket; half a dozen carters blindfolded and armed with their cart whips are placed round it, who, after being turned thrice about, begin to whip the cock, which if any one strikes so as to make it cry out, it becomes his property; the joke is, that instead of whipping the cock, they flog each other heartily.'[55]

Cocks occasionally featured in grotesque fairground attractions. In May 1758, the *Newcastle Courant* gave notice of traditional amusements at Swalwell's popular fair: 'For the entertainment of some of our readers, we mention the annual diversions at Swalwell, on the 22d of May, viz. dancing for ribbons, grinning for tobacco, women running for smocks, ass races, foot courses by men, with an odd whim of a man eating a cock alive, feathers, entrails, &c.'[56] Brutal pastimes were not restricted to cocks, and one popular game was 'mumbling a sparrow', which also took place at fairs and wakes. Grose described it in his dictionary: 'a cock sparrow whose wings are clipped, is put into the crown of a hat, a man having his arms tied behind him, attempts to bite off the sparrow's head, but is generally obliged to desist, by the many pecks and pinches he receives from the enraged bird'.[57]

One newspaper piece in 1840 on customs in Derbyshire's Peak District mentioned what they considered to be another barbarous diversion:

Two posts are set up at a convenient distance from each other, and a communication by means of a cord is made between them:

a goose is then tied by the legs and suspended from the cord, having previously had the head and neck sufficiently greased to prevent much hold being taken. A number of men riding at full speed, in passing the goose attempt to pull off the head, when he who succeeds in so doing is declared the winner, and becomes the possessor of the mutilated fowl. Some years ago a similar custom was practised at Edinburgh, and was called 'riding the goose.'[58]

A Christmas pastime was hunting was for owls. Towards the end of the eighteenth century, John Brand revised his *Observations on Popular Antiquities*, and in a discussion on 'The Christmas Carol', he gave a version that contained the lines:

> *The wenches, with their wussell-bowles,*
> *About the streets are singing;*
> *The boyes are come to catch the owles.*

He added an explanatory note. 'A credible person born and brought up in a village not far from Bury St Edmunds in the county of Suffolk, informed me that, when he was a boy, there was a rural Custom there among the youths of hunting Owls and Squirrels on Christmas Day.'[59]

Much more common was the winter custom of 'hunting the wren', and yet to kill a wren was believed to bring bad luck, so catching and killing one on Christmas Day, Boxing Day or New Year's Day was a rather perverse custom. It occurred widely across England, Wales, Ireland and the Isle of Man, as well as France and America.[60] William Buller was a naturalist in Sussex, who in 1890 wrote: 'I think the persecution of the Wren, in Sussex, is a thing of the past; but in my younger days [he was born in 1814] it was a regular institution to hunt it at Christmas time, when numbers of boys, on both sides of the hedges, amused themselves by beating the bushes and throwing at the Wren whenever it showed itself, with knobbed sticks about eighteen inches long called "libbets".'[61] In Essex, the slaughtered wren was carried around in furze foliage and used as an excuse by boys ('wren-boys') to chant for money:

The wren! the wren! the king of birds,
St. Stephen's Day was killed in the furze;
Although he be little, his honour is great,
And so, good people, pray give us a treat.[62]

The origin of this tradition is uncertain, perhaps based on fairy myths or religious beliefs, and countless ideas developed. According to the seventeenth-century antiquary John Aubrey, the last battle in Ireland between Protestants and Catholics was at Glinsuly in Donegal, near to which a Catholic force was about to surprise a group of Protestants who were sleeping, but wrens woke them up by dancing and pecking on the drums. 'For this reason,' Aubrey said, 'the wild Irish mortally hate these birds, to this day, calling them the Devil's servants, and killing them wherever they catch them; they teach their children to thrust them full of thorns: you will see sometimes on holidays, a whole parish running like mad from hedge to hedge a wren-hunting.'[63]

In a book that was first published on the Isle of Man in 1726, the antiquary George Waldron mentioned how the wren was hunted there:

On the 24th of *December*, towards evening, all the servants in general have a holiday, they go not to bed all night, but ramble about till the bells ring in all the churches, which is at twelve a-clock; Prayers being over, they go to hunt the Wren, and after having found one of these poor birds, they kill her, and lay her on a bier, bringing her to the Parish-Church, and burying her with a whimsical kind of solemnity, singing dirges over her in the *Manks* language, which they call her Knell; after which Christmas begins.[64]

The wren is often assumed to be female, and in 1816 Hannah Ann Bullock wrote a history of the Isle of Man, in which she said that the hunting of the wren was founded on a tradition that a beautiful fairy once exerted undue control over the male population, and while they were trying to destroy her, she took the form of a wren:

though she evaded instant annihilation, a spell was cast upon her, by which she was condemned on every succeeding New Year's Day, to reanimate the same form [and become a wren] ... every man and boy in the island ... devote the hours between sun-rise and sun-set, to the hope of extirpating the fairy, and woe be to the individual birds of this species, who shew themselves on this fatal day to the active enemies of the race: they are pursued, pelted, fired at, and destroyed, without mercy.[65]

Such pastimes involving ill-treatment towards wild birds and cocks were eventually suppressed, though long after the first humane voices spoke out. Keeping a cockpit became illegal in London in 1833 and two years later in England and Wales, but the legislation had little effect, and so cockfighting itself was banned in 1849, though not until 1895 in Scotland.[66] Cockfighting may have been prohibited, but it was difficult to suppress, being so ingrained in society. For at least three decades, there were countless instances of illegal fights and many prosecutions, and across the world it is still legal in a few countries and tolerated in many others. As a boy in Scotland, Hugh Miller heard a neighbour tell his Uncle James that excellent clergymen had approved cockfighting over the ages, to which his uncle retorted: 'Yes, excellent men! but the excellent men of a rude and barbarous age; and in some parts of their character, tinged by its barbarity.'[67]

One important element of the annual wren massacre on the Isle of Man was to provide the Manx fishermen with feathers to protect them from shipwreck, and so this cruelty was justified as being for the good of humanity and the fight against evil. Bullock described the superstition: 'their feathers [were] preserved with religious care, it being an article of belief, that every one of the relics gathered in this laudable pursuit, is an effectual preservative from shipwreck for one year; and that fishermen would be considered as extremely foolhardy, who should enter upon his occupation without such a safeguard'.[68]

Another Manx historian, William Harrison, said in 1869: '"Hunting the Wren" has been a pastime in the Isle of Man from time immemorial, and is still kept up on St. Stephen's Day, chiefly by boys.' After catching a wren, he said that they suspended it in garlands and evergreens and then went on a procession, calling at each house, where they gave a feather for luck: 'these are considered an effectual preservative from shipwreck, and some fishermen will not yet venture out to sea without having first provided themselves with a few of these feathers to insure their safe return. The "dreain," or wren's feathers, are considered an effectual preservative against witchcraft.'[69] In the evening, the plucked bird was solemnly buried in a corner of the churchyard, after which wrestling and all manner of sports took place.

CHAPTER FOURTEEN

———— •◆• ————

BY-PRODUCTS

Cackle, cackle, Madam Goose,
Have you any feathers loose?
Truly have I, little fellow,
Half enough to fill a pillow;
And here are quills, take one or ten,
And make from each, pop-gun or pen.

Traditional nursery rhyme[1]

Birds have thousands of feathers that are used for flight, display and insulation. Formed from keratin, fully grown flight feathers have a hollow shaft or quill with vanes of interlocking barbs. Feathers for insulation are less complex. When birds moult, the entire plumage is replaced in stages, usually once a year after breeding and lasting up to five weeks. Some birds moult twice a year, allowing their appearance to change between seasons.

For centuries, birds were slaughtered or harvested for two key types of feather – the down feathers and the long, rigid flight feathers of the wings and tail.[2] Thomas Pennant watched live geese being plucked in 1769 as he travelled through the extensive low-lying fenlands of Lincolnshire, where immense numbers of geese were reared:

The geese are plucked five times in the year; the first plucking is at *Lady-Day*, for feathers and quills, and the same is renewed, for feathers only, four times more between that and *Michaelmas*. The old geese submit quietly to the operation, but the young ones are very noisy and unruly. I once saw this performed, and observed that goslins [goslings] of six weeks old were not spared, for their tails were plucked, as I was told, to habituate them early to what they were to come to. If the season proves cold, numbers of geese die by this barbarous custom.[3]

Most of the feathers were sold for quills, bedding and upholstery, but the trade declined after the wetlands were drained and the commons enclosed, especially in Lincolnshire and the Somerset Levels, as the immense flocks of geese could no longer freely graze. In April 1797, the dissenter and historian William Hutton wrote 'The Cottager', a poem about Hodge (the generic name for an uneducated farm labourer), who had managed to support his family until the common was enclosed. He was brought before the very judge who had allowed its enclosure, accused of stealing a goose out of hunger. When sentenced to be hanged, Hodge meekly replied:

> *The crime is small in man or woman,*
> *Should they a goose steal from a common;*
> *But what can plead that man's excuse*
> *Who steals a common from a goose?*[4]

One particular reason for plucking geese was to obtain quills for pens. Such pens had been used for writing with ink from about the sixth century, initially on parchment and later on hand-made paper. The word 'pen' was derived from the Latin *penna*, meaning a feather or wing. Instead of 'feather', the word 'plume' might be used, which is French for a pen, itself derived from the Latin *pluma*, also meaning a feather. The French word *plumer*, to pluck, led to the word *plumassier* being used for someone who prepared and sold decorative feathers.[5]

The most robust quill pens were made from the flight feathers of geese, swans and occasionally bustards, with those from older,

plucked geese considered as better quality than ones from juve-
niles or slaughtered birds. Feathers from the left wing suited
right-handed people, and vice-versa.[6] The feathers of other birds
were generally not so good, although William MacGillivray, in
1837, rated rook quills highly: 'Their quills are often collected by
the herds [herd boys], and although small, are superior to goose
quills for writing, provided the fingers of the person using them be
sufficiently flexible and delicate to adapt themselves to so slender
an instrument.'[7]

Pens were formed by cutting the tip of the quill and making a
small slit to aid the flow of ink when writing, a design that metal
nibs later copied. As the pen wore down, the tip of the shaft was
reshaped with a penknife. Because ink rose part-way up the hollow
shaft, it was possible to write several words before having to dip
the quill back into the inkwell. In his poem 'The Pen', written in
June 1796, Hutton addressed his quill pen, comparing the slit with
slitting a magpie's tongue to make it talk:

> *I own, I could not let thee rest,*
> *But rudely stripp'd thy downy vest;*
> *Ere with impurity wert ting'd,*
> *I scrap'd thee as a pig that's sing'd;*
> *And, as the cook-maid serves a trout,*
> *Have drawn thy tender entrails out;*
> *Often, as folks the magpye tweak,*
> *I've slit thy tongue to make thee speak;*
> *Pursuing still the rude attack,*
> *I've dy'd thy slender limbs in black.*

He then imagined the pen wearing out:

> *And when, dear pen, thou'st had thy day,*
> *Like me, worn out, art thrown away.*
> *Our end the same, we're neither free;*
> *A knife cuts thee up – Time cuts me.*[8]

Although quill pens could be purchased commercially, they were normally fashioned at home, and instructions in a 1677 manual for maidservants showed that it was a skilled task:

> Having a penknife with a smooth, thin, sharp edge, take the first or second quill of a goose wing and scrape it, then hold it in your left hand with the feather end from you, beginning even in the back, cut a small piece off sloping, then to make a slit enter the knife in the midst of the first cut, put in a quill and force it up, so far as you desire the slit should be in length, which done cut a piece sloping away from the other side above the slit, and fashion the nib by cutting off both the sides equally down.[9]

Writing from Wreyland in Devon, Cecil Torr said that, in the early 1860s, 'I was taught to make a pen (that is, to cut a quill into a point) as one of the things that every child must learn. Metal pens did not come into common use until after 1840, though introduced some years before.'[10] Quills for handwriting were used for well over a thousand years, before being replaced by metal nibs, and they persisted in traditional professions, such as for legal documents, as Charles Dixon mentioned: 'the quill-pen makers ... earn a livelihood by fashioning the instruments which men of the law and bankers deem necessary in compiling their mystic documents. We verily believe the lawyers would consider a document illegal which had been written with a vulgar steel pen – nor are the bankers far behind in their prejudice. There is an amount of pliability about a quill pen never to be found in one of steel.'[11]

During the financial year 1870–1, it was recorded that clerks in public offices went through 10,344 steel pens and 430,087 quill pens.[12] A secondhand trade existed, because thousands of quill pens were discarded when broken: 'time is too valuable to be frittered away in pen-mending, and therefore it is more economical to use new pens than to be frequently repairing old ones ... In some cases the office pens go to certain dealers by contract, and in other cases they lapse as perquisites into the hands of somebody or other who has a prescriptive claim to them. What seems most likely is that very few of them are wasted.'[13]

For those with pen-cutting skills, they were a wise purchase:

The man whose monotonous trade it is to drive the quill, and whose fate it is to scribble away from morn to eve for a living, is well aware of the merits of the office pen. It is his oldest acquaintance and most tried friend. He rarely buys a new pen or an uncut quill – the new article would not be half so well seasoned, or so tough, or of half so good a quality in relation to the price, and would not last half so long. The office pen cost originally double the sum per thousand at which it is retailed to the public at the close of its official life, and instead of being worse for its past career, is, for all practical purposes, when deftly mended, as good as new.[14]

Being of superior quality, secondhand pens were counterfeited:

These impositions are manufactured from the cheapest refuse quills – are cut into pens by means of a hand machine, which does the business in a moment, and, being first tied in bundles, are dipped into ink hundreds at a time. They have never entered any office, or seen any service at all, and it is very little service that is to be got out of them beyond the practice they afford to the tyro in the art of pen-making ... The professional quill-driver is never taken in by these rubbishy things.[15]

While clerks and hack writers were called quill-drivers, outstanding uses of the quill have included Magna Carta, the novels of Jane Austen, the musical scores of Beethoven and the writing and signing of America's Declaration of Independence. The list is endless and the number of quills used throughout history incalculable.

Quills were not just for writing. They were extremely versatile, though the range of uses has been largely forgotten. Born in 1782, John Dudeney eventually became a schoolteacher, but when twenty years old, he was a shepherd on the South Downs between

Brighton and Rottingdean, where he would catch wheatears: 'From what I have heard from old shepherds, it cannot be doubted that they were caught in much greater numbers a century ago than of late.' One shepherd from East Dean, he said, took so many 'that he could not thread them on crow-quills in the usual manner, but took off his round frock and made a sack of it'.[16]

Hutton ended his poem 'The Pen' with the lines:

You'd better be a tooth-pick made
Than follow the poetic trade;
Unless you can, with powers alert,
Instruct the reader, or divert.
But you'll attain a double worth,
If ever you accomplish both.[17]

Toothpicks were also mentioned by Charles Dixon at the end of the nineteenth century: 'An almost equally important use for quills is the manufacture of toothpicks, thousands of gross of this useful little article being consumed every year in England alone. The largest toothpick factory in the world is near Paris, which is said to turn out twenty millions each year.'[18] An Australian journalist was told about these toothpicks from a buyer for a large drug house:

We buy ours from a broker in Paris, who obtains them from a large manufactory near that city. M. Bardin, at Joinville le Pont near Paris, had the largest manufactory in the world engaged in the quill industry. He has two million geese, and produces annually twenty million quills. Formerly this factory made quill pens, but when these went out of general use the quills were used to make brushes for artists and toothpicks ... The price is so low that there is very little margin of profit in the business. The wooden tooth-pick has taken the place of the quill.[19]

Artists' brushes were often of feathers, and in his 1816 *Instructions to Young Sportsmen*, Peter Hawker discussed how to obtain prized paintbrushes from woodcocks: 'The feather of the woodcocks, which is so very acceptable to miniature painters, is that very small one, under the outside quill of each wing; to be sure of finding which, draw out the extreme feather of the wing, and this little one will appear conspicuous from its sharp white point.'[20]

Some musical instruments were operated with quills. When a key of a harpsichord was struck, it caused a quill plectrum to pluck a string and make a sound. The rector of Fyfield in Hampshire in the late eighteenth century was Henry White, younger brother of the naturalist Gilbert White, and he was an accomplished musician who owned a harpsichord, noting in his diary on 23 December 1784: 'Began tuning ye Harpsichord and quilling it.' The next day, he wrote: 'Harpsichord completely tuned and quilled, and ye keys perfectly rectified, so that it is now in better condition than ever.' One of the biggest fairs in England took place at nearby Weyhill, where three years earlier White bought quills: '1781, October 9th. – Quills 1500 b^t from ye Sedgemoor Marchant on Weyhill at 1s. per Hund., the cheapest market by far. At Andover they are full three times as dear.'[21] These were goose quills, as the merchant was from Sedgemoor in the Somerset Levels, renowned for its flocks of geese.

Another use of quills was for fishing floats, which were attached to lines, enabling the baited hook to be suspended at a set depth. They were also a visual aid, because they were pulled beneath the surface of the water whenever a fish was hooked. *The Sporting Magazine* in 1794 said that anglers valued quills of the bustard for floats, because, 'being spotted with black, these spots are supposed to appear as flies to the fish, and are consequently alluring to them'. Many early floats comprised a cork body fitted on a crow's quill. William Palmer, just after World War Two, described a typical scene along the River Mole at Dorking: 'Late February must be a good time for coarse fishing in these waters. All day anglers ply with rods, lines and red-topped quill-floats in every sort of nook not swept by the thin breeze.'[22]

Quills were invaluable as tubes, and in 1802 William Holland in Somerset said of his young son: 'Little William quite a hero, being able to shoot out of a quill pop gun I made him. He is an expert boy at learning all kind of play.'[23] They made ideal containers for gold dust or priming gunpowder for cannons and handguns, and fireworks might incorporate quills. An 1826 encyclopedia gave instructions on how to make 'bearded rockets': 'Fill the barrels of some goose-quills with the composition of flying rockets [gunpowder, saltpetre and charcoal], and place upon the mouth of each a little moist gunpowder, both to keep in the composition, and to serve as a match. If a flying rocket be then loaded with these quills, they will produce at the end a beautiful shower of fire, which is denominated golden rain.'[24]

A fashion developed for quills painted with pictures, and in 1836 a Liverpool newspaper advertised a sale of Chinese rarities, including 'Feather Fans, Painted Fans or Fire Screens, curiously Painted Quills'. Feathers often adorned fans, and a 'Ladies' Gossip' column in a late Victorian Glasgow newspaper shared the latest trend: 'Feather fans have regained their waning popularity, but the most chic are composed of hand-painted swan quills mounted on fragile ivory or mother-of-pearl, and finished with a border of curling ostrich tips, or a cloud of softest marabout.' An example of a design was mentioned in another Scottish newspaper: 'Painted quills are much in vogue, a couple of brown ones bearing the most natural-looking ears of corn painted on them.' They were also fixed to hats, as a Cardiff newspaper informed its readers: 'Dame Fashion is showing us her prettiest fantasies in the hats and toques of the moment. Everything that pleases finds a place in the newest winter millinery – soft velvets, panne, pliable felts, feathers, painted quills, and flowers.' One pretty hat was of black felt, 'trimmed with generous scarves of crimson velvet, and two black and white hand-painted quills in front'.[25]

A fund-raising use for painted quills was recorded by Hugh Gladstone in his 1919 book on birds in World War One: 'the New Zealand Government sent a quantity of SWANS' quills with appropriate paintings thereon, to be sold for the benefit of the Belgian Relief Fund at 1s. and 2s. 6d. each'.[26]

Goose wings with all the feathers were handy brushes, and in late Victorian times an industry evolved for feather dusters, particularly with imported ostrich feathers. Dixon spoke of feathers in 'the manufacture of brooms, chiefly from the long narrow hackles of the barn-door fowl and from the broad feathers of turkeys'. In June 1821, a spur-winged goose from sub-Saharan Africa was shot by John Bickford near St Germans in Cornwall, and Henry Mewburn from the village said that although Bickford knew he was a bird stuffer, his wife thought she could stuff it, 'but being soon convinced of her inability, she cut off the wings for dusters and threw the skin away'.[27]

Feathers were also useful for cleaning guns. After each firing of a flintlock gun, William Daniel advised that the flint, pan and hammer should be wiped, 'and a partridge or woodcock's *wing feather* introduced into the *touch-hole*'.[28] This was also recommended by George Markland in his 1727 poem *Pteryplegia*. While preparing to fire a gun, he said,

> *Yet cleanse the Touch-hole first: A Partridge Wing*
> *Most to the Field that wise purpose bring.*[29]

Massive quantities of feathers were sold for upholstery and bedding, including mattresses, pillows and quilts. Mattresses of straw and flock (discarded wool and shredded rags and carpets) were a cheaper, harder alternative, but feather beds – mattresses of ticking filled with feathers – were manufactured from medieval times and were common until recent decades. Because geese are water birds, their feathers and down were preferred, being fluffy, light, warm and elastic, and of superior quality to the harsher feathers of fowl such as hens.[30] During the investigation of children's employment in London in 1841, the manager of a feather, horsehair and flock manufacturer said: 'In the feather department boys are employed in picking out the quills or stronger feathers from the soft feathers.' Dr James Mitchell saw the factory conditions: 'The work of the boys in the feather department is very light ... Their clothes were covered over with downy feathers, which are brushed off before they go home. The breath must inhale some of the down, and therefore the work cannot be very healthy. Still, so long as mankind will have feather beds, this work must be done, and boys are as suitable as any other description of persons to do it.'[31]

A domestic cookery book of 1808 gave advice on collecting feathers:

> In towns, poultry being usually sold ready picked, the feathers, which may occasionally come in small quantities, are neglected; but orders should be given to put them into a tub free from damp, and as they dry to change them into paper bags, a few in each; they should hang in a dry kitchen to season; fresh ones must not be added to those in part dried, or they will occasion a musty smell, but they should go through the same process. In a few months they will be fit to add to beds, or to make pillows, without the usual mode of drying them in a cool oven, which may be pursued if they are wanted before five or six months.[32]

One way of reducing costs for bedding was to mix goose with other feathers, especially pigeons or chicken, but in the early nineteenth century Thomas Wilkie in southern Scotland said: 'For a newly married pair to sleep on beds or pillows filled with the feathers of wild fowls, or that such bed should be used by any person, were always reckoned unlucky, as all those who slept on them would soon die.'[33] Elsewhere, a widely held superstition existed that it was difficult for anyone to die on a bed containing pigeon or game feathers, since it caused much suffering. A correspondent to *Notes and Queries* in 1852 said it was a common belief among the poor in Sussex, though the dialect feels more 'north country':

> A friend of mine a little time back was talking to a labourer on the absurdity of such a belief; but he failed to convince the good man, who, as a proof of the correctness of his belief, brought forward the case of a poor man who had lately died after a lingering illness. 'Look at poor Muster S——, how hard he were a dying; poor soul, he could not die ony way, till neighbour Puttick found out how it wer, – "Muster S——," says he, "ye be lying on geame feathers, mon, surely;" and so he wur. So we took'n out o'bed, and laid'n on the floore, and he *pretty soon died then!*'[34]

The superstition was so well known that it found its way into novels. In Emily Brontë's *Wuthering Heights*, set in the late eighteenth century, Cathy Linton is in bed, mentally deranged and slowly starving. In a fit of madness, she tore her pillow: 'she seemed to find childish diversion in pulling the feathers from the rents she had just made, and ranging them on the sheet according to their different species: her mind had strayed to other associations. "That's a turkey's," she murmured to herself; "and this is a wild duck's, and this is a pigeon's. Ah, they put pigeons' feathers in the pillows – no wonder I couldn't die! Let me take care to throw it on the floor when I lie down." '[35]

Writing from Exeter in 1871, the retired clergyman William Grey said: 'When I lived on Salisbury Plain some thirty years ago [as a curate], I used to observe that when a pigeon-pie was being concocted, my housekeeper invariably burned the feathers. On my asking a reason for this waste, she assured me that if a single feather found its way into a bed or pillow, nobody could die upon it, but would be "dying hard" until the obnoxious feather was removed.'[36] Such feather superstitions long persisted in East Anglia, according to a paper read at the 1926 International Poultry Congress, where it was said that such mixed feathers caused restlessness at night, insomnia and enuresis, and that pillows with feathers of gamebirds or pigeons were removed from under the heads of the very sick so that they could die. A related belief from Sussex was that 'You must not turn a feather-bed on a Sunday, or the person who sleeps on it will have fearful dreams for the rest of the week.'[37]

A considerable industry involved feathers from seabirds. A history of the Isle of Wight published in 1781 referred to the chalk cliffs at Freshwater: 'The country people take the birds that harbour in these rocks, by the perilous expedient of descending by ropes fixed to iron crows, driven into the ground: thus suspended, they with sticks beat down the birds as they fly out of their holes; a dozen birds generally yield one pound weight of soft feathers, for which the merchants give eight pence.'[38] When James Wilson was at St Kilda in 1841, he watched as a boat brought back several of the male population who had spent ten days on the neighbouring island of Boreray: 'we had an opportunity of inspecting their harvest. The large boat was filled with huge bundles of feathers.' He explained their purpose: 'The people pay their rent (about £60, as we were told) chiefly by means of feathers, which they collect from both the young and old birds.'[39]

Over half a century later, in June 1896, Richard Kearton spent a fortnight on St Kilda with his brother Cherry, who was a photographer, and their friend John Mackenzie, on his annual visit as factor to the island. The fulmar looks like a gull, but is actually a petrel, and Kearton said: 'The oil gives off such a strong odour that everything in St Kilda smells of it ... My friend Mackenzie told me that the feathers plucked from the Fulmars are mixed with those of other birds and sold to the Government for stuffing soldiers' pillows. Before being used they are thoroughly fumigated, but in about three years the smell returns to them so strongly that Tommy Atkins refuses to rest his sleeping head on them until they have been again roasted.'[40]

The most expensive products were filled with down from eider ducks – coastal ducks that feed on mussels and other molluscs. In a process that is unique to wildfowl, the females pluck their own down to line their nests and insulate and camouflage the eggs. In the first half of the twentieth century, this down was removed from the nests, and much was exported from Scandinavia and Russia to Britain, especially from eider duck farms.[41] Being extremely warm and lightweight, the down was used mainly for bed quilts, and in 1850 Heal and Son's Bedding Factory in London's Tottenham Court Road advertised: 'THE EIDER DOWN QUILT is the warmest, the lightest, and the most elegant covering; it is suitable for the bed, the couch, or the carriage; and its comfort to invalids cannot be too highly appreciated.'[42]

John Keast Lord had been a naturalist to the British North American Boundary Commission, but returned to England in 1862 and shortly afterwards wrote a popular magazine piece on migrating waterfowl, including eider ducks: '"Eider-down," though a household word, is nevertheless, a material too costly for any but the wealthy to enjoy ... As an article of commerce eider-down is a very important one, being used for making quilts, stuffing cushions, and in the manufacture of tippets and muffs. Swan's down and the down from different species of water-fowl are often mixed with it, the adulterated article being vended as real eider.'[43]

Before working in North America, Lord had served as a veterinary surgeon in the Crimean War, and at Scutari (modern Üsküdar) he saw a substantial eider-down industry:

> The great market for eider-down is Turkey, where it is in great demand for cushions and coverings of various sorts, for the ladies of the harems to luxuriate on, and under. Any one who has ever wandered round that quaintest of all quaint places, Lower Scutari, must have observed curious tumble-down shops, with curious dilapidated-looking people working in them, busy amidst heaps of white down; cushions, pillows, and beds of every shape and size are piled together, and gorgeous quilts that you could well-nigh blow away, warm, yet so astoundingly light, hang like tapestry against the dirty walls, everything and everybody appearing as if just out of a snow-storm: these are the manufactories that supply the 'City of the Sultan' with luxuries in eider down. A good quilt costs in Scutari from seven to ten pounds sterling.[44]

The word 'eiderdown' entered the English language to mean a padded bed covering, so that in 1908 a north London retailer offered for sale eiderdown quilts and cot eiderdowns. The following year, in a column 'For the Ladies', a Sheffield newspaper gave instructions on how to make eiderdowns from scrap materials:

> USEFUL 'EIDERDOWNS.' A very useful nursery quilt for a child's bed can be made from an old blanket. A large one should be chosen, or two smaller ones, and they should be folded till the required size is obtained, allowing two or three thicknesses of blanket. This should then be covered with a pretty washing silk, cretonne or sateen, and be quilted. The effect will be much the same as that of an eiderdown, and a few buttons covered

with the material and caught into the quilt will greatly improve the appearance.[45]

Feathers have also long been critical for arrows used against people and animals, because until gunpowder the arrow was the most important weapon. As Puck left on an errand for Oberon in Shakespeare's *A Midsummer Night's Dream*, he cried:

> *I go, I go; look, how I go,*
> *Swifter than arrow from the Tartar's bow.*

At that time, no weapon moved faster, further or more accurately than an arrow fired from a bow, and the Tartars of central Asia were renowned archers. An arrow had three parts – a lethal arrowhead, fixed to a wooden shaft, and feather flights (or fletching) at the end of the shaft. The feather flights counterbalanced the arrowhead and made the shaft rotate, giving the arrow stability in flight and increasing its range and accuracy. An injury on a human skull found in a medieval cemetery at Exeter in Devon showed that it was caused by an arrow that was fletched to rotate clockwise.[46]

Simple bows and arrows are depicted in prehistoric cave paintings of hunting scenes. These were an improvement on hand-thrown spears, stones and throwing sticks, and in Britain arrowheads of flint date back at least ten thousand years. The very first arrows may not have had feathered flights, but Ötzi, the man preserved in Alpine ice, was carrying arrows with flights when he perished around 3000 BC on what is now the border between Austria and Italy. A few centuries later, arrows with flights were being depicted in ancient Egyptian art.

By the Middle Ages, longbows and crossbows were key weapons in warfare and hunting, and the longbow was the supreme weapon of war in Europe for more than a century, so that a considerable arrow industry was needed. The arrowheads were made by an arrowsmith, while a fletcher crafted the wooden shafts and fitted the feather flights, giving rise to the surnames Arrowsmith and Fletcher. Feathers from geese were preferred, along with swans, pigeons, ducks and even peacocks. Ballads about the legendary archer Robin Hood were collected and published as 'garlands', mainly from the mid-eighteenth century, and one included a verse referring to arrows with goose feathers:

Bend all your bows, said Robin Hood,
And with the grey goose wing,
Such sport now show, as you would do,
In the presence of the king.[47]

Fletchers needed flight feathers from the wings, which were cut to shape and either glued on the arrow shafts or secured with fine thread. The best arrows had feathers that were glued, bound on and then waterproofed to prevent the glue deteriorating in damp conditions. Over three thousand arrows were recovered from the Tudor warship *Mary Rose* that sank off Portsmouth in 1545, although 9,600 were recorded in the ship's original inventory. The feathered flights and most of the bindings had been destroyed by sea-water, but traces of feathers were preserved in the glue still clinging to the shafts.

The most famous use of arrows occurred more than a century earlier, at the battle between the English and French at Agincourt in 1415. By then, artillery guns were beginning to feature in battle, and Henry V actually took guns on this French campaign, which proved effective at the siege of Harfleur, but were not used at Agincourt. Only a few years afterwards, the writer of *Gesta Henrici Quinti* ('The Deeds of Henry the Fifth') described the role of the archers in the battle:

The French cavalry that was posted on the flanks made charges against our archers on both sides of our army. But, by the will of God, they were quickly forced to fall back and flee under a rain of arrows ... As the fighting grew more intense and fierce, our archers fired their sharp-pointed arrows right into the enemy's flanks, constantly keeping up the fighting. And when the arrows were exhausted, they seized axes, stakes, swords, lances and spears that were lying about and struck down, stabbed and destroyed the enemy.[48]

The demand for feathers to make flights for arrows diminished as guns and gunpowder gradually improved. While archery declined in warfare, it grew as a sport and pastime, and by the end of the nineteenth century feathers were needed for darts, which became popular as a pub game. In battledore, which was a forerunner of badminton, racquets or bats were used to hit a shuttlecock made from a cork ball fitted with overlapping trimmed goose feathers, similar to flights in arrows and darts. In a description of 'battledore and shuttlecock', a boy's handbook of 1863 offered some tips: 'A good shuttlecock may be made, where there are no toy-shops to supply it, by cutting off the projecting ends of a common cotton reel, trimming one end with a knife, and drilling holes in the flat surface left at the other, in which holes the feathers of quill pens are to be inserted.'[49] Plastic is now the main material for all these sport functions.

Fly fishing was another sport that needed feathers, and according to Charles Dixon, some birds fetched 'very high prices indeed for this peculiar purpose'. Reusable artificial flies made largely from feathers were attached to a hook and were intended to deceive the fish. Cornwall Simeon in 1860 commented: 'I am persuaded that for fly-fishing colour, not form, is the principle thing to be looked to in the selection of flies.' It all depended on the weather, geology and the actual insects and bugs in circulation, and so feathers of different-coloured birds were obtained for making the often exotically named artificial flies. Peter Hawker in April 1814 went out trout fishing in his local chalk stream: 'notwithstanding a bright sun, I in a few hours killed 36 trout. N.B. – My flies were (what I always use) the yellow dun at bottom, and red palmer bob.'[50]

Izaak Walton is best known as the author of *The Compleat Angler*, first published in 1653. In the dialogue between the traveller Viator and the fisherman Piscator, advice was included on how to make artificial flies for trout fishing: 'I confess, no direction can be given to make a man of dull capacity able to make a flye well; and yet I know, this, with a little practice, will help an ingenous Angler in a good degree.' Piscator suggested carrying in a bag some hooks ready equipped with flies:

> First you must arm your hook, with the line in the inside of it; then take your Scissers and cut so much of a browne *Malards* feather as in your own reason will make the wings of it, you having withall regard to the bigness or littleness of your hook, then lay the out-most part of your feather next to your hook, then the point of your feather next the shank of your hook; and having done so, whip it three or four times about the hook with the same Silk, with which your hook was armed, and having made the Silk fast, take the hackel of a *Cock* or *Capons* neck, or a *Plovers* top, which is usually better; take off the one side of the feather, and then take the hackel, Silk or Crewel, Gold or Silver thred, make these fast at the bent of the hook.[51]

And so he continued, giving alternative ways of devising flies and recommending one in particular: 'a *fly* made with a peacocks feather, is excellent in a bright day: you must be sure you want not in your *Magazin* bag, the Peacocks feather'.[52] Making flies was time-consuming and skilled, and many fishermen would later purchase them ready-made.

The feathers of the dotterel, a species of plover, were prized for artificial flies. During its spring migration from north Africa, this beautiful wader stops off at places across England before heading to its breeding grounds, mainly in Scotland. A traveller to the Lake District in the summer of 1803 remarked: 'my guide told me the week before he saw twenty brace ... The bird itself when stripped of its plumage, sells only for four-pence; but its feathers at Keswick are always worth sixpence, for the purpose of making artificial flies, for the numerous fishermen who live in this neighbourhood.'[53] In about 1835 Thomas Heysham, a naturalist from Carlisle, was concerned: 'These birds, I understand, are getting every year more and more scarce in the neighbourhood of the lakes; and from the numbers that are annually killed by the anglers at Keswick and the vicinity, – their feathers having long been held in high estimation for dressing artificial flies, – it is extremely probable that in a few years they will become so exceedingly rare that specimens will be procured with considerable difficulty.'[54]

In 1907 Thomas Nelson described their persecution in Yorkshire:

As is generally known, the breast feathers of this bird were formerly, and are still, in great request by fly fishers . . . from the Holderness coast right up to the high grounds about Bempton and Specton, the shooting of Dotterel was a regular occupation in spring: for some coast gunners and old shooters boast that in former days – fifty years ago – they have taken as many as fifty couple in a season. The destruction was carried on with equally disastrous effects on the Wolds, moors, and commons inland.[55]

In June 1888 a prosecution was brought against a local man in the Lake District, James Gilpin, 'charged under the Wild Birds Protection Act [of 1880 and 1881] with shooting a dotterel during the close season . . . Police-constable Hogg said he heard of defendant offering dotterels for sale . . . Gilpin said that it had taken him four days to get the birds, and that they were for a gentleman at Milnthorpe. The Bench imposed a penalty of £1 and 14s. 6d. costs, or in default fourteen days' imprisonment.'[56]

The blue patch on the wings of jays was likewise valued for fishing flies and was also desirable for craftwork, as John Knapp mentioned in 1829: 'the beautiful blue-barred feathers, that form the greater coverts of the wings, distinguish it from every other bird, and, in the days when featherwork was in favour with our fair countrywomen, were in such request, that every gamekeeper, and schoolboy brother with his Christmas gun, persecuted the poor jay through all his retirements, to obtain his wings'.[57]

Wearing feathers to adorn clothing, especially hats, undoubtedly dates back to prehistoric times. A feather placed in someone's headgear for every person slain was adopted as a universal sign of military prowess, and so the term 'a feather in your cap' has become one of achievement.[58] From medieval times, feathers embellished the headgear of the military and civilians, at times adorned with jewels. An inventory of numerous richly ornamented plumes of many colours for helmets and horses was compiled in 1560 by Elizabeth I's new Master of the Armoury, including 'oone hundreth and nynetene toppes for hedde peces for men all of red ffethers; oone hundreth and seventene like toppes for horses of red ffethers; three score and eightene toppes for hed peces for men all of red and yellow ffethers'.[59] Another traditional use of feathers was in morris dancing, and in the mid-nineteenth century Thomas Sternberg related: 'The clown, or "Tom Fool," has generally an old quilt thrown over him, plentifully hung with rabbit-skins; his cap is ornamented with a feather, and in his hand he holds a stick with an inflated bladder attached to the end by a cord.'[60]

The Victorian era saw a huge demand for ornamental feathers for women's hats, causing the massive slaughter of birds in Britain and around the world, something that Frank Buckland witnessed in March 1864 when exploring a stretch of the River Thames:

> I met a man in a punt on the Thames, whose special mission on that day was to destroy kingfishers. He had one (a beauty), and he had two shots at others. They were going, he told me, to London, to be made into ornaments for ladies' hats. It seems a very great pity to destroy these little birds, who are just now building their nests; but ladies' fashions rule the day ... Ladies, if you wish to do service to your husbands and brothers, make the white swan of the Thames fashionable; for they are useless and spawn-eating brutes. If I knew who the individual was who *sets* the fashions, I would certainly do my best to cause this really modern semi-god to make swans' plumes fashionable.[61]

Two years afterwards, with his intimate knowledge of the fur and skin trade, John Keast Lord took his readers on a ramble round London's fur stores for *The Leisure Hour*, at the end of which he pointed out the numerous imported bird skins and feathers:

One storey higher, and now we are amidst large deal boxes filled with birds, skins, and feathers. 15,196 grebes bear melancholy evidence of the terrible slaughter a capricious taste or a so-styled fashion may cause amidst the harmless denizens of the land or water. Grebe-skins are worth all sorts of prices, from 7s. 6d. to £1 and upwards. I need not specify their uses: a peep into any clothing-shop, or a stroll up Regent Street on a cold day, will amply suffice. The best grebes are procured from the Lake of Geneva, but a great many are obtained from Norway, Sweden, and, I may say, from all high latitudes. Next to these are 100 scarlet ibis-skins. Ladies, ladies! why will you wear these and other poor birds' bright wings and plumes in your hat? Surely art could supply a decoration quite as becoming.[62]

He was also appalled by the trade in ostrich and egret feathers:

In 1850 3988lbs. of ostrich feathers were imported into the United Kingdom. In 1856 ladies' riding-hats, ornamented with ostrich feathers, became the fashion, and now-a-day every variety of plumage, from the body of a humming bird to a duck's wing or an owl's face, may be seen in the hats of ladies and children … but when living things are wantonly sacrificed, and actual cruelty perpetrated, simply to gratify a stupid fashion, then I feel justified in pleading for the victims.[63]

He found the new fashion of ostrich feathers in funeral processions equally grotesque:

Ostrich feathers are extensively used – I cannot say to decorate the horses and equipages of the dead; the black tufts of feathers, like huge dusting-brushes, ranged round the sombre hearse, are surely not ornamental at any time … Are dirty white feathers, crisp and yellow from age and London smoke, emblems of youth and innocence? Surely clean white drapery, or black, if it be the customary emblem of mourning for maturer years, would be as effective, and, at the same time, save many a poor bird's life.[64]

Kittiwakes were popular for decorating Victorian hats, and in October 1867 the ornithologist John Cordeaux encountered their slaughter at Flamborough in Yorkshire:

> The London and provincial dealers *now* give one shilling per head for every 'white gull' forwarded; and the slaughter of the poor birds during the season affords almost constant and profitable employment to three or four guns. One man, a recent arrival at Flamborough, boasted to me that he had in this year killed with his own gun four thousand of these gulls; and I was told that another of these sea-fowl shooters had an order from a London house for ten thousand ... The young kittiwake is particularly in request by these feather-makers, as the rich black markings on the plumage contrast favourably with the pure white of the under parts and pearl-gray of the back.[65]

Two years later, the Act for the Preservation of Sea Birds was passed, protecting named seabirds between 1 April and 1 August, after which the killing would promptly resume, when many young birds were still in the nest or could hardly fly. The ornithologist Howard Saunders described the situation:

> Some years ago, when the plumes of birds were much worn in Ladies' hats ... the barred wing of the young Kittiwake was in great demand for this purpose, and vast numbers were slaughtered at their breeding-haunts. At Clovelly, opposite Lundy Island, there was a regular staff for preparing the plumes, and fishing smacks with extra boats and crews to commence their work of destruction at Lundy Island by daybreak on the 1st of August, continuing this proceeding for upwards of a fortnight. In many cases the wings were torn off the wounded birds before they were dead, the mangled victims being tossed back into the water.[66]

On Lundy, Saunders himself saw hundreds of young kittiwakes dead or dying of starvation in their nests: 'On one day 700 birds were sent back to Clovelly, on another 500, and so on; and, allow-ing for the starved nestlings, it is well within the mark to say that at least 9,000 of these inoffensive birds were destroyed during the fortnight.'[67]

Not only were loopholes exploited in the legislation, but the plumage (or fancy feather) trade expanded massively around the world, massacring birds on a vast scale to supply decorative plumes for millinery in Britain, Europe and America. Millions of feathers and skins were brought to London because it was legal to do so, and most people were in ignorance of the worldwide devastation. In 1885 the Plumage League was founded, which urged members not to wear feathers. Various related bodies were formed and merged, mainly by women, out of which came the Society for the Protection of Birds (SPB) in 1889. Awarded a royal charter in 1904, the main aim of the RSPB at that time was to stop the devastating plumage trade.[68]

With increased criticism, the feather industry became more secretive and deceptive, even claiming that egret plumes were artificially made, as the RSPB revealed in a 1911 booklet:

> The Egret feather was no longer to be labelled as 'real'; milliners' and drapers' assistants were instructed to assure lady customers that these delicate sprays were manufactured by the million out of quills and other material, by an army of factory-workers, who earned their living by this pleasant and artistic work ... Ladies were told that these things were manufactured from quills, ivorine, silk, wood, the feathers of poultry, etc.; that they could not be sold so cheaply if they were 'real' feathers.[69]

Once this fraud was exposed, the trade instead claimed that they dealt only in moulted feathers that had been collected, even though they tend to be worn out, brittle and soiled. After failing to get a bill through parliament in 1908 to stop the trade, World War One temporarily halted it, but it was finally stopped in 1921 with the Importation of Plumage (Prohibition) Act, also known as the Plumage Act, though it did not control the sale or wearing of feathers.[70]

Caged canaries and white mice from World War One carved on the Scottish
National War Memorial at Edinburgh Castle

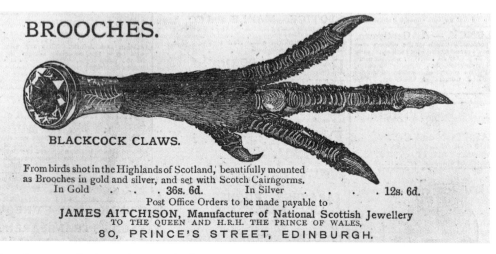

Advertisement for brooches made from the claws of black grouse shot in the
Highlands

The guano workings on the Chincha Islands off Peru

The vast accumulation of guano on the Chincha Islands off Peru

Unlucky peacock feathers

Pigeon, peacock and turkey (left to right)

Pheasant (left) and two partridges

Young cormorants

A gamekeeper shooting
magpies

A boy robbing a nest

Eggs from a hedge sparrow's nest were sought after
by boys

Trying to shoot birds with an early flintlock gun

Tawny owls, often treated as vermin

Wildfowler Snowden Sleights from East Cottingwith, Yorkshire, in his punt on the River Derwent

A native American bald eagle on a 1942 quarter dollar

Homing pigeons trained for racing

An 1883 advertisement for goose feather beds

Swallows (left) and a kingfisher (right)

Magpies (left) and rooks (right)

Queen Elizabeth I medal of c.1574 linking her to the legendary phoenix

A Victorian caged bird c.1876

A Victorian taxidermist with an owl c.1897

Sometimes feathers were simply for decoration in houses. Mary Frampton's parents were well acquainted with the literary hostess Mrs Elizabeth Montagu, and Mary visited her magnificent house at the corner of Portman Square in London on several occasions, recalling that 'one room was entirely hung with peacock's feathers, which made by no means a pretty ornament'.[71] Mrs Montagu was by then widowed and had in 1781 moved into this new property, constantly begging her friends to give her feathers, such as pheasant, cock, swan and peacock, to decorate one room. In 1788, when William Cowper heard of it, he wrote 'On Mrs Montagu's Feather-Hangings', which began:

> *The Birds put off their ev'ry hue,*
> *To dress a room for MONTAGU.*
> *The Peacock sends his heavenly dyes,*
> *His* Rainbows *and his* Starry eyes.[72]

Others were afraid of peacock feathers, and several decades later *Notes and Queries* published a comment from a reader who signed himself as P.P.: 'A correspondent speaks of Peacocks' feathers as being unlucky, and considers this may refer to the evil eye. I never heard of such a superstition, and cannot think it general. I only know that, between servants, labourers, young ladies, and little children, I find it difficult to secure any of mine. A group of these feathers, stuck behind a picture-frame or a looking-glass, is a very common cottage or farm-house ornament in the north of England.'[73]

The editor of the *Derby Telegraph*, Llewellyn Jewitt, disagreed:

P.P., I perceive, doubts the fact of there being a general superstitious feeling regarding peacocks' feathers. I can vouch for such feeling being general in Derbyshire and the surrounding counties. It is considered extremely unlucky to have them in the house, and they are believed to bring losses and various misfortunes, including illness and death, to the inmates. I have seen people perfectly horrified when a child or other person has unwittingly brought into the house one of these feathers.[74]

In East Yorkshire, John Nicholson also knew about such a superstition:

> Though peacock feathers are now [1890] fashionable and aesthetic, they are looked upon with disfavour by those of the old school, for these feathers were always deemed unlucky. Some aesthetic friends, in choosing a 'fairing,' selected a lovely fire-screen, made of peacock's feathers. The person to whom it was sent was greatly troubled about it, and, after spending several days in great mental distress, actually burnt the costly gift, being fearful of the ill-luck its possession would entail.[75]

According to the *Newcastle Weekly Chronicle* a few years later, it was 'lucky to have a live peacock on a farm, but very unlucky to have peacock feathers in the house'.[76]

Nicholson described one ceremony with feathers that was intended to bring good luck to newly weds: 'Near the house in which the bridal feast is spread, stand three or four men with guns crammed to the muzzle with feathers, and, as the party passes them, the guns are discharged, and the air is filled with falling feathers, thereby betokening a wish that nothing harder may ever fall on the happy pair.'[77]

With such conflicting beliefs about feathers, it is not always possible to explain their presence in strange places, such as the swans' feathers and skins that were discovered lining pits at Saveock in Cornwall. In February 1673, sixteen-year-old Thomas Isham of Lamport Hall in Northamptonshire wrote in his diary that witches had been much talked about and that two women at nearby Broughton were suspected of witchcraft. When a little girl was frightened by one of them and the rest of the family became 'vexed with enchantments', Thomas's father 'ordered them to examine the pillows, when they found feathers stuck together; which they threw into the fire and burnt'.[78] The discovery of feathers woven into a rope at Wellington in Somerset still generates discussion about whether or not it was associated with witchcraft. Abraham Colles was a local doctor who, in 1886, lived in the High Street, and on a call to a new house just outside the town, he was told that it was constructed on the site of a dilapidated ancient farmhouse, formerly the property of the Popham family. While being demolished a few years earlier, builders had found a small room just below the thatched roof, which Colles described:

> In pulling down the upper storey there was found in a space which separated the roof from the upper room, and to which there was no means of access from below – First: six brooms. Second: an old arm-chair. Third: a rope with feathers woven into it ... composed of a piece of rope about five feet in length, and about half-an-inch in diameter. It is made with three strands, and has at one end a loop, as if for the purpose of suspending it. Inserted into the rope cross-ways are a number of feathers – mostly goose, but some crow or rook – not placed in any determinate order or at any regular intervals, but sticking out on all sides of the rope ... It was a piece of *new* rope with feathers woven into it, only that now the feathers are in a very imperfect condition.[79]

Colles sought advice from local women who practised or believed in witchcraft and gained the impression that the rope was connected with casting spells, something that has been recorded elsewhere in Europe.[80] It was donated by Edward Tylor, a mill-owner and anthropologist at Wellington, to the Pitt Rivers Museum in Oxford, where it is described as a witch's ladder, but one contrary opinion is that it was a sewel, set up to frighten deer when hunting. In September 1887, the Reverend Evan Daniel wrote to the *Manchester Guardian*, saying that this interpretation was likely considering how close Wellington was to the stag-hunting district of Exmoor, but added a note of doubt: 'On the other hand, Somersetshire is, I am afraid, one of the most superstitious counties in England.'[81]

As well as feathers, other parts of birds were put to practical use. Many bird bones are hollow, making them light to aid flight, and so they were rarely turned into artefacts, but some of the sturdier hollow bones were shaped into musical whistles and other items.[82] The Ward family became renowned for artistic taxidermy in Victorian England, and in his 1880 handbook for sportsmen, Rowland Ward gave advice on what to do with certain bird bones: 'It may sometimes fall to the lot of the traveller to secure an albatross, and at the same time he may not always know what to do best with his unwieldy and not very rare specimen. If he does not care to preserve it whole, he may well be reminded that there are some parts of it which may be profitably saved. The long tubular wing bones are prized for pipe stems, to which they can be adapted well.' Ward was referring to pipes for smoking tobacco, for which he also recommended flamingo bones: 'The leg-bones of the Flamingo are long and have an elegant curve – they will form admirable pipe-stems.'[83]

He had further ideas for the albatross: 'the great web-feet will make beautiful tobacco pouches when properly prepared, or a curious small work-bag for a lady can be formed of the same trophies'. After giving instructions on how to prepare the bones and the feet, he commented: 'Of course the utilisation of parts of birds and animals in this way is mainly a question of inventiveness and ingenuity, for many natural features may be adapted to the most useful and ornamental purposes, while at the same time they retain the character of trophies.'[84] Hugh Gladstone in 1910 also said of albatrosses: 'Sailors often bring back from their voyages in the Southern Ocean the feet of these birds as tobacco pouches and their beaks as curios.'[85]

Martin Martin, a Scottish traveller and author, visited St Kilda in 1697 and documented that the women wore no shoes in the summer: 'their ordinary and only shoes are made of the necks of *Solan* Geese [gannets], which they cut above the eye, the crown of the head serves for the heel, the whole skin being cut close at the breast, which end being sowed, the foot enters into it, as into a piece of narrow stocking; this shoe does not last above five days, and if the downy side be next the ground, then not above three or four; however, there are plenty of them.'[86]

The manure, dung or droppings of birds was even put to various uses, from medicinal applications to the supernatural. In East Yorkshire, Nicholson said that goose dung was a cure for baldness, while pigeon and poultry manure was employed by tanners to help soften animal skins and remove the hairs.[87] On the Isle of Man in about the mid-nineteenth century, hen's dung was believed to counter witchcraft: 'The old farmers – now very long ago – the first day they were taking the horses out to plough [usually in March], used to get a bucket of Chemerly (Chamberlye or mooin), and put some hen's dung in it, and stir it for some time, and then sprinkle the horses and the plough with it. That was the charm against witchcraft and the evil-eye.'[88]

Dung became invaluable for the manufacture of gunpowder, whose ideal ingredients were 10 per cent sulphur (brimstone), 15 per cent charcoal and 75 per cent saltpetre. Saltpetre was critical, because when set alight, oxygen is given off that enables burning to occur in a confined space like a cannon. With the increase in weapons needing gunpowder in Tudor and Elizabethan England, the restricted supplies of both foreign gunpowder and saltpetre were problematic.[89] From the mid-sixteenth century information reached England about producing it artificially by mixing suitable earth or animal manure with lime and ashes, constantly turning the heaps and watering with urine. The liquid with the saltpetre ran off, which was then refined by boiling and spread out to dry in order to collect the saltpetre crystals.[90]

Initially, gunpowder makers made their own saltpetre, but it became the responsibility of saltpetremen, who were permitted, under a royal prerogative, to enter anywhere suitable to dig out soil, such as inside dovecotes (pigeon houses), stables, barns, cellars and even church crypts. From around 1590 for half a century, the intrusion was immense, and it was illegal for owners to put down paving or other floor surfaces.[91] Dovecotes proved one of the best sources, but damage was invariably done to the eggs and pigeons, as well as the buildings. Much resentment was caused, and in order to obstruct the saltpetremen, some dovecote owners removed their own soil. On 15 November 1634, Richard Bagnall, a saltpetreman, sent to the Admiralty secretary Edward Nicholas a list of names of those in Oxfordshire and Warwickshire 'who have lately carried forth their earth in their pigeon houses. If some course be not taken others will do the same, and it will be impossible for the saltpetremen to supply their great proportions.'[92]

In 1637 Christopher Wren, who was the rector at East Knoyle in Wiltshire and father of four-year-old Christopher, the future architect, complained directly to the king about the damage caused by the workmen of Thomas Thornhill, the saltpetreman, to the rectory's pigeon house:

> There had been two diggings in this pigeon-house, one by Helyar [Hilliard], whom Thornhill then served, about eight years ago, the other by Thornhill in March 1636–7. On the first occasion the pigeon-house, built of massy stone walls 20 feet high, was so shaken that the rector was forced to buttress up the east side thereof. On the last occasion the foundation was undermined, and the north wall fell in. The loss to the rector had been that of three breeds, whereof the least never yielded fewer than 30 or 40 dozen, and of the whole flight, which forsook the house.[93]

From the early seventeenth century, the East India Company brought back shipments of saltpetre to England from India, mostly as ballast, and by the end of the century they were supplying far more saltpetre than the saltpetremen could ever manufacture, of much better quality, so dovecotes ceased being plundered.[94]

Poultry manure and pigeon droppings from dovecotes also proved valuable as a fertiliser in agriculture and market gardening due to the high levels of nutrients such as nitrogen, phosphate and potassium. Occasionally, other sources of bird dung might be available, such as from seabird colonies. The historian John Campbell in a 1774 survey of Britain talked about the small islands close to Barrow-in-Furness, including Foulney: 'Amongst these is one called Fowlney, i.e. Fowls Island, from amazing numbers of wild-fowl resorting thither; the dung of which collected, and spread on the meadows nearest to it on the Continent [mainland], makes them so rich that they commonly let from fifty shillings to three pounds an acre.'[95] The adoption of seabird dung as a fertiliser was not widespread in Britain, because the rainfall reduced its efficacy, but that changed in the nineteenth century when guano was imported.

The word 'guano' is Spanish for any manure derived from animal droppings, especially bird droppings from the islands off Peru. For thousands of years, it was used by the indigenous people, but was little known in Europe until the Prussian naturalist Alexander von Humboldt visited the coast in 1802 and took some back to France for analysis. In Britain the scientist Humphry Davy also analysed guano samples: 'It exists abundantly, as we are informed by M. Humboldt, on the small islands in the south sea . . . 50 vessels are laden with it annually at Chinche, each of which carries from 1500 to 2000 cubical feet . . . I made some experiments on specimens of guano sent from South America to the Board of Agriculture in 1805. It appeared as a fine brown powder; it blackened by heat, and gave off strong ammoniacal fumes.'[96]

In 1808, on the Atlantic island of St Helena, the new governor Major-General Alexander Beatson began experiments with different manures, including guano from Egg Island, a mile offshore, after being alerted to its properties by Sir Joseph Banks, President of the Royal Society. Beatson found the guano incredibly effective and in 1811 wrote: 'I have been informed that guana [guano] is sold at Lima, and other towns on the coast of Peru, for a dollar a bag, of 50 pounds weight, and that it is much in use there for manuring fruit trees and gardens.'[97] Other scientists also undertook research, but the political upheavals in South America following the Napoleonic Wars in Europe meant that guano was not commercially exploited in Peru until 1840. The Peruvian government nationalised the industry, which became a great source of revenue.[98]

Cuthbert Johnson, a barrister and expert on manures and fertilisers, published a small book in 1843 on the use of guano, including impressive results of trials in Britain, and a huge demand developed.[99] In 1850, writing in the *Morning Chronicle* about London's merchant shipping, Henry Mayhew said: '*The South American traders* are generally large ships, averaging from 500 to 800 tons, and carrying from twenty to thirty men ... They sail at all times of the year, taking out cargoes of iron and general merchandise, and bringing back home hides, skins, tallow, horns, hoofs, bones, and guano.' He spoke to one intelligent and reliable seaman about working conditions and learned that on his last voyage, the heavy seas around Cape Horn drenched everything. The captain therefore let them sleep on the guano that they had loaded as ballast at the Chincha Islands off Peru: 'The guano is in cliffs ... From the sea it looks like a rocky mountain ... They say it's the ordure of birds, but I have my doubts about that, as there could never be birds, I fancy, to make that quantity. Why, I have seen as much guano on the Chinqua Islands ... as would take thousands of ships twenty years to bring away. There are great flocks of birds about the guano places now, chiefly small web-footed birds.'[100]

The scale and origin of the deposits appeared extraordinary to him, but they did actually represent centuries of bird droppings. The seaman explained how the guano was loaded:

> We have two 'shoots.' A shoot is made of canvas, equally square on all sides. The diggers bring the guano a quarter of a mile, to the shore ... There the diggers empty their bags, through an open place into the shoot which is spread below, and held by ropes. The shoot is then lowered down from the guano mountain by the diggers, and the seamen who hold the ropes to regulate it must keep the lines a moving, to keep the guano from choking (going foul) in the shoots. We must regulate it by the pitch of the ship. The ship is moved alongside, and so the shoot is emptied down the hatchways at a favourable moment.[101]

Guano was difficult to cope with: 'There is a very strong smell about the guano mountain. It oft makes people's noses bleed. The diggers on board and trimmers in the ship have to keep handkerchiefs round their noses ... It affects the eyes, too; no trimmer can work more than fifteen or twenty minutes at a time ... The diggers and trimmers are labourers who live on the guano mountain ...

There were a few women there, but no bad women. The diggers are chiefly foreigners.'[102]

Thousands of coolie labourers, mainly from south China, were employed, especially after 1854 when slavery ended in Peru. A correspondent of *The Leisure Hour* was shocked to discover the situation when in 1860 he sailed from Hong Kong to the port of Callao on board a ship carrying 384 young coolies who 'had signed an agreement to be taken to Peru, where they were to work for seven years for the sum of two hundred and forty Mexican dollars each, and to be returned to the port from whence they sailed . . . A coolie is a slave for the time he has volunteered to serve.'[103] Conditions on board the ship were so terrible that forty-two of the young men died 'and several others were not much better than dead men when they were sent ashore a few miles from Callao'.

Mayhew's seaman told him that the crew was not allowed on the island, and on board ship their cargo of guano was very unpleasant, though they became accustomed to it: 'At first the ship smelt as if she was laden with hartshorn [ammonium carbonate, used for smelling salts].' Mayhew himself experienced the stench: 'I may add that a cargo of guano was being unladen at the West India Dock [London] on one of my visits there. It was hoisted out of the hold in bags, and had altogether the smell of very strong and unsound cheese. The whole atmosphere of the ship was cheesy. Some of the guano, which had been spoiled by salt water, had the appearance of yellow mud or slime.'[104] By the 1870s, the industry dwindled due to many factors, including the depletion of the guano in Peru.

Dead birds were also used for fertiliser. The scientific name for the Manx shearwater is *Puffinus puffinus*, though perversely it is not related to the bird commonly known as a puffin, with its colourful beak, and whose own name is *Fratercula arctica*. The shearwater breeds in colonies on offshore islands, including Skomer Island off the Pembrokeshire coast of Wales, where they were known as 'Cockles'. At the end of May 1884, Murray Mathew and a friend spent some time there, watching the shearwaters at night after they emerged from their burrows. There were so many, he said, that with a lantern it would have been possible to knock them down with a stick: 'One night we were told that the farm servants actually destroyed a multitude in this manner, and that the bodies of the birds were ploughed into the ground as a dressing for wheat. Alas, poor "cockles"! to what vile uses did they come.'[105]

Oil was another major by-product of seabirds, and ones like the fulmar produced stomach oil that coastal communities relied on for lighting, amongst many other uses. When James Wilson visited St Kilda in 1841, he said that the men collected oil as well as feathers on neighbouring Boreray:

> They ... brought numerous long distended bags (the stomachs of old solan geese) filled with oil, which they extract nearly pure from the stomach of the fulmar ... This article, as may easily be imagined, is one of essential service to the people throughout the darkness of their long-enduring winter. It is extracted from both the young and old birds, which ... they must seize suddenly and strangle, else as a defensive movement the desired (and pungent) oil is immediately squirted in the face and eyes.[106]

Edward Stanford, a prominent industrial chemist and pharmacist in Scotland, also observed how the oil was collected and later analysed it: 'The oil is a good deal mixed with a rougher sort from the solan geese, and realises a poor price as an ordinary rough fish-oil. When genuine, it is of a clear, dark, slightly reddish sherry colour, and has a powerful and peculiar odour – an odour of which the whole island and all the inhabitants smell. It is certainly a fish-oil, and it possesses nearly all the properties of cod-liver oil.'[107] On his trip to Skomer, Mathew said that it was too early to find the storm petrel nesting, though its smell lingered: 'we detected ... the unmistakable Petrel scent which clings to eggs of the true Petrels ... and also to those of the Fulmar for many years after they have been blown, as it also does to their skins. Like all the Petrels, this species is so full of oil that it is said that the inhabitants of the Faeroe Islands, by merely passing a wick through the bodies of the birds and lighting the end, supply themselves with a light.'[108]

Frank Buckland, when talking of herons, wrote about one belief: 'Herons stand motionless in the water when fishing; they do not go after the fish, but wait for the fish to come to them. By some it is supposed that there is a peculiar scent in the heron's legs which attracts the fish, and with this idea it is the custom in some countries (I heard of it in Oxfordshire) to cut off the legs of a heron and obtain from them the oil, which is used by the fisherman to anoint his bait. I should be inclined to doubt its efficacy.'[109] On the Isle of Wight in the late eighteenth century, intrepid men would climb down the chalk cliffs, kill the seabirds and sell the feathers, while the carcasses were bought by the fishermen at sixpence a dozen to bait their crab pots.[110]

Anyone with imagination or need could make use of countless parts of birds. In February 1826, when at Keyhaven opposite the Isle of Wight, Peter Hawker's manservant Reade rushed outside in a deluge of rain and shot a cormorant (or, as it was described, 'lowered a parson', the derogatory term used there for shooting these large, black seabirds). Hawker noted in his diary: 'He was disposed of as follows: the skin to make a dandy collar for a coat; the feathers to make me drawing pens; and his carcase begged by my boatman Williams ... for a delicate Sunday's dinner.'[111]

Even claws found a function. One high-class jeweller in Edinburgh in 1883 went upmarket and advertised macabre brooches made from black grouse claws, which sold at £1 16s. 6d. in gold and 12s. 6d. in silver: 'Blackcock claws. From birds shot in the Highlands of Scotland, beautifully mounted as Brooches in gold and silver, and set with Scotch Cairngorms.'[112]

CHAPTER FIFTEEN

———— •·• ————

AT WORK

In one company a certain canary became a regular
'old soldier.' On entering a mine he would topple
off his perch immediately, and pretend to be dead.
On being taken out of the mine he would 'recover'
at once and hop about the cage and chirp merrily
as though he enjoyed the joke.

Nottingham Evening Post May 1936[1]

Ingenious ways have long been devised to make practical use of
live birds, which include potentially life-saving roles such as carrying messages and detecting gases underground. Birds have even
inadvertently provided an invaluable role by their behaviour in the
wild, most notably by clearing up detritus on the battlefield.

Birds have an incredible ability to navigate, and none more so
than domesticated or homing pigeons. Originally derived from wild
rock doves, they are capable of returning home from wherever they
are released. This gift has enabled them to provide an immense
service throughout history as carrier or messenger pigeons, taking
or carrying messages back to the 'home' from which they had been
temporarily removed, or else competing in races. Such homing
pigeons were easily bred and, with good care, lived up to twenty
years. They are often loosely referred to as carrier pigeons, but
are not the same as passenger pigeons or the Carrier (or English
Carrier) breed of fancy pigeon.[2]

Young homing pigeons are trained to regard their loft, cote or coop as home, initially by releasing them a short distance away, then gradually increasing the distance, until they are accustomed to finding their way back, probably using magnetic fields to navigate by. If adult pigeons were moved to a different loft, they often returned to their old home if not sufficiently acclimatised. When Edmund Dixon wrote *The Dovecote and the Aviary* in 1851, he said that 'many a time, when a boy, have I gone out for a country walk, with two or three Pigeons in my pocket, or wrapped up in a silk handkerchief tucked under my arm, to be tossed off at the furthest point of the excursion, and to be found at home on my arrival there.'[3] At that time, it was incorrectly assumed they could only return home if the entire route was familiar, perhaps by recognising particular landmarks, causing Dixon to state: 'The Pigeon which will traverse with practised ease the space from London to Birmingham may be unable to find its way from Bangor or Glasgow to the same town.'[4]

The earliest record of pigeons bearing messages is in a poem by the ancient Greek poet Anacreon, writing in the sixth century BC, and from the Roman period pigeons are mentioned as a method of communication between merchants in different cities or carrying messages in times of war. In April 43 BC, the troops of Decimus Brutus were besieged inside the town of Mutina (Modena) in Italy by Mark Antony's troops. Running out of supplies, Brutus sent messages by pigeons to allied troops further away, outwitting Mark Antony, as Pliny the Elder explained: 'at the siege of Mutina, Decimus Brutus sent to the consuls' camp dispatches tied to their feet. What use to Mark Antony were his rampart and watchful besieging force, and even the barriers of nets that he stretched in the river, when the message went by air?'[5]

By the sixteenth century, specially bred carrier pigeons had been introduced to Europe from the Middle East. Alexander Russell, a physician to the English factory at Aleppo in Syria from 1740, related that the Europeans at the Mediterranean port of Scanderoon (modern-day Iskanderun) used pigeons to alert the merchants at Aleppo of a ship's arrival:

A small piece of paper, with the ship's name, day of arrival, and what else material could be contained in a very narrow compass, was fixed so as to be under the wing, to prevent its being destroyed by wet. They also used to bathe their feet in vinegar, with a view to keeping them cool, so as they might not settle to drink or wash themselves, which would have destroyed the paper. And I have heard an *English* gentleman, who remembered that practice, say, that he has known them arrive in two hours and a half.[6]

After aeronauts in nineteenth-century France improved the art of hot-air ballooning, it made possible, during the Franco-Prussian War of 1870–1, one of the most famous uses of homing pigeons.[7] After the Prussian chancellor Otto von Bismarck had provoked an over-confident France to declare war, the defeat of the French army at the small town of Sedan a few weeks later proved shocking, especially as Napoleon III was captured and sent into exile in England. As the Prussian forces headed towards Paris, the French government shifted to Tours and later Bordeaux.[8]

Paris was completely surrounded, and all means of communication with the outside world, such as the telegraph lines, were destroyed. In late September 1870, a manned balloon using coal gas was successfully launched, landing 20 miles away with letters and dispatches that were forwarded to Tours. Private messages were subsequently printed in newspapers across France and in *The Times* at London, and this success led to many balloons being manufactured in Paris that could accommodate two men, baskets of homing pigeons and mailbags. A few balloons were captured, lost or went astray, but about sixty-seven of them reached safety, enabling a regular pigeon post back into Paris to be established.[9]

During a talk at the United Services Institution in London a few years later, Captain Henry Allatt explained that a pigeon loft was created at Tours, and whenever pigeons were needed for sending messages to Paris, about five were taken by train as close to the city as possible, before being released.[10] The first messages had to be brief, so in order to carry longer messages, microphotography was used, with images taken of hundreds of letters, dispatches and newspapers that were processed as tiny negatives on sheets or leaves of collodion film about 2 by 3 inches. Felix Whitehurst, the Paris correspondent for the *Daily Telegraph*, was trapped in the city throughout the siege, and he was shown how the system worked: 'From fourteen to eighteen of these tiny leaves were put into a quill [at Tours] and tied to a pigeon's tail, several copies of the same leaves being sent by different pigeons, so as to diminish the risk of loss.'[11]

Allatt mentioned that the 'birds were stamped on the wing feathers with numbers, the first indicating the number of birds sent, the second the number of the series of messages, and the third the number of pigeons remaining.'[12] Whitehurst also said that when a pigeon reached Paris,

> the quill was immediately forwarded to the telegraph station, where the leaves were read through a microscope to a clerk, who wrote out the despatches for each person. But this was a terribly slow process; it permitted the employment of only one reader and only one writer, which was insufficient ... So, after a few days, the leaves were successively placed in a large microscope, to which electric light was adapted; and the magnified image of each leaf was projected on a white board, from which it was copied by as many clerks, taking a column each, as could manage to get sight of it from a writing table.[13]

Any private messages were delivered to the intended recipients, so that the pigeon post proved an invaluable means of transmitting news and maintaining morale. The Prussians did try to shoot the pigeons or bring them down with hawks and falcons, but about seventy reached Paris.[14] During the siege, Whitehurst noted how the system worked:

> Here is a rapid act of transmission of news, A.D. 1871: – On the 9th of December, 1870, I sent a letter by 'mounted balloon' to London. It arrived on the 13th of December, on which day a

telegraphic answer was sent, and which I received by pigeon from Tours on the 23rd of January, 1871. Before the Prussian era, I used to send at 7.40 A.M., Monday, a letter to the 'Daily Telegraph,' and buy the paper containing it on the boulevards at 7 P.M. on Tuesday![15]

Food became desperately short, so that rats, cats, dogs, horses and rare specimens in the zoo were eaten. Whitehurst heard that when one lady was offered some roast pigeon, she replied: 'Not any, thank you; I should think I was eating a postman.'[16] After heavy bombardment, Paris capitulated towards the end of January 1871.

Homing pigeons were not solely used for carrying messages. Pigeon breeding, 'the fancy', became increasingly popular in nineteenth-century Britain, France and Belgium, especially with the working classes, who reared pigeons as a hobby. Most of these pigeons took part in races, and a journalist in Yorkshire remarked in 1916: 'Racing pigeons have provided as much pure sport to the working men of Yorkshire and Lancashire as any other recreation, and the soul of the sport is that the birds will as soon fly home to a small box cote in a dingy coal house of a Lancashire miner as the most palatial of lofts such as those of His Majesty the King [George V] of Sandringham.'[17]

Breeding fine specimens had grown fashionable from the mid-eighteenth century, and they were often exhibited at shows. The innumerable breeds have names as colourful as tumblers, jacobins, magpies, fantails, homers, rollers, trumpeters, archangels, pouters and English carriers.[18] The show birds tend not to have a homing instinct, as highlighted by an incident in 1897 when some were being transported by steam train along with racing pigeons: 'Pigeon clubs are in the habit of sending off baskets of homers to country stations with a request to the stationmasters to set them free at a certain hour. By mistake a basket of show homers was let loose with the others. These birds were worth from £2 to £5 apiece. The railway company offered £30 for the recovery of ten of them, describing them. None returned to their homes, and very few were caught.'[19]

Thousands of domestic homing pigeons were transported

by railway each Saturday during the summer, and the railway's involvement in racing was well known to the Yorkshire journalist: 'The homing pigeon traffic is now an important branch of railway work, and although a quarter of a century ago [1890] no organised arrangements for conveying homing pigeons away to the race points were ever dreamed of, it is now no unusual occurrence for from sixty to eighty special vans at a construction cost of £800 to be utilised for the conveyance of racing pigeons alone, from Lancashire on a particular night.' Racing pigeons, he further explained, were once used by newspaper offices across the country to obtain information rapidly, such as the results of football matches:

> It is within the recollection of the writer how a bird, which was known to be the bearer of some very important news from a country district where no telegraph office was convenient, arrived at the newspaper office in good time but would not enter the loft, though some precious time was spent in trying to coax the bird in. The bird had to be straightway shot owing to the importance of the news ... Such use of pigeons, however, has been rendered obsolete by the telephone.[20]

Although many European states had begun to invest greatly in pigeons for military use, in Britain the army was less enthusiastic. Captain Hornby of the Royal Navy talked in 1886 about his experiments with homing pigeons on men-of-war and the fact that 'it would be quite possible to train pigeons to maintain communications between ship and shore'.[21] Unfortunately, he pointed out, government spending departments were too inept to take up such new ideas. Four years later, Charles Dixon added his views: 'The uses to which Homing Pigeons have been applied have considerably increased of late years. The gentle Dove has been made a messenger of war; and numerous experiments have been made to determine its fitness for carrying messages rapidly from one district to another, where the telegraph or railway is unavailable ... there can be little doubt that in the warfare of the future, the Homing Pigeon will play an important part.'[22]

At the end of the nineteenth century, James Wilson, a correspondent for *The Navy and Army Illustrated*, disclosed that the
Admiralty was setting up pigeon lofts, albeit on a shoestring:

> The whole amount is only a few thousand pounds ... but the
> training of the birds will have to be thorough, or the money,
> small as the amount is, will have been wasted, for France,
> Germany, Russia, and other foreign Powers are all exerting
> themselves to obtain the best service of messenger pigeons it is
> possible to get ... Although foreign Naval and Military authori-
> ties have made use of pigeons to carry Government despatches
> backwards and forwards from one Naval or Military station to
> another for many years past, it is only within the last eighteen
> months that the British Admiralty has done so, while, as for the
> War Office authorities, they are still unbelievers in the useful-
> ness of the pigeon as a messenger of war.[23]

Wilson said that bright red-and-white pigeon lofts could be seen
in some seaside towns, as well as naval pigeon-cotes at Portsmouth,
Devonport and Sheerness: 'At the first-mentioned place the cote
has been erected on Rat Island ... At Devonport the loft has been
erected on Mount Wise, in the near vicinity of the admiral commanding at that port. Mount Wise makes an excellent landmark
for the birds when returning with a message ... At Sheerness the
pigeon-post has only been in existence a year or so, the loft being
erected on the Naval Recreation Ground.'[24] About one thousand
birds were being trained:

> Compared with the thousands of messenger pigeons owned by
> France, Germany, and one or two other Powers, the number is
> not very large. France, for example, has more than 1,000 birds
> at Toulon alone, while at each of the towns of Cherbourg, Brest,
> Rochefort, and Paris from 500 to 600 birds are kept under train-
> ing. The distances the British birds can fly are short compared
> with those covered by some foreign pigeons. In Germany 300
> and 500 miles are often flown by the Government pigeons.

There may be some British Naval pigeons capable of flying this distance, but so little information is allowed to transpire about them that nothing very definite is known.[25]

During their training, the distance each bird travelled was gradually increased, until they were ready to be taken to sea in torpedo-boats, cruisers or battleships and released about 10 to 12 miles from land. Wilson thought they were indispensable:

> it will be as a means of communication between ship and shore that the pigeon will be found most valuable in war time, as far as the Navy is concerned; and the time will probably come when each of our battle-ships and cruisers will carry a pigeon cote ... during the trials of the first-class cruiser 'Terrible,' and the third-class cruiser 'Pelorus,' pigeons were taken out for the purpose of sending them back with messages, and in this manner the trials were known on shore long before the vessel returned to port.[26]

He added that American Government pigeons had more experience, 'and a bird which was recently released from the flag-ship "New York," 350 miles out at sea, reached the Navy yard at that city in seven hours'.

Alfred Osman in London was devoted to pigeon racing and breeding. In 1898 he founded *The Racing Pigeon*, a weekly magazine, and he also promoted competitive races. In 1907 he published *The Pigeon Book*, in which he remarked: 'I shall be probably underestimating the number of fanciers, when I say that there are from 50,000 to 60,000 keen enthusiasts racing each week.'[27] That same year, the War Department abolished the fledgling pigeon service, an ill-judged and reckless decision according to the newspapers: 'It cannot be that the advent of wireless telegraphy has made the pigeon post obsolete, for it will be many years before small scattered sections of an army can carry their own Marconi apparatus. It could not be on the score of economy, for the cost of keeping the pigeons was infinitesimal. A large number of the birds were presented to it by patriotic pigeon owners. All these birds have now been sold and the lofts destroyed.'[28]

At the outbreak of World War One, German spies were so feared throughout Britain that civilian-owned pigeons were ordered to have their wings clipped, be destroyed or released into the wild, and their pigeon lofts closed. Osman intervened and became involved in re-establishing pigeons as part of the war effort, initially on the home front, calling on all his contacts for help. A series of pigeon lofts along the east coast was created, and Osman provided pigeons for trawlers that were engaged in minesweeping, as well as for the seaplanes of the Royal Naval Air Service and for those naval vessels operating in the North Sea, so they could send back intelligence or call for assistance in emergencies.[29]

On 5 September 1917 Flight Lieutenant Robert Leckie, a Canadian airman serving with the Royal Naval Air Service based at Great Yarmouth, was piloting an American 8666 flying boat, accompanied by a modified de Havilland DH 4 aircraft. After attempting to intercept two Zeppelins – the huge airships used by the Germans to drop bombs – and being fired at by German warships, the DH 4 was forced to turn back, escorted by the flying boat with its crew of four. The DH 4 ended up in the North Sea and sank, 50 miles from land, so the flying boat landed on the sea, pulled the two crewmen on board, but was then unable to take off in the difficult conditions. When their petrol ran out and the engines stopped, it was a constant struggle to try to prevent the vessel from sinking. What they did have was the standard number of homing pigeons for their craft, four in all, and the local Yarmouth newspaper related what Leckie supposedly told them:

> I ... released a pigeon at once, carrying a message to the base, giving our position and cause of troubles. On Thursday a second bird was released, carrying a similar message. On Friday morning a third bird was sent, and knowing that we could not last much longer our remaining pigeon was released on Friday afternoon, carrying an S.O.S. signal. The first three pigeons failed to reach England, but the fourth was a winner. When I released

him there was a strong west wind blowing; also fog. In spite of these adverse conditions, and also in spite of the fact that he must have been suffering from hunger like ourselves, and thirst, the pigeon struggled over fifty odd miles of sea without a landmark or without a rest. He failed to reach his loft, but reaching the English coast about twenty miles north of our base, picked out a coastguard station, and fluttering into the courtyard fell dead from exhaustion. At the cost of his own life he delivered his message, and thus saved ours – we were picked up by the U.S. – the next day.[30]

He added that they were incredibly thankful to the pigeon: 'although it died its memory will live. The grateful airmen have had it stuffed, and a little plate bears the inscription: "A very gallant gentleman." '

The reality was a little different, because two pigeons were actually released on the first day, another the next morning and the final one in the afternoon with an SOS, which was received at Yarmouth air station the day after. That led to their rescue by the gunboat HMS *Halcyon*. Two pigeons were never seen again, but one of those from the first day reached land and dropped dead while the rescue mission was taking place. After being picked up by chance, it was taken to a nearby signal station, and the message was phoned through to Yarmouth. The six rescued men were so moved by this gallantry that they had the pigeon stuffed and put in a glass case with a brass plaque, which was displayed in the wardroom at Yarmouth and is now in the Royal Air Force Museum.[31]

After Belgium and northern France had been overrun by German forces, about a million pigeons were ordered to be destroyed in this pigeon-fancying region. Alec Waley of the Intelligence Corps was in France from 1914, and because of difficulties with communications in the trenches, he borrowed pigeons from the French and set up a Carrier Pigeon Service, which gradually expanded to thousands of pigeons in mobile and static lofts. He built up a dog service as well, because pigeons tended to need anything from a week to a fortnight to acclimatise to a new mobile loft. Osman also supplied pigeons to the Allied forces, and it was found that messages could be relayed by pigeons much more quickly and efficiently than any other method, as so many things could go wrong with wireless telegraphy. Pigeons were even used in the new British weapon – tanks.[32]

Philip Gosse in the Army Medical Corps heard that mobile pigeon lofts were being sent out, needing expert hands to look after the birds. The battalion he was then with had been recruited in a Yorkshire town where pigeon racing and fancying were a way of life. According to him, when the men paraded, 'the order was read out, and any man who knew about the care of carrier pigeons and how to fly them was instructed at the word "Advance" to take two steps forward'.[33] When the order rang out, the entire battalion moved forward two paces and halted.

The use of pigeons during World War One proved hugely successful, and batches were even dropped by parachute in occupied areas in the hope that knowledgeable sympathisers would return them with completed questionnaires and messages giving intelligence. The results far exceeded expectations. In the Lord Mayor's Show in London in 1918, at the end of the war, a travelling pigeon loft was included in the procession to remind the public of the contribution made by pigeons in the war effort.[34] This did not stop the Admiralty deciding that wireless telegraphy was the future, and they therefore disposed of all the pigeons.

In World War Two the same fear developed about pigeons being used in espionage, and so in August 1940 new rules were brought in: 'The Home Office state that people concerned with homing pigeons are warned that a new provision has been added to the Defence Regulations making it illegal for any person, without official authority, to send a message by means of a homing pigeon. The regulation does not apply to the ordinary identification rings or rubber race rings attached to the homing pigeons.'[35]

Pigeons yet again were proved to surpass new technology in particular circumstances. In 1944 David Gunston, who was a wildlife writer at Portsmouth, sang the praises of carrier pigeons, especially in relation to the Royal Air Force:

It is in the R.A.F. that carrier-pigeons have proved their worth in this war. Many an airman can honestly say that he owes his life to their never-failing homing instinct. Nearly every aircraft on reconnaissance or offensive patrol or on a bombing raid carries two pigeons in a square watertight container. This is also buoyant so that if the aircraft is forced down into the sea, it will float and so allow the birds to be released with an S.O.S. message in the tiny bakelite container that is fastened to one leg.[36]

Gunston also described their racing qualities, saying that some of the pigeons had set up amazing records of endurance, with one bird flying through terrible weather across the North Sea from Stavanger, Norway, some 250 miles, while another returned to the Midlands from Holland in just over four hours: 'To date, the speed record for one of these "feathered Spitfires," as the R.A.F. call them, is 68 miles an hour.'[37] As in World War One, pigeons were also dropped by parachute into enemy territory to try to obtain important information from sympathisers.[38]

The pigeons faced various hazards, including the menace of powerful peregrine falcons, as their talons can instantly kill a bird the size of a pigeon. Although protected by law, in 1940 it was decreed that they could be destroyed in certain areas, and a Falcon Destruction Unit was established. Gunston said that Cumberland and Westmorland were the first counties affected, as they covered the western approaches, but the chief danger from falcons was when the tired pigeons reached the coast with vital messages.[39] One journalist, Ignatius Phayre, pointed out that pigeons also risked being shot for food by the hungry population: 'No wonder our fighting Services warn thoughtless persons: "Do not harm the homing pigeon. His pot-value – even in rationing time – may be only a shilling, while the message it may carry can mean life or death to our air-pilots." As an added warning the peace-time fine was hugely increased, and even a gaol sentence added to it!'[40]

In the past, other birds have been used to carry messages, even between ships. The French explorer and naturalist François Levaillant left a vivid description of the strategy when, in mid-July

1784, he set sail from False Bay in South Africa on board the Dutch East Indiaman *Ganges*. His vessel managed to keep close to three other Company vessels on their long journey back to Europe, and when the weather was calm, they visited each other by boat, but in stormy seas, birds were employed:

> we had recourse to another [method], that of mutually writing letters, of which the gulls and terns were the carriers. These birds, beaten by the winds and tired with their flight, would pitch upon our yards to rest themselves, where the sailors easily caught them. Having fastened our little epistles to their legs, we then let them fly, and, making a noise to prevent their alighting again on our vessel, obliged them to wing their course to the next. There they were caught again by the crew, and sent back to us in the same manner with answers to our letters. This curious stratagem has something kind and affectionate about it ... one of the circumstances of my travels which I always recollect with additional pleasure.[41]

A few years later, during the French Revolutionary Wars, a merchant on board a vessel that had called at Lindisfarne, off the Northumberland coast, caught a hawk. Tied to its neck was a letter with the following message: 'On board the Lion, September 4 [1795]. I sent this from on board the Lion of 64 guns 25 leagues off the Texel, in chace of a frigate and a sloop of war – He that gets this letter will put it in the Newspapers. RICHARD WILKINSON, Midshipman.'[42] The *Lion* was a naval warship on the North Sea station in 1795, and after the midshipman had attached the note, the hawk obviously flew across the North Sea, a distance of about 250 miles, before being caught the very same day. A fortnight later, the story duly appeared in the newspapers.

Amongst other birds that have given warnings through history, geese are renowned for making a noise and being aggressive towards strangers or when sensing some danger, and they are still recommended for home security today. The most famous legend about geese is when the Gauls attacked Rome in about

387 BC, pillaging and burning the city, but when the invaders tried to enter the citadel – the Capitol – the geese in the temple of Juno Moneta rose up and stopped them.[43] In 1948, just before Christmas, a Scottish newspaper reported how festive poultry was being guarded: 'Some Fife farmers have taken a tip from the Romans. They're keeping geese in their poultry yards as a guard against thieves. The vigorous cackling of these birds gives warning of intruders ... But maybe it's better not to rely too much on the ever-wakeful geese. Two were reported stolen in Fife last year. Many North Fife farmers are taking more practical steps by arranging all-night gun guards over their stock.'[44]

During World War One, gulls were watched carefully at sea, as their behaviour at times indicated the presence of German submarines, something that one naval seaman in early 1915 said had saved his battleship and the crew of eight hundred: 'I was standing ... at my twelve-pounder when a large flock of seagulls, which followed the ship for food, suddenly rose from the surface of the water. They had by their sharp eyes detected the periscope of a submarine. But for the seagulls we should all be in Davy Jones' locker. Their alarm saved us.'[45] In early January 1918, after guiding a vessel from the English Channel into an unnamed south-eastern port, the pilot mentioned that the presence of gulls had likewise stopped them being blown up: 'Whilst in the Channel ... I noticed some seagulls sitting upon a floating object. Upon closer investigation I saw it was a mine with five prongs. On top of each prong was perched a seagull. I just had time to alter the ship's course and thus averted disaster.'[46]

A similar bird sighting in October 1914 saved a naval vessel on patrol in the North Sea, a few miles off the west coast of Norway. Lieutenant John Noble Kennedy was a keen birdwatcher and kept a constant record of the birds he observed while on patrol, including many that followed in the ship's wake. On 9 October, he noted: 'While watching a Puffin through my glasses, I suddenly saw the periscope of a German submarine appear above the water close to the bird. We altered course just in time to evade, by a few feet, two torpedoes which were fired at us! We tried to ram the submarine, but unfortunately she dived too quickly.'[47]

Also during World War One, parrots were installed at the Eiffel Tower in Paris to give warning of approaching aircraft. Built between 1887 and 1888, the tower was the tallest structure in the city and became involved with radio transmissions. It also housed parrots, as one newspaper reported in February 1915:

The domestic parrot is a great feature in many Paris homes, particularly in the poorer quarters. It has been noticed ... that parrots have an extraordinary faculty for discerning the approach of an aeroplane. They do not detect them by sight ... as a rule, the birds commence to draw attention to the airships coming long before the human eye can see anything of them ... They work themselves up into bristling excitement, and yell and screech ... A number of birds are at present on trial in the outposts of the camp of Paris, while others are posted on the summit of the Eiffel Tower.[48]

A few weeks later another newspaper gave more details:

If parrots could state the nationality of any aeroplane they hear there would be no need for men to be continually on the look out for hostile aircraft. The parrots would give warning quick enough. The French authorities have had a number of parrots kept in the outposts of Paris, as well as on the summit of Eiffel Tower. The birds have shown a remarkable power of heralding the approach of an aeroplane when the latter had been quite invisible to trained observers stationed near. Warning is given by the birds in a peculiar way. Their feathers literally bristle with excitement, and they yell and screech until they are pacified ... This peculiar power of parrots was discovered quite accidentally by the excitement they showed whenever the Paris air patrol were flying, or a raid was made by the German aircraft.[49]

Other birds were equally alert to unexpected noises, which made them draw attention to unusual activity. Pheasants especially picked up strange sounds, and Arthur Savory mentioned what happened after the death of Queen Victoria, while her coffin was being transported from Cowes on the Isle of Wight to Portsmouth: 'When the vessel was crossing the Solent, in 1901, some very heavy salutes were fired from the battleships, and, the day being still and the air clear, the detonations carried to an immense distance ... The reports were so loud at Woodstock, near Oxford, that the pheasants began crowing in the Blenheim preserves.'[50]

Pheasants were especially sensitive to noise that was not apparent to the human ear. In World War One, they often gave warning of approaching Zeppelins, and *The Mail* in 1916 commented: 'During the first year of war the approach of a Zeppelin was heralded by the birds and animals. Pheasants, partridges, sheep, horses, and cows made a tremendous din long before the engines of the airship could be heard. But now, when a Zeppelin approaches, they make no noise, they restrain themselves until the bombs begin dropping. With so many of our own aircraft nowadays, both bird and beast are becoming accustomed to noises in the air.'[51]

There were some false warnings when birds used their powers of mimicry, as on the Western Front in 1916. One artillery officer wrote: 'The Starlings out here have acquired the trick of giving three shrill taxi whistles, in imitation of the call for enemy aeroplanes. It is great fun to see everyone diving for cover; I was nearly taken in myself the other day.'[52]

The sound of birds was once equally valuable as a guide to time, an alarm call, particularly for workers needing to rise early in the morning. Those in towns relied on night watchmen doing their rounds and calling out the time, but for those in the countryside it might mean looking at the stars, moon or sun, checking sundials and church clocks, or listening out for church bells and birds. One song in Shakespeare's *Cymbeline* begins: 'Hark, hark! the lark at heaven's gate sings, And Phoebus 'gins arise'. This is a reference to the skylark starting to sing before the sun rises, which made the skylark invaluable as a timepiece, a sign for the ploughman to set off across the fields, as in Shakespeare's *Love's Labour's Lost*: 'When shepherds pipe on oaten straws, And merry larks are ploughmen's clocks.'[53] It gave rise to the phrase 'up with the lark'.

In mid-Victorian Norfolk, Henry Stevenson pointed out that early risers do not actually hear the first notes of the skylark:

> Late as these birds are during the light summer evenings in retiring to rest, their song may be heard again by two o'clock the next morning, whilst the stars are still shining brightly in the cold grey sky, and scarce a streak of light yet indicates the approach of dawn. I have often, at such times, when out on the broads, heard the sky-lark's notes high over-head, when far too dark to distinguish the bird; or from the neighbouring fields, not 'poised in the air,' but warbling from the ground, several have simultaneously burst into song.[54]

At the height of spring, the dawn chorus would have made a significant noise, even in towns and cities, with several birds starting to sing well before sunrise. Geoffrey Egerton-Warburton said that song thrushes in Cheshire were depleted in the hard winter of 1895, but their numbers increased after a couple of years, 'and I heard people complain of their night's rest being spoilt, there were so many and they sang so early and loud. From April to June they sing almost incessantly, from earliest light to quite dark. They begin at three in the morning, or even earlier ... In 1905, on the longest day of the year, I woke at 2.30 A.M. to hear a throstle in full song just outside my window.'[55]

In the winter months, time could be tracked by observing rooks going to and from their roosts. Every winter morning in Wiltshire, Richard Jefferies related, one group of rooks would leave the elm trees in which they roosted and fly a mile or so to the meadows, where they spent the day feeding. 'Before watches became so common a possession,' he wrote, 'the labouring people used, they say, to note the passage overhead of the rooks in the morning in winter as one of their signs of time, so regular was their appearance; and if the fog hid them, the noise from a thousand black wings and throats could not be missed.'[56]

Cock crowing for many was a useful means of waking up, a sign that dawn was imminent. During World War One, Philip Gosse in the Army Medical Corps made good use of the Army Postal Service, and when his mother died he found that she had maintained a correspondence with several other young soldier friends, including Siegfried Sassoon. On 6 January 1916, in a letter to Mrs Gosse, Sassoon wrote: 'We are having a rest at a quiet village miles from the Somme valley ... A cock crows discordantly at 6.30 A.M. within a few feet of my slumbering head. He crows with a French accent, I think.'[57]

Birds have, throughout history, performed another beneficial service by clearing up the detritus produced by human activity and ridding society of dead human and animal bodies before they turned into a decomposing, pestilential and stinking nuisance. Although not trained to do such work, the birds involved were generally tolerated and therefore had some degree of protection from persecution. In prehistoric times in Britain, during the Neolithic period, there is evidence that dead human bodies were placed on raised wooden mortuary platforms, to keep them out of reach of predatory animals. As the bodies decomposed, the skeletons were picked clean by carrion-eating birds. From time to time the disarticulated bones were gathered up and interred in burial chambers or within the ditches of long barrows.[58]

Vultures are especially useful in dealing with decomposing flesh, and so they acquired a reputation of being able to sense death. In the late nineteenth century, the ornithologist Charles Dixon wrote:

The King of scavenger birds is the Vulture. In all hot countries these birds of various species abound. They scent the offal and the garbage, and decaying animals from afar, and on ample wings speed to the uncleanly feast, picking up every morsel, and ridding the tainted air of its unpleasantness. These birds are everywhere protected – in fact it is an offence in some countries to shoot a Vulture, so highly are they esteemed ... We have often admired the huge Griffon Vultures in Northern Africa, tame as barn-door fowls, either in the dirty towns, or on the dreary expanse of the Great Desert, where the dead camel or gazelle were the objects of their hungry quest.[59]

On battlefields, vultures were of great benefit, because they consumed both human and animal corpses. It is uncertain how the dead were treated in most battles, in particular the dead of the losing side, but the numbers killed could be immense. Dead soldiers were likely to be stripped of useful clothing, armour, weapons and possessions, before being buried in mass graves, burned or simply left to the mercy of scavenging birds and animals. Early in the Peninsular War, the British troops under Sir John Moore were in danger of being overwhelmed by Napoleon's advancing army in northern Spain, and so an evacuation began in January 1808. It meant marching in the most wretched conditions towards the ports of Vigo and Corunna. Many soldiers and camp-followers were in a desperate plight and fell by the wayside, as did the horses, with bloated carcasses strewn along the route from Lugo to Corunna. Adam Neale, an army physician, took it for granted that vultures would be on the scene:

> When the horses reached Lugo many fell dead, and others were mercifully shot. Above four hundred carcasses were lying in the streets and market-places, which it was impossible for the army to bury; and the town's-people were in too great a state of terror . . . the firing of muskets in all directions gave notice of the slaughter of these poor animals, whose bodies lay, swelling with the rain, putrifying, bursting, and poisoning the atmosphere, faster than the dogs and vultures could devour them.[60]

Several species of vulture were present across the Iberian Peninsula, but black vultures, with a wingspan approaching 10 feet, would have been adept at tearing open horse carcasses with their huge bills.

In March the following year, the French tried to take control of southern Spain and achieved a resounding victory at the Battle of Medellín, killing some ten thousand Spanish infantry. Lieutenant de Rocca of Napoleon's 2nd Regiment of Hussars described the grotesque sights that must have been repeated on many battlefields:

We lived among carcasses, and often saw from them thick black vapours, which the winds bore away to spread contagion and disease through the surrounding country ... Thousands of huge vultures collected from all parts of Spain in that vast lonely valley of death. Perched on the heights ... they seemed as large as men ... These birds would not leave their human repast on our approach, until we came within a few yards of them; then the beating of their vast pinions above our heads resounded far and near, like the funereal echoes of the tomb.[61]

Later on in the campaign, when the French were being driven out of Spain, John Malcolm of the 42nd regiment was horrified by what he witnessed at the stronghold of San Sebastian, captured by Wellington's forces in September 1813: 'The recollection of St Sebastian will haunt me as long as I live ... the dead in some instances had been buried – but so partially, that their feet and hands were frequently to be seen ... and innumerable heads, legs, and arms were strewn around, in the various stages of decay, and mangled and half-devoured by birds of prey.'[62]

Such horrors of the Napoleonic Wars were rarely reported back home, but in the 1880s outrage was expressed in many newspapers about vultures at battlefields during Britain's involvement in the Sudan. For much of the nineteenth century, Sudan was under the oppressive rule of Ottoman Egypt, but in the 1870s a Muslim cleric declared himself as the Mahdi or saviour, and an uprising began. With Britain increasingly involved in Egypt, Lieutenant-Colonel Hicks was put in charge of an Egyptian force to suppress the revolt. In early November 1883, on their way from Khartoum to relieve the city of Al-Ubayyid, his army was ambushed and slaughtered.[63]

The following year, 1884, General Gordon was sent to the Sudan to withdraw the Egyptian garrisons, but came under siege for ten months at Khartoum and was killed at the end of January 1885 when the city was taken. Two weeks earlier, the Gordon Relief Expedition had clashed with Mahdist forces at Abu Klea, and passing the battlefield a month later, a medical officer from the Light Camel Force wrote to a friend: 'It is on the side of a hill, and the air is putrid; heaps of ravenous vultures about.'[64] A combined British and Indian force heavily defeated the Mahdists in March 1885 at the Battle of Tofrek, 5 miles from the Red Sea port of Suakin, and an officer of the Scots Guards described that battlefield only a few days later: 'Strewed around, thicker and thicker, as we neared the scene of that Sunday's fight, lay the festering bodies of camels and mules; and round them hopped and fluttered, scarcely moving when our column passed, hundreds of kites and vultures. The ground was also thickly sown with hands and feet dragged from their graves by the hyaenas, and the awful stench and reek of carrion which loaded the air will never be forgotten.'[65]

Once commonplace in Europe and Africa, vultures were to the British soldiers such an unusual sight that they often commented on them, so that reports of this battlefield were met with horror. Stung by the expressions of outrage, the *Army and Navy Gazette* responded, giving an insight into centuries of grim battlefield scenes:

> the public shuddered, the friends of humanity were shocked, and the newspapers uttered exclamations of horror in type, as though they expected all the graves to bloom with flowers, and the air to be perfumed with sweets ... War is a dreadfully serious business ... As to bad smells and ugly sights, the public may be quite sure that battle-fields under the best conditions are disagreeable. They must ... accept the results, with the 'trimmings' of gorged vultures and prowling hyaenas, pariah dogs, and the like, which may be served up as a dish of horrors as long as the war and liberty of pen, ink, paper, and press endures.[66]

At the end of the nineteenth century, Charles Dixon said that the black kite was almost as important as vultures for scavenging, 'a positive boon throughout Oriental countries, where it picks up all the scraps of offensive matter, and shares the larger feasts with the Vultures and the various kings of Eagles'.[67] He was expanding on what the naturalist Osbert Salvin had said, after spending several months bird-nesting in Tunisia and Algeria in 1857: 'The Black

Kite plays the part of a scavenger in the districts where it abounds; and over every French settlement and Arab village several may be seen flying boldly round, on the look-out for any fragment of carrion that may be lying about.'[68]

Such scenes would have been familiar in medieval and later towns and cities in Britain, with kites and ravens in the unpaved streets clearing offal (waste animal parts, usually from butchery), carrion (dead bodies, usually putrefying) and other filth (including animal dung). Both these birds were protected by law as scavengers in urban areas, though in Britain black kites were rare. It was red kites that were once a common species, and their habit of hovering, gliding and swooping in search of offal or wounded prey was so distinctive that the word 'kite' was adopted for children's paper kites.[69]

When Baron Leo von Rozmital from Bohemia visited England from 1465 to 1467, his secretary mentioned that he had nowhere seen so many kites as London.[70] Some thirty years later, a Venetian nobleman in London recorded: 'the raven may croak at his pleasure, for no one cares for [heeds] the omen [of death]; there is even a penalty attached to destroying them, as they say that they keep the streets of the towns free from all filth. It is the same case with the kites, which are so tame, that they often take out of the hands of little children, the bread smeared with butter ... given to them by their mothers.'[71]

William Turner, originally from Morpeth in Northumberland, travelled widely in Europe and was a pioneer of botany and ornithology. In about 1544 he wrote a description of kites: 'I know of two sorts of Kites, the greater [red] and the less [black]; the greater is in colour nearly rufous, and in England is abundant and remarkably rapacious. This kind is wont to snatch food out of children's hands, in our cities and towns. The other kind is smaller, blacker, and more rarely haunts cities. This I do not remember to have seen in England, though in Germany most frequently.'[72]

Kites also tended to take ducklings, goslings, chickens and rabbits and so were unpopular in rural areas. They came to be persecuted, which was a complete change in their fortunes, with rewards being offered for their destruction from the sixteenth century, as at Tenterden in Kent, where payments were made from 1654, including tenpence to Joseph Page for five kites' heads in 1681–2 and fourpence to John Tompkinds for two heads in 1682–3.[73] Once sanitation was improved in urban areas, it was no longer necessary to protect kites, and along with enclosures, the rise of shooting estates and persecution, kites were pushed to the edge of extinction.[74]

Robert Smith, who called himself 'rat-catcher to the Princess Amelia', published a book in 1772 on destroying vermin, including ravens: 'This is the largest bird that feeds on carrion … in some places it is very serviceable, in eating up the stinking flesh or carcasses of dead beasts and other carrion, but in many other places very mischievous.' Where ravens were taking rabbits or lambs, he was paid to destroy them: 'I have often caught the London Ravens near twenty miles from home in warrens, where they will sometimes come after the young rabbits; by the London Ravens I mean those that generally frequent the outskirts of the metropolis, and live upon the filth lying there, grubbing up the dirt, in order to get at their food, from whence the tops of their wings become of a nasty, dusky brown colour.'[75]

The ornithologist William MacGillivray was a prodigious field-worker and researcher and writer, and in the early nineteenth century he wrote down his observations of ravens in the Outer Hebrides devouring the carcasses of dead animals, including dolphins cast ashore. He added an acerbic swipe at armchair naturalists:

> I well remember standing when a boy for a long time to observe the proceedings of about a dozen ravens devouring a dead cow that had been dragged to the sand banks on the farm of Northtown [Northton, Isle of Harris]. Some were tearing up the flesh of the external parts, others dragging out the intestines, and two or three had made their way into the cavity of the abdomen. It was amusing, and perhaps might be disgusting to a delicately organised snuff-taking and clean-fingered gentleman-inspector of birds'-skins, to see them drag out the intestine to the distance of several feet.[76]

He also mentioned how rooks and jackdaws helped to clean the streets of Scottish towns. In the early morning, the rooks either went to the fields, or else 'you may find it in the streets of the populous city, carefully searching for whatever is applicable to its wants among the garbage that waits the scavenger's cart ... Jackdaws often obtain a large proportion of their food in the streets, which they frequent more especially in the mornings, along with Pigeons, and sometimes Rooks. On these occasions they pick up the refuse of whatever serves as food to man.'[77]

Gulls are also very adaptable and will eat anything, dead or alive, inland or at sea. Their ability to scavenge human refuse has seen their numbers increase significantly. Writing in 1899 about Devon, Charles Dixon was reminded of scavenging vultures:

> When fields have been manured with fish or other offal, the Gulls congregate in force, coming from various parts of the coast ... The Herring Gulls in Tor Bay generally assemble on the fields near a slaughter-house midway between Torquay and Paignton, more especially on 'killing days,' when an abundance of scraps and offal may be obtained. Sometimes we have seen them soaring at a great height above this gruesome spot as if waiting for the feast, reminding us of the flocks of Vultures we met with years ago in Northern Africa.[78]

He also noted: 'Favourite feeding places, especially for the Herring Gulls, are at the various sewage outfalls round Tor.'[79]

Three decades earlier, Harry Blake-Knox wrote an account of gulls and described what he had observed off the Irish coast. The gull, he said, seemed intent on one purpose:

> that of restlessly scavengering the sea. In fact, a good feed of floating excrement, bread, fat, garbage, or carrion, seems more congenial to its taste than the pure silver fry. At all seasons we meet this gull in Kingston Harbour [Dun Laoghaire] ... Their food here is chiefly pieces of biscuit, bread, fat mess pork, oil, tallow, &c., thrown from the Queen's ships and the other craft

in harbour. So cunning are the old frequenters that they know dinner-hour on board the coast-guard frigate as well as the oldest tar, and at times when not a bird is seen, some twenty or thirty will make their appearance round the vessel at the dinner-hour to feed on the pieces thrown overboard. Its strangest food is floating oil and grease.[80]

Writing about birds in Scotland a few years later, Robert Gray wryly commented that the white stork was rarely seen but sorely needed: 'It is said . . . that in some towns in Holland and Flanders the species is of great use in devouring all kinds of garbage that might accumulate and putrify in the streets; and bearing in mind this convenient habit, it might not be an unwelcome invasion were a few hundred Storks to diverge from their line of flight, and pass an hour or two in some of our coast towns and villages.'[81]

The canary is arguably the bird that has been of the greatest use and saved most lives, because for several decades it was used to detect carbon monoxide in coal mines and other underground situations. Coal mining was a perilous occupation, but the invention of the safety lamp in the early nineteenth century reduced the number of explosions, because miners could tell from the lamp's flame if dangerous gases were present, particularly methane, which they called firedamp (from the German 'dampf', meaning 'vapour'). Being odourless, carbon monoxide (CO, 'white damp' or 'sweet damp') was one gas that could not be detected, and miners were liable to die from inhaling even small amounts.

In spite of safety lamps, underground disasters still occurred, and on 27 January 1896 there was a huge explosion at Tylorstown Colliery in the Rhondda Valley. John Haldane, a Scottish lecturer in physiology at the University of Oxford, travelled to Wales straightaway. On his arrival, he learned that fifty-seven men and eighty pit ponies had been killed, but on examining several corpses, he was astonished to see that carbon monoxide was the main killer. Amongst the notes he compiled were descriptions giving the tell-tale appearance of those who had died, including: 'Collier. No injuries or burns. Face red. Nails pink. Carbon monoxide poisoning.'[82]

The afterdamp – the gases resulting from an explosion – included lethal amounts of carbon monoxide, and more men died of carbon monoxide poisoning than from catastrophic injuries. Rescuers were also at risk, rapidly becoming dizzy, drowsy and even dying. Haldane had already been researching the effects of carbon monoxide, but had never imagined that it was the main cause of death in pit explosions. Instead, he was convinced that miners in explosions perished mainly through suffocation due to lack of oxygen.[83] In 1894, he had conducted experiments on himself and on white mice and discovered that mice were good indicators of the invisible gas being present, as they were affected far more quickly than humans, even taking mice with him as a safety precaution while collecting samples from pits in the Staffordshire collieries.[84]

In July 1895, a few months before Tylorstown, five workers died down a sewer at East Ham in London, overcome by sewer gas, which is a mixture of gases. At their inquest, it was stated that the district council had since introduced safety measures and that 'acting on the advice of Dr. Haldane, a mouse or bird had been lowered each time the men went down',[85] so mice and birds were already beginning to save lives in underground locations. In April 1896, a few weeks after Tylorstown, an explosion at Brancepeth Colliery near Durham caused the loss of twenty men, and Haldane went there to seek similarities. He concluded that all but one had died of carbon monoxide poisoning. Barely a week later, an explosion at the Peckfield pit at Micklefield in West Yorkshire killed sixty-three men and boys, and here he found that every single fatality was due to carbon monoxide.[86]

In his report on the 1896 pit explosions, Haldane recommended that miners should take mice into the mines: 'The mouse may be carried in a small cage, or a lamp chimney closed at the ends with wire gauze ... A few white mice might easily be kept in the engine room of the winding engine, or in stables or other places in the pit.' As for the rescuers, he suggested that they take mice 'or other equally small warm-blooded animal. A few white mice might easily be kept for this purpose in the engine room at the top of the downcast shaft, and be taken down in small cages by the rescuers.'[87] Although canaries were not specifically mentioned, miners were already unofficially taking them and other small birds down the pits, aware of their usefulness but not the science.[88]

It is likely that miners had for centuries tried unofficial ways of keeping safe underground. In the coalfields around Newcastle, cockfighting was a very important pastime, and in about 1795, a century before scientific research on carbon monoxide, the Northumberland curate John Brand wrote: 'In the North, before any Collier ventures down a pit, which is suspected to contain foul air, a Cock is let down.'[89]

Canaries were actually more responsive than mice and had the added advantage of being regarded as pets, and in February 1901 canaries were reported as aiding rescuers in the Engine Pit at Hill of Beath in Fifeshire. Suspicions that an old fire was spreading had led the manager and inspector to check it out, but they never returned. A rescue party went to investigate, but only one man came back, and so Henry Rowan from a nearby colliery descended into the pit and 'took with him a canary, which one of the miners had kindly given him on loan. No sooner was the white damp reached than the bird dropped helpless to the bottom of its cage, although Mr Rowan had not felt any of the evil effects himself.' It was decided to alter the air currents, 'and the system of testing the air by two canaries was adopted. Again and again the canaries dropped to the bottom of the cage, but they always revived on being taken to purer air.'[90] Finally, the bodies of seven men who had perished were recovered.

Four men died of carbon monoxide poisoning in an explosion at Urpeth Colliery, south of Gateshead, in mid-December 1906. This was a few months after Europe's worst coal mining disaster, at Courrières in northern France, when 1,099 miners lost their lives in a massive explosion. Although Haldane still preferred the use of mice, at the Urpeth inquest the mining engineer William Blackett explained that he had tried to reach the scene of the accident, 'taking with him a canary in a cage, a safety lamp, and an electric hand lamp'.[91] He and other rescuers advanced some way, left the bird and rapidly retreated. When observing that the canary remained on its perch, they kept repeating the process so as to locate the bodies. Blackett then reported what suddenly happened next to one dead man:

> he saw the bird fall from its perch, and, incautiously taking a breath of the air himself, his knees gave way to a small extent, but he managed to scramble out into the better air, taking the bird with him. The bird recovered within three or four minutes, and again got on to its perch. It was most extraordinary to see the rapid effect which the carbonic monoxide had on the bird, and he was quite satisfied, after the experience with the bird in this way, that a bird was a comparatively safe guide, and much to be preferred to using mice, as the fall of the bird from the perch could be easily seen.[92]

The *Nottingham Evening Post* commented: 'This is a new use for the canary, and it is satisfactory to know that the bird recovered from the effects of the after-damp, and was produced at the inquest. As an aid to the exploration of a coal pit after an explosion has occurred, birds may be of very great value, as is shown in this case, but humane people will hope that they will be used if at all with due consideration to their well being.'[93] Both mice and canaries could often be revived in the fresh air or by small oxygen devices, and yet this newspaper within a coal mining community was curiously wary of these popular cage-birds taking on a role that might help the miners.

Following the Coal Mines Act of 1911, it became mandatory to use canaries in mines in order to detect carbon monoxide, and rescue stations had to be established with equipment that included canaries. When the rescue station at Chesterfield in Derbyshire was opened in 1917, it was reported in the *Sheffield Daily Telegraph*: 'Most people will be familiar with the part played by white mice in testing for gas in mines, but the employment of canaries for the same object will not be so generally known ... Canaries are bred at the rescue station, and just now there is a nest containing several young ones.'[94]

In the United States on 9 December 1911, a huge explosion ripped through the Cross Mountain coal mine near Briceville, Tennessee, and a rescue team from the newly formed Bureau of Mines attended with gas masks and oxygen tanks, as well as canaries in cages. They searched the labyrinth of tunnels and managed to bring out a few men alive and recover the bodies of eighty-four miners who had perished. Only a few weeks later, popular accounts published in magazines explained that this 'was the first chance for the government rescue workers to test the value of canary birds in connection with their work, and for the first time the rescue cars [old Pullman railway carriages] each had from one to three dozen birds on board as part of their equipment.'[95]

George Burrell, who worked for the Bureau of Mines, now published the first research comparing canaries and mice, developing Haldane's experiments:

> These tests show that canaries may be better than mice as indicators of the presence of noxious gases in the atmosphere of mines, since they more quickly show signs of distress in the presence of small quantities of carbon monoxide. In addition the symptoms of poisoning in birds are much more clearly defined. A bird sways noticeably on its perch before falling and its fall is a better indication of danger ... Consequently birds not only give more timely warnings of the presence of small quantities of carbon monoxide, but exhibit symptoms that are more easily noticed by exploring parties.[96]

He added: 'The experiment shows that small birds are much more susceptible to the action of carbon monoxide than are men, and demonstrates the desirability of using small birds, such as canaries, rather than larger ones, such as pigeons.'[97]

The worst mining disaster in Britain occurred at Senghennydd, near Cardiff, when a pit explosion in October 1913 killed 439 miners and a rescuer. The use of canaries was still such a novelty that the *Daily Mirror* explained their role: 'Each [rescue] party goes down with a canary in a little wooden cage. The canary is the danger signal. Highly susceptible to "afterdamp," canaries fall off their perch on encountering any of the fumes, and as soon as a bird falls down the spot is marked as "danger" and no one is allowed to go beyond that mark unless provided with breathing apparatus.'[98]

The following year saw the outbreak of World War One, and Haldane was consulted over German poison gas attacks during the war and helped to develop respirators. His earlier research into carbon monoxide also proved invaluable for the underground or trench warfare in Europe, which involved the construction of a vast network of tunnels and mines for detonating explosives beneath German positions. This work was done by Royal Engineer tunnelling companies, whose men included hundreds of miners from Britain. Carbon monoxide was a constant problem, and so canaries and mice were essential, and they also saved lives in sensing poison gas attacks. Each company initially bred its own canaries, though that role was later transferred to the main Calais stores of the Royal Engineers.[99]

One newspaper was struck by a story that a 'considerate company commander kept a full record of his canaries. After a canary had been gassed three times, he classed it as "P.B." [Permanent Base, excused from frontline duties] and promoted him to the headquarters dugout, where his only duty was to sing to the commanding officer.'[100] Less happy was the fate of a canary that deserted:

> He escaped, or was liberated by some kindly disposed person, and fled to a bush in No Man's Land. A sniper was called upon to dispatch the renegade. He failed, and the bird flitted across to the German wire. Then the trench mortars were told off to deal with the situation. After their first round no more could be seen of the truant. A lot of fuss to make about a canary, apparently, but there is more in it than meets the eye. The reason was that mining was being carried on at that particular spot unknown (we hoped) to the Germans. Had they spotted the canary, the secret would have been revealed.[101]

Even after the war and the successful use of canaries, Haldane admitted that he still favoured mice: 'This test [using a bird or mouse] has now come into very general use, and was, for instance, largely used during the war by the tunneling companies. It is easier to see the signs of CO poisoning in a bird in a small cage, as it becomes unsteady on its perch, and finally drops, while a mouse only becomes more and more sluggish; but the mouse is easier to handle, and less apt to die suddenly and thus leave the miner without any test.'[102]

In British coal mines, canaries were bred for detecting carbon monoxide until 1986–7, and they have prevented the deaths of countless men worldwide. So important has been the role of the canary that the expression 'canary in the coal mine' has come to mean an indicator of an invisible hazard or looming disaster, as well as a reminder of how people's lives owe so much to birds for their well-being.

POSTSCRIPT

———•◆•———

The skylark has proved to have had more emotional and artistic appeal than any other bird. Over two centuries ago, in the summer of 1820, Percy Bysshe Shelley was staying near Livorno in Italy with his wife Mary Shelley, who related what led to his poem 'To A Skylark': 'It was on a beautiful summer evening while wandering among the lanes whose myrtle-hedges were the bowers of fireflies, that we heard the carolling of the skylark, which inspired one of the most beautiful of his poems.'[1] Shelley celebrated the stream of song produced by this skylark while flying so high as to be virtually invisible:

> *Hail to thee, blithe spirit!*
> *Bird thou never wert,*
> *That from heaven, or near it,*
> *Pourest thy full heart*
> *In profuse strains of unpremeditated art.*[2]

When Thomas Hardy was at Livorno in 1887, almost seven decades later, he took a more down-to-earth approach in his own poem 'Shelley's Skylark', describing how the same skylark had inevitably perished and turned to dust:

> *Somewhere afield here something lies*
> *In Earth's oblivious eyeless trust*
> *That moved a poet to prophecies–*
> *A pinch of unseen, unguarded dust:*

> *The dust of the lark that Shelley heard,*
> *And made immortal through times to be;–*

Though it only lived like another bird,
And knew not its immortality.[3]

Skylarks were then an everyday sight and sound, and their sheer numbers were memorable, as the naturalist Richard Kearton mentioned in 1898: 'One of the grandest sights, from an ornithological point of view, to be seen on the south coast of England in wintertime, is undoubtedly the immense flights of larks, fieldfares, and redwings ... there is never any lack of larks in England.'[4]

Another English novelist and poet, George Meredith, also wrote about skylarks, in a lengthy poem published in 1881, called 'The Lark Ascending', which began:

He rises and begins to round,
He drops the silver chain of sound.[5]

Meredith's work was then very popular, and this poem inspired Ralph Vaughan Williams to compose a piece of music, bearing the same title, shortly before World War One. Time and again, 'The Lark Ascending' has been voted Britain's favourite piece of classical music, and it is also popular worldwide. The sequences of rising notes echo the song of the bird and depict its upward flight – a quarter of an hour of the most exquisite and nostalgic sound, reflecting the tranquillity and beauty of the pre-war landscape.[6]

The outbreak of the war delayed the premiere of his masterpiece, but actual skylarks were comforting to the soldiers in the bleakest of times. 'I was sitting ... in a trench one spring-like morning, having a rest and listening to the incessant shriek of bullets,' one soldier wrote, 'when, amid the noise of firing I heard the note of a skylark. Anyone who has heard the din of a battlefield can perhaps appreciate the wonderful contrast of that skylark's song. I sat and listened for some time and shortly saw the bird soaring and singing only twenty-yards beyond our trench.'[7]

In little over a hundred years, skylarks have become an endangered species in Britain and are in steep decline across Europe as well. Everyone has heard of skylarks, but very few have seen or heard them, yet they are part of the nation's soul. The disappearance of the skylark has become a symbol, 'a canary in the coal mine', for the wholesale decline of British birds, and the erosion and destruction of nature.

World War One highlighted the country's dependence on imported food, a problem that became critical during World War Two, leading to a substantial increase in farming during the post-war years. In Europe, the Common Agricultural Policy (CAP) was set up in 1962, and Britain came under those rules in 1973. Since then, industrialised agricultural practices, along with unfettered development, have caused an immense degradation of the land and a massive collapse in the numbers of insects and birds. The British Trust for Ornithology has shown that, since the 1960s, the skylark population alone has shrunk by three-quarters, with some other bird species suffering even more.[8]

While the countryside can look superficially attractive, the true picture is rather unsettling, hidden but in full view, such as the industrial-sized barns holding huge herds of dairy cows that are kept indoors all year, the vast, empty fields with no hedges, the dreary monoculture with little biodiversity and the diversification that too often means something sinister.[9] Now that Britain has left the European Union, the government can no longer hide behind the rules of the Common Agricultural Policy, but must take full responsibility for the transformation of agriculture and the environment for the benefit of everyone. Local authorities must also cease preaching a green message, while at the same time giving the green light to damaging developments. With factors such as global warming, flooding, pollution and a collapse in wildlife, all inextricably linked, it is time for drastic action. Thankfully, it is starting to happen, and a widespread concern for birds in particular is leading the way.

Although the dwindling number of birds was noticed over two centuries ago, such messages tended to be ignored then. We now live in such a connected world that it is proving possible to bring about real change. An awareness of the natural world is growing rapidly, alongside a huge groundswell of opinion in favour of reclaiming what has been lost in our rural, maritime and urban environments. Birdwatching as a pursuit has never been more popular, and many wildlife, conservation and heritage bodies are involved with birds, such as the Royal Society for the Protection of Birds (RSPB), the Council for the Protection of Rural England (CPRE), the National Trust, the National Trust for Scotland and numerous local natural history societies and wildlife trusts.[10] At an armchair level, superb television and radio programmes are stimulating the interest in wildlife, especially birds, while astonishing webcams of nesting birds provide an unexpected glimpse into their world.

One of the most exciting projects is the rewilding of the Knepp estate in East Sussex, an extraordinary ecological restoration project of over 3,500 acres that includes a regenerative farming project, river restoration and the reintroduction of white storks. It has been an inspiration, highlighting the potential to improve other local environments. This can be achieved at an individual level, from window boxes filled with flowers to setting aside wild areas in gardens to attract birds and other wildlife. Small-scale measures such as bug hotels, bird feeders, water features and ponds can make a significant difference overall and lead to a noticeable increase in the bird population.

In our own garden of a few acres, we have planted (and continue to plant) various kinds of trees, including fruit trees. Brambles and nettles are left to flourish, and benign neglect is the order of the day. Our small, scruffy-looking fields turn each year into flowering meadows full of butterflies, bees, dragonflies and other insects. With the insects, come the birds. Pushing our way into a meadow that is waist high, on a sunny day, then watching and listening to all the insects and birds, is an enchanting experience. While writing this book, the birds have been a constant distraction, and over the past year we have seen more than thirty different species, from wrens and long-tailed tits to crows and huge, soaring buzzards – and we watch, wait and listen for those elusive skylarks to arrive.

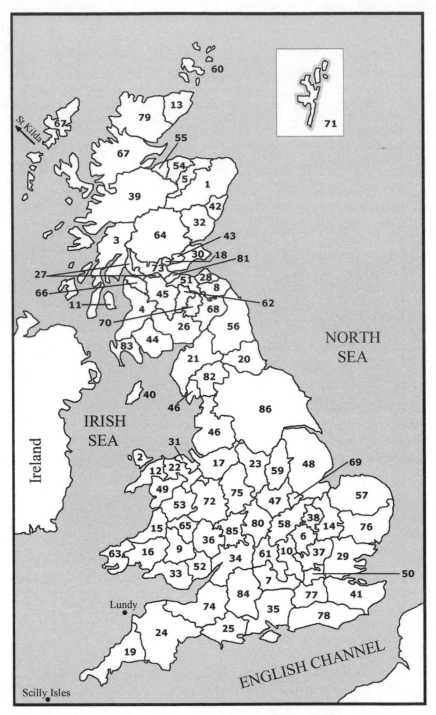

Map 1: Key to the old counties of Great Britain

1. Aberdeenshire
2. Anglesey
3. Argyllshire
4. Ayrshire
5. Banffshire
6. Bedfordshire
7. Berkshire
8. Berwickshire
9. Breconshire
10. Buckinghamshire
11. Buteshire
12. Caernarvonshire
13. Caithness
14. Cambridgeshire
15. Cardiganshire
16. Carmarthenshire
17. Cheshire
18. Clackmannanshire
19. Cornwall
20. County Durham
21. Cumberland
22. Denbighshire
23. Derbyshire
24. Devon
25. Dorset
26. Dumfriesshire
27. Dunbartonshire
28. East Lothian
29. Essex
30. Fife
31. Flintshire
32. Forfarshire
33. Glamorgan
34. Gloucestershire
35. Hampshire and Isle of Wight
36. Herefordshire
37. Hertfordshire
38. Huntingdonshire
39. Inverness-shire
40. Isle of Man
41. Kent
42. Kincardineshire
43. Kinross-shire
44. Kirkcudbrightshire
45. Lanarkshire
46. Lancashire
47. Leicestershire
48. Lincolnshire
49. Merionethshire
50. Middlesex
51. Midlothian
52. Monmouthshire
53. Montgomeryshire
54. Morayshire
55. Nairnshire
56. Northumberland
57. Norfolk
58. Northamptonshire
59. Nottinghamshire
60. Orkney Islands
61. Oxfordshire
62. Peeblesshire
63. Pembrokeshire
64. Perthshire
65. Radnorshire
66. Renfrewshire
67. Ross & Cromarty
68. Roxburghshire
69. Rutland
70. Selkirkshire
71. Shetland Islands
72. Shropshire
73. Stirlingshire
74. Somerset
75. Staffordshire
76. Suffolk
77. Surrey
78. Sussex
79. Sutherland
80. Warwickshire
81. West Lothian
82. Westmorland
83. Wigtownshire
84. Wiltshire
85. Worcestershire
86. Yorkshire

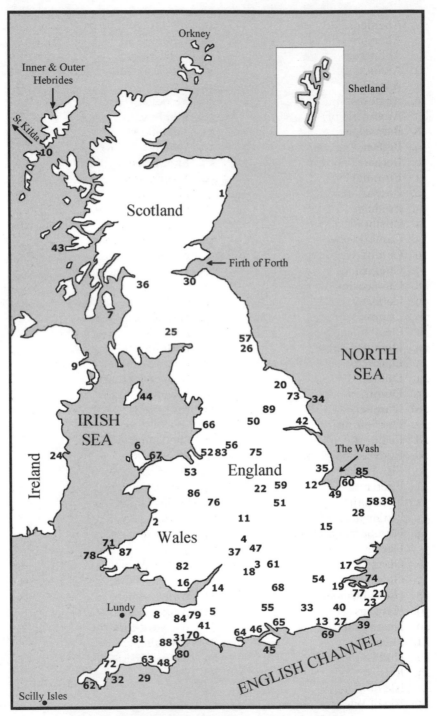

Map 2: Key to selected place-names of Great Britain

WEIGHTS, MEASUREMENTS AND MONEY

——•◆•——

Most weights and measurements are given as imperial ones, and so for those better acquainted with the metric system, the following may be useful:

Linear

12 inches (in.) = 1 foot (ft)
3 feet = 1 yard (yd)
1,760 yards = 1 mile

Capacity

2 pints = 1 quart
4 quarts = 1 gallon
36 gallons = 1 barrel
1½ barrels = 1 hogshead

Weight

18 pennyweights (approx.) = 1 ounce (oz)
16 ounces (oz) = 1 pound (lb)
14 pounds = 1 stone
8 stone = 1 hundredweight (cwt)
20 hundredweight = 1 ton

Some metric equivalents:

1 inch = 2.54 centimetres
1 foot = 30.48 centimetres
1 yard = 0.91 metres
1 mile = 1.61 kilometres
1 pint = 0.568 litres
1 gallon = 4.54 litres
1 ounce = 28.35 grams
1 pound = 0.45 kilograms

Money

¼ penny = 1 farthing
½ penny = 1 halfpenny
12 pence or pennies (12d.) = 1 shilling (1s.)
20 shillings = £1 (one pound)

Approximate equivalents in decimal coinage are one pound = £1.00; one shilling (1s.) = 5 pence (5p); sixpence (6d.) = 2.5p; and one penny (1d.) = 0.416p. In terms of purchasing power, a penny (1d.) in 1800 was worth about 18p today, a shilling (1s.) was worth about £2.20, and one pound (£1) about £44.07 (using The National Archives currency converter, which translates prices in 1800 to values in 2017).

CHRONOLOGICAL OVERVIEW

———— •◆• ————

This list of dates provides a summary of some of the events mentioned in *When There Were Birds*, mainly associated with Britain or England:

387 BC	Geese on the Capitol in Rome warned off an attack by Gauls.
249 BC	Augury of sacred chickens before the naval Battle of Drepana.
104 BC	Gaius Marius chose the eagle as the symbol of Roman legionary standards.
43 BC	At the siege of Mutina, Decimus Brutus sent messages by pigeons.
1348–9	Black Death in England.
1360	An Act made it illegal to conceal or keep stray falcons (34 Edw III cap. 22).
1363	An Act made it a felony to conceal or keep stray falcons (37 Edw III cap. 19).
1390	An Act restricted the hunting and killing of game to landowners and the clergy (13 Ric II Stat. I cap. 13).
1415	The Battle of Agincourt, with the famous deployment of archers.
1438	A mallard was found in the foundations of All Souls College, Oxford.
1533	Henry VIII married Anne Boleyn (January).
1533	An Act to destroy rooks, crows and choughs, the earliest reference to eliminating vermin (24 Hen VIII cap. 10).
1533–4	An Act against the Destruction of Wild-Fowl (25 Hen VIII cap. 11).

1545   The shipwreck of the *Mary Rose* containing thousands of arrows.

1558   Elizabeth I became monarch.

1566   Legislation to kill vermin (8 Eliz cap. 15).

1575   An aviary was constructed at Kenilworth Castle for the visit of Elizabeth I.

1583   A witch, by boiling eggs, was believed to have drowned fourteen men who were buried at Wells-next-the-Sea.

1584   Reginald Scot published *The Discoverie of Witchcraft.*

1602   Richard Carew published his *Survey of Cornwall.*

1615   Gervase Markham published *Countrey Contentements* (including *The English Hus-wife*).

1621   Gervase Markham published *Hungers Prevention.*

1621   Robert Burton published *The Anatomy of Melancholy.*

1645–6   A windfall of skylarks prevented starvation in the besieged city of Exeter (probably February 1646).

1653   Izaak Walton's *The Compleat Angler* was published.

1660   Restoration of the monarchy (Charles II).

1663   Pigeons were used on Catherine of Braganza to remove illness.

1671   The Game Act formed the basis of game laws until 1831 (22–3 Car II cap. 25).

1672   The ornithologist Francis Willughby died.

1673   Nicholas Cox published *The Gentleman's Recreation.*

1686   Charles Morton, who thought that birds migrated to the moon, went to America.

1697   Martin Martin visited St Kilda.

1727   George Markland wrote *Pteryplegia: Or, the Art of Shooting-Flying.*

1728   César de Saussure was in England.

1737   Eleazar Albin published a book on English songbirds.

1745	The Battle of Culloden in 1746 was foretold by a fight between sparrows one year earlier.
1748	The Swedish naturalist Peter Kalm was in England.
1748	The Royal Society published a paper by Réné-Antoine de Réaumur on preserving birds.
1758	Peter Collinson read a paper to the Royal Society on the migration of swallows.
1762	Friedrich von Kielsmansegge was in London.
1769	Thomas Pennant toured the Highlands of Scotland.
1776	America declared independence (July).
1782	A bantam-cock applauded the success of the Battle of the Saintes (April).
1782	An eagle at Gibraltar was taken as a good omen during the Great Siege (September).
1782	The United States used for the first time its Great Seal with the native American bald eagle symbol (September).
1788	Cowper wrote 'On Mrs Montagu's Feather-Hangings'.
1790	John Byng toured Cambridgeshire.
1791	The tax for an annual stamped certificate or licence for shooting game increased to 3 guineas (31 Geo III cap. 21).
1798	William Bingley toured north Wales.
1798	A lucky bird was at the Battle of the Nile (August).
1798	A ringed woodcock was shot in Dorset (December).
1801	William Bingley toured north Wales again.
1801	England had a population of about 8,300,000.
1802	The brass eagle lectern from Bristol cathedral was sold.
1804	Napoleon decided on eagle standards for his regiments (July).
1805	Charles Fothergill toured Yorkshire.
1805	Humphry Davy did experiments on Peruvian guano.

1805   A cock on board the *Colossus* cheered at the Battle of Trafalgar (October).

1806   A lighthouse was built at Flamborough Head.

1807   Forsyth patented his 'scent bottle' lock.

1808   Vultures were present during the army's retreat to Corunna.

1808   Major-General Beatson did experiments with guano on St Helena.

1813   Vultures were present after the capture of San Sebastian.

1820   Shelley composed 'To A Skylark'.

1822   A white stork impaled by an African spear was shot near Klütz in Germany.

1827   Mantraps and spring guns became illegal.

1830   Swing Riots; peacock was eaten at Christmas in Moreton House, Dorset.

1831   The Game Act replaced most earlier statutes and ended the property qualification.

1833   Keeping a cockpit in London became illegal (3 Wm IV cap. 19).

1834   Richard Leech died at Longparish House amidst strange bird omens.

1835   The Cruelty of Animals Act made it illegal to keep a cockpit in England and Wales (5 & 6 Wm IV cap. 59).

1835   Joseph Arch began crow-scaring, aged nine.

1841   William Eley was killed in a fulminating powder explosion.

1841   The zoologist James Wilson toured the Scottish coasts and islands, including St Kilda.

1843   William Yarrell published *The History of Birds* in three volumes.

1843   Cuthbert Johnson published a book on the use of guano.

1848   Henry Davenport Graham moved to Iona.

1848   Revolution in France.

1849   Cockfighting was prohibited in England and Wales (12 & 13 Vict cap. 92).

1849–50   Henry Mayhew investigated the lives of London's labouring classes.

1851   Thomas Sternberg published *The Dialect and Folk-Lore of Northamptonshire.*

1851   The Great Exhibition in London included stuffed birds in glass cases.

1854   The Plomley collection was donated to the museum at Dover.

1854   The first acclimatisation society, in France, was founded.

1859   A huge bird wreck occurred on the west coast of Scotland and Ireland.

1860   Walter White did a walking tour of eastern England.

1860   Isabella Bird toured the Outer Hebrides.

1862   The first dinner of the Acclimatisation Society of Great Britain.

1863   Poisoned grain for killing birds became illegal.

1863   Elihu Burritt did a walking tour of Britain.

1863   A major irruption of Pallas's sandgrouse.

1865   James Gardner published a book on taxidermy for amateurs.

1867   Wirt Sikes was appointed the United States consul for Wales at Cardiff.

1869   An Act for the Preservation of Sea Birds (32 & 33 Vict cap. 17).

1870–1   Homing pigeons conveyed messages into Paris during the Franco-Prussian War.

1872   The start of a succession of Wild Birds Protection Acts.

1873   The Lombe collection of birds was acquired by Norwich Castle Museum.

1875   John Gurney junior stayed six months in Egypt.

1876   'Uncle Toby' set up the Dicky Bird Society for children.

1879   Lighthouse migration surveys were begun.

1880   An Education Act led to the virtual cessation of child labour.

1880   Murray Mathew became curate of Wolf's Castle, Pembrokeshire (to 1888).

1880   The taxidermist Rowland Ward published *The Sportsman's Handbook*.

1880   Wirt Sikes published a book on Welsh folklore.

1880   Wild Birds Protection Act.

1880–1   The severe winter killed huge numbers of birds.

1881   'The Lark Ascending' by George Meredith was published.

1881   Wild Birds Protection Act.

1881   William Greener published a book on the history of guns.

1884   Andrew Carnegie took a party of tourists to the West Country.

1885   Numerous vultures and kites were present after the Battle of Tofrek in Sudan (March).

1885   The Plumage League was founded.

1888   Another major irruption of Pallas's sandgrouse.

1890   Hans Mortensen in Denmark began to ring birds.

1890   The Forth Bridge opened, with lark pie for the luncheon.

1891   Sarah Morton, an American journalist, was the first woman to dine at London's Cheshire Cheese inn.

1895   Cockfighting was prohibited in Scotland.

1895   Five men died of sewer gas at East Ham (July).

1896   An explosion at Tylorstown colliery (January).

1896   An explosion at Brancepeth Colliery (April).

1896   An explosion at Peckfield pit (April).

1896   Richard and Cherry Kearton visited St Kilda.

1898   Alfred Osman founded *The Racing Pigeon* magazine.

1899   Joseph Arthur Gibbs of Ablington died, aged thirty-one.

1899	The poem 'Sympathy' by Paul Laurence Dunbar was published.
1900	The music-hall ballad 'A bird in a gilded cage' was published as sheet music.
1900	The World Fair at Paris, with the Olympic Games, included live pigeon competitions.
1901	England had a population of 30½ million.
1901	Canaries were used in a rescue at Engine Pit, Hill of Beath (February).
1901	William Eagle Clarke spent a month at Eddystone lighthouse to watch migrating birds.
1903	William Eagle Clarke spent five weeks on board the Kentish Knock lightship to watch migrating birds.
1904	A trial at Scarborough of Mr and Mrs Cooper, whose defence was being victims of witchcraft.
1906	An explosion at Urpeth Colliery; canaries were used in the rescue (December)
1906	Europe's worst coal mining disaster at Courrières, France.
1907	The War Department abolished the pigeon service.
1911	The Coal Mines Act; it became mandatory to use canaries to detect carbon monoxide.
1911	An explosion at Cross Mountain coal mine, United States; canaries were used in the rescue and recovery (December).
1913	Cecil Torr returned to Wreyland in Devon from London.
1913	The worst mining disaster in Britain at Senghennydd colliery; canaries were used in the rescue (October).
1914	'The Lark Ascending' was composed by Vaughan Williams.
1914	The outbreak of World War One. Canaries were used to detect carbon monoxide in tunnelling operations.
1917	The Battle of Messines (June).
1917	Pigeons sent from Leckie's flying boat in the North Sea saved the crew (September).

1918   The hospital ship *Llandovery Castle* was torpedoed; a pet
          canary was saved (January).

1921   The Importation of Plumage (Prohibition) Act.

1925   The use of birdlime was made illegal in Britain.

1928   The Protection of Lapwings Bill.

1933   The Protection of Birds Act.

1940   The Rodd collection of birds from Trebartha Hall was donated
          to the Royal Institution of Cornwall.

1940   The museum at Dover was wrecked.

1942–3   Further shelling of Dover destroyed the Plomley collection.

1953   The Protection of Birds Act replaced all earlier laws.

1983   A huge bird wreck occurred on the north and east coasts
          of Scotland.

1986–7   Canaries ceased being used down coal mines.

2003   The much-valued Walter Potter collection was sold off.

2011   England had a population of 53 million.

CAST OF CHARACTERS

———— •◆• ————

The key characters are listed below. Further information about other people is contained in the Notes.

Albin, Eleazar (c.1690–c.1742), possibly German, a watercolour painter and naturalist in London. His works included *A Natural History of English Song-Birds*.

Atkinson, John (1814–1900), grew up in Essex, a Cambridge graduate; became an antiquarian and vicar of Danby in Yorkshire from 1847.

Blundell, Nicholas (1669–1737), a Catholic landowner and diarist from Little Crosby near Liverpool.

Browne, Sir Thomas (1605–82), an author, physician and Oxford graduate, who spent time in Ireland and Europe before settling in Norwich. His works included *Vulgar Errors* (*Pseudodoxia Epidemica*), published in 1646.

Buckland, Francis Trevelyan (known as Frank) (1826–80), a surgeon, naturalist, writer and specialist in fish and fishing, son of William Buckland (geologist and clergyman).

Burritt, Elihu (1810–79), from Connecticut, USA; a blacksmith and self-educated polymath. He lived in Britain for a decade from 1846 and set up the League of Universal Brotherhood. He made another trip in 1863 and became US consular agent in Birmingham.

Byng, John (1743–1813), after an army career, he was a commissioner of stamps in London, from 1782, but is known mainly as a diarist. Towards the end of his life, he became 5th Viscount Torrington.

Cornish, James George (1860–1938), born in Suffolk, was a curate at Faringdon, then rector of East Lockinge and vicar of Sunningdale, all in Berkshire. In 1919 he retired to Sidmouth in Devon.

Daniel, William (1754–1833), clergyman and writer, best known for his *Rural Sports*. He was a Cambridge graduate and knew East Anglia well.

Dixon, Charles (1858–1926), grew up in Sheffield, an ornithologist and writer of many books and papers who lived for several years in the Torquay area.

Dixon, Edmund Saul (1809–93), born in Norwich, a Cambridge graduate and, from 1842 to his death, rector of Intwood with Keswick, near Norwich. He was an expert on poultry and pigeons, wrote numerous magazine articles and spent much time in France.

D'Urban, William Stewart Mitchell (1836–1934), grew up in South Africa and Canada; a hearing impairment prevented a military career, but as a keen naturalist, especially interested in birds, he joined the Royal Albert Memorial Museum in Exeter as its first curator, leaving in 1884. He co-authored *The Birds of Devon* with Murray Mathew in 1892 and was active up to his death at the age of ninety-seven.

Eagle Clarke, William (1853–1938), an engineer and surveyor from Leeds who turned to natural history and worked at the Royal Scottish Museum in Edinburgh from 1888. He was much involved with studying bird migration.

Fowler, William Warde (1847–1921), renowned worldwide as a historian of ancient Rome, he was an Oxford graduate and tutor who was also passionate about birds. He lived at Kingham with his sisters.

Gardner, James (1827–1920), a leading taxidermist in London and naturalist to the Royal Family.

Gibbs, Joseph Arthur (1867–99), first-class cricketer, educated at Eton and Oxford. In 1892, he bought Ablington Manor in the Cotswolds, but died suddenly of heart failure at the age of thirty-one.

Gladstone, Hugh Steuart (1877–1949), a Cambridge graduate, country gentleman, landowner, ornithologist and author, who lived at Capenoch near Dumfries. He served in the military and fulfilled many public roles.

Graham, Henry Davenport (1825–72), born in London, served in the Royal Navy from 1839, lived on Iona 1848–54, studying birds and antiquities, then married, went to Canada, returned to Scotland, and in 1866 went to southern England.

Gray, Robert (1825–87), born at Dunbar, he was a Glasgow banker before moving to the Bank of Scotland in Edinburgh in 1874. He was an extremely knowledgeable ornithologist, best known for his book *The Birds of the West of Scotland* (1871).

Gurney, John Henry (1848–1922), a landowner and internationally recognised ornithologist. He lived at Keswick Hall near Norwich. His father was John Henry Gurney senior, a banker, politician and ornithologist.

Haggard, Henry Rider (1856–1925), he spent time in Africa before becoming a barrister (briefly), adventure novelist and an authority on farming. He farmed some 350 acres in Norfolk, at Home Farm, Ditchingham, and Moat Farm, Bedingham, close to Bungay in Suffolk.

Harting, James Edmund (1841–1928), solicitor, librarian, ornithologist, keen hawker, writer and editor for *The Field*, who lived at Weybridge in Surrey. His works included *The Ornithology of Shakespeare*. He was apparently known as Edmund.

Harvie-Brown, John Alexander (1844–1916), a naturalist, landowner and prolific writer who lived at Dunipace near Falkirk in Scotland.

Hawker, Peter (1786–1853), he served in the army in the Peninsular War, until injured. He was an inventor, musician and game-shooter, especially of birds, with an unsurpassed knowledge of guns. He owned the Longparish estate in Hampshire and a large shooting cottage at Keyhaven.

Hayes, Richard (1725–90), yeoman farmer of Cobham in Kent who came to own Owletts Farm (now National Trust) and kept a diary from 1760 to 1778.

Henderson, William (1813–91), a woollen carpet manufacturer from Durham with many interests including horticulture and folklore, who in 1866 published a book on the folklore of northern England and southern Scotland.

Holland, William (1746–1819), vicar of Over Stowey in Somerset, an Oxford graduate and diarist.

Housman, Henry (1832–1912), clergyman, author and uncle of A.E. Housman. He and his brother Francis created a museum when growing up at Woodchester House, Gloucestershire.

Howitt, William (1792–1879), a prolific rural writer from Heanor, Derbyshire, who later moved to Nottingham and then Esher in Surrey.

Jefferies, Richard (1848–87), a writer about rural England who came from Coate near Swindon.

Johns, Charles (1811–74), born in Plymouth, a Dublin graduate, ordained priest, author of popular books on natural history, and a schoolmaster and headteacher, initially at Helston in Cornwall and latterly his own school in Winchester.

Knapp, John Leonard (1767–1845), a naturalist who was born in Buckinghamshire, joined the navy and then the militia. He lived near Worcester, then in Wales and finally Alveston in Gloucestershire. He is best known for his *Journal of a Naturalist*.

Latham, Charlotte (1801/2–83), born in Sussex, she lived at Llanynys in Wales until her first husband (the vicar) died in 1844. She married Henry Latham, vicar of Fittleworth in West Sussex, but moved to Torquay after his death in 1866.

Leadbeater, Benjamin (1760–1837), a renowned naturalist, ornithologist and taxidermist based at Brewer Street in London.

Lilford, Lord (1833–96), Thomas Littleton Powys, 4th Baron Lilford. He was a landowner based in Northamptonshire and an avid ornithologist who travelled widely in the Mediterranean and Spain.

Linnaeus, Carl (1707–78), a renowned Swedish scientist who devised the naming system for living organisms. He did not believe in bird migration. The Linnaean Society of London took his name.

MacGillivray, William (1796–1852), an Aberdeen graduate who worked in Edinburgh in museum-type roles until he became a professor at Aberdeen in 1841. As a naturalist and ornithologist, he was a prodigious fieldworker, researcher, teacher and writer, best known for his volumes of *A History of British Birds*.

Markham, Gervase (c.1568–1637), he was from a gentry family in Nottinghamshire and wrote many works including *Countrey Contentments* and *Hungers Prevention*.

Mayhew, Henry (1812–87), a London journalist and social reformer who in 1841 founded the magazine *Punch*.

Mathew, Murray Alexander (1838–1908), a clergyman, writer and ornithologist, he grew up in Barnstaple, was an Oxford graduate and became vicar of Bishop's Lydeard, curate at Wolf's Castle and vicar at Buckland Dinham in Somerset.

Morris, Francis Orpen (1810–93), the son of a Royal Navy admiral, an Oxford graduate, clergyman and naturalist, who wrote works on numerous topics, especially religion and birds. He was a curate in Cheshire, Nottinghamshire and Yorkshire, then became vicar of Nafferton in 1844 and of Nunburnholme in 1854.

Nicholson, John (c.1853–1927), a Hull headmaster, honorary librarian and lecturer of the Hull Literary Club, member of the Yorkshire Dialect Society and Hull Geological Society, he wrote several books, including *Folk Lore of East Yorkshire*.

Payne-Gallwey, Sir Ralph William Frankland (1848–1916), an expert on military and sporting history, a knowledgeable sportsman and author of several books, who owned Thirkleby Hall (demolished 1927). His only son was killed in 1914 in World War One.

Pennant, Thomas (1726–98), a Welsh naturalist and landowner who lived at Downing Hall in Flintshire (demolished 1953). He corresponded with Gilbert White and published numerous books, including his successful volumes of *British Zoology* and accounts of his travels.

Rodd, Edward Hearle (1810–80), Penzance town clerk and ornithologist whose bird collection passed to his nephew Francis Rashleigh Rodd of Trebartha Hall.

Salvin, Francis (1817–1904), he joined the army in 1839, retiring in 1864 with the rank of captain. He was dedicated to field sports, especially falconry, on which he wrote books and articles.

Saunders, Howard (1835–1907), he worked for Anthony Gibbs & Sons, bankers and South American merchants, and lived at Reigate on his return from Peru in 1862, devoting himself to ornithology in Britain and Europe. He wrote many well-received works and held positions in several leading societies.

Savory, Arthur Herbert (1851–1921), he went to agricultural college, married at Alton in Hampshire in 1873, farmed at Aldington in Worcestershire until 1906, then retired to Burley in the New Forest.

Simeon, Cornwall (1820–80), an Oxford graduate, barrister and naturalist who lived on the Isle of Wight. He became County Treasurer at Winchester and moved to Bishopstoke. He was a good friend of Charles Johns and a keen fisherman and shooter.

Sterland, William John (1815–81), grew up in Ollerton, Nottinghamshire, where his father was a grocer, then worked for twenty-one years for the Home and Colonial School in Middlesex, wrote for *The Field* and is best known for his book *The Birds of Sherwood Forest*. He died in Ramsgate.

Sternberg, Vincent Thomas (1831–80), the son of a Northampton wine merchant and music teacher, he published *The Dialect and Folk-Lore of Northamptonshire*. He became the librarian of the Leeds Library and his ghost apparently haunts the building today.

Stevenson, Henry (1833–88), lived in Norwich, he was the editor and proprietor of the *Norfolk Chronicle*. He wrote three volumes of *The Birds of Norfolk*, and his obituary said that 'ornithology possessed his soul'.

Swainson, Charles (1841–1913), a clergyman, naturalist and folklorist who published *Provincial Names and Folk Lore of British Birds*. An Oxford graduate, he became rector of High Hurst Wood in Sussex and then of St Luke's at Charlton near Greenwich.

White, Gilbert (1720–93), brought up in Selborne, Hampshire, he was an Oxford graduate, naturalist and clergyman who inherited the family home, The Wakes, and became increasingly interested in natural history. His correspondents included Thomas Pennant. He is famous for his *Natural History and Antiquities of Selborne*.

White, Walter (1811–93), a self-educated son of an upholsterer in Reading, he worked for a while in America, but returned to England and became a library attendant at the Royal Society in London and, in 1866, its assistant secretary. He published several books about his walking tours.

Wilkie, Thomas (c.1789–1838), he was a ship's surgeon for the East India Company and in 1821 set up a medical practice at Innerleithen in his native Scottish Borders, all the while collecting ballads, songs and traditions that he shared with Sir Walter Scott.

Willughby, Francis (1635–72), also spelled Willoughby, he was born at Middleton Hall in Warwickshire, of a prosperous family whose main seat was Wollaton Hall (Nottingham). As a Cambridge graduate, he was an early ornithologist and scientist who travelled widely. His friend John Ray published his works after his premature death.

Yarrell, William (1784–1856), he took over his father's newsagent and bookselling business in London. Self-educated, he acquired an incredible knowledge of birds and fishes and published the influential *The History of Birds* in three volumes.

NOTES

———•◆•———

PROLOGUE: THE DAWN CHORUS

1 *The Times* 8 June 1917, p. 8. See also Barton et al 2004. Harry Perry Robinson lived 1859–1930.
2 MacGillivray 1837, p. 473.
3 See Moss 2018 and Swainson 1885 for folk names, while Cocker and Mabey 2020 is an excellent guide to individual birds. Amongst the many websites, see rspb.org.uk/birds-and-wildlife/wildlife-guides/birds-a-z/ and bto.org/understanding-birds/birdfacts, which gives fascinating details about the numbers of birds and their decline (or occasionally growth) over the years. An example of a small snippet of information is that one Bewick's swan was counted as having 25,216 feathers.
4 Hudson 1909, p. 118. He was at the hamlet of Crux Easton.
5 *Burnley News* 8 May 1915, p. 11.
6 *Dumfries and Galloway Standard* 7 August 1918 p. 5. This was part of a report on a memorial service for local soldiers, including John Bryden, who enlisted in 1915 aged twenty-three. He was a corporal in the 6th battalion Cameron Highlanders. He died on 30 May 1918.

ONE: ABUNDANCE

1 Nicholson 1890, p. 129. He was probably living in Berkeley Street in Hull.
2 Cobbett 1830, pp. 466–7. This was 11 September 1826. Cobbett actually wrote Ocksey. Somerford is in Gloucestershire and Oaksey in Wiltshire.
3 Hawker 1893a, p. 20. Diary entry for 7 November 1810.
4 Simeon 1860, p. 196.
5 Lubbock 1845, p.92. Lubbock was rector from 1837 to 1876, when he died.
6 Gibbs 1898, p. 392. He was Joseph Arthur Gibbs, but as an author was known as Arthur.
7 Bingley 1804, p. 350.
8 Albin 1737, preface.
9 Lubbock 1845, pp. 93–4.
10 Harvie-Brown 1890, pp. 240–1.
11 Harvie-Brown 1890, pp. 226–7.
12 Torr 1918, pp. 4–5.
13 Torr 1918, p. 5.
14 Burritt 1864, pp. 45–6. Burritt set out on this tour on 15 July 1863. In 1865 he became the US consular agent at Birmingham.
15 John Denison Champlin from Connecticut, USA, lived 1834–1915. He visited Europe in 1884. Another account (but only as far as Salisbury) was

published in *Harper's New Monthly Magazine* 70, December 1884, pp. 21–36, called 'A few Days' More Driving' (anonymous, but by William Black).

16 Champlin 1886, pp. 191–2. This was Tuesday 17 June 1884. The meadow was by the trout stream called the River Tale, on the east side of the hamlet, and they could see the cottages at Fairmile and the church of St Philip and St James.

17 Champlin 1886, p. 193.

18 Tunnels, bunkers and tree-houses were built by the protestors, who included the famous 'Swampy' (Daniel Hooper).

19 Massingham 1936, p. 27 (the location is given on pp. 26–7).

20 Massingham 1936, p. 99. The season is not given, but it was autumn or winter, because the rooks were roosting, and Massingham had earlier watched ploughing, suggestive of autumn. Alfred's Castle is 30 miles north-east of Stonehenge. Close by, King Alfred conquered the Danes in AD 871 at the Battle of Ashdown. Ashdown House is now in the care of the National Trust.

21 *Corvus* is the scientific name of a genus in the crow family, containing the 'true' crows, which include crows, rooks, jackdaws and ravens. It is derived from the Latin word *corvus*, meaning a black bird or specifically a raven. Choughs are distinguished by red legs and bills. Hooded crows are closely related to carrion crows but are mixed grey and black.

22 Cornish 1939, p. 107. The Grove now seems to be Grove Wood, which was oak wood, about 1 mile north-east of the town. Childrey is 6 miles north-east of the hillfort. Cornish's father moved the family to Childrey in 1882, which was then in Berkshire, but has since changed to Oxfordshire. Before that the family spent many years in Suffolk, and James was born and brought up in Suffolk.

23 C.W. Bingham 'Flock of Starlings' *Notes and Queries* 9 (2nd ser.), April 1860, p. 303. Bingham was by then rector of Bingham's Melcombe, a big manor house and church next to a deserted medieval village. Half a century later, he was immortalised as Parson Tringham in Hardy's *Tess of the D'Urbervilles*. Bingham founded the Dorchester Museum and the William Barnes Society. He lived 1810–81 and is buried at St Andrew's, Bingham's Melcombe. The word 'murmuration' does not refer to normal daytime flocks of birds.

24 Dixon 1899, pp. 43–4.

25 Dixon 1899, p. 44.

26 Gibbs 1898, pp. 391–2. By 'thrashing machine', he meant a steam-powered threshing machine.

27 *Pall Mall Gazette* 23 October 1912, p. 2. NB it says 'or few', not 'or a few'. Nelson's Column was used from about 1897. For urban starling roosts, see Fitter 1942 and Potts 1967.

28 *The Times* 4 October 1928, p. 17. The British Museum roosts began in 1919.

29 Dixon 1899, p. 196. Whimbrels breed in northern Scandinavia and northern Russia.

30 Gosse 1934, p. 50. The letter was written on 26 October 1915 as a reply to Philip Gosse who was serving in the Army Medical Corps on the Western Front. He was sent books to read by his mother, and his favourite ones were by Hudson on natural history, which is why he had written to him.

31 Mill 1848, p. 311. John Stuart Mill lived 1806–73.

32 Possibly as many as 8,600,000 if uncounted children and members of the armed forces are added. By the 1811 census, the population had increased to almost 9,500,000.

33 In 2019, England had a population of over 56,000,000. A useful discussion of population is in Wood 1995, pp. 18–21. The reasons for the expanding population are various, including the cessation of prolonged wars, immigration from Ireland into the new industrial towns like Manchester, reduced infant mortality and people living longer. It is also speculated that there was increased fertility due to improvements in food supply and sanitation.

34 MacGillivray 1840, p. 568. The village is known now as East Linton.

35 See Macdonald 2020, in particular pp. 52–3. He points out that the European Union's Common Agricultural Policy has proved far more damaging in Britain than the Tudor vermin acts, because the landscape has been turned into a monoculture factory, destroying everything in its path.

36 A/BTL/2/4, Somerset Archives and Local Studies Service, 15 January 1810.

37 Knapp 1829a, pp. 197–8.

38 Knapp 1829a, p. 198.

39 D'Urban and Mathew 1892, pp. 325–6.

40 Dutt 1903, pp. 160–1. See also Shrubb 2003, pp. 127–55.

41 Rackham 1986, Hollowell 2000, p. 1, Williamson 2013, pp. 95–101, Shrubb 2003, pp. 66–113.

42 *The Journal of the House of Commons* 47 (1803), p. 318; M'Culloch and Reid 1881, pp. 651–64.

43 Holland 1808, p. 121. The author was Sir Henry Holland, a physician who was born at Knutsford in Cheshire and lived 1788–1873. The correspondent was Mr Fenna from Blackhurst in north-east Cheshire. See also Williamson 2013, pp. 103–7, and Shrubb 2003, pp. 114–21.

44 Burritt 1864, p. 58.

45 Burritt 1864, p. 60.

46 Burritt 1864, pp. 56–7.

47 White 1865, pp. 210, 213. He set out on his trip in July 1860.

48 White 1865, pp. 217–18. He went to this farm in error (a mistake on his letter of introduction).

49 'The Border-Land of London' *The Leisure Hour* 15 (1866), pp. 78–80.

50 Mathew 1872, p. 2917.

51 Torr 1923, p. 12. He does not specify the date, but Lustleigh station opened in July 1866, closed to passengers in 1959 and completely closed in 1964.

52 The first telegraph poles were used by the Great Western Railway in 1843.

53 For the term 'telegraphed', see Nelson 1907a, p. 41.

54 *The Zoologist* 1 (2nd series), 1866, pp. 94–5.

55 Gray and Anderson 1869, p. 21. They spent fifteen years doing excursions related to natural history.

56 Coues 1876, pp. 734–5.

57 Stevenson 1866, pp. 337–8.

58 Mathew 1894, p. xxxvi. He was curate of St Laurence, Wolf's Castle, from 1880 to 1888. For his obituary, see *Somerset Standard* 17 July 1908, p. 6. See also Love 2015, pp. 256–8.

59 Stevenson 1870, p. 277. Abel Towler was a magistrate and head of the firm Towler, Allen & Co, cloth makers and dyers. He lived at 1 Grove Terrace, Unthank Road, Norwich.
60 Stevenson 1870, p. 71.

TWO: VERMIN

1 Baker 1854, p. 50.
2 Fowler 1913, p. 84.
3 The Act was 24 Hen VIII cap. 10. See Amos 1859, pp. 80–2, Anderson 2005, p. 209, Brushfield 1897, Lovegrove 2007, pp. 79–80, and Raithby 1811, p. 133. An original transcription is given in *Statutes of the Realm* 3 (1817), pp. 425–6. Kernelling is the formation of the edible part (kernel) of the grain.
4 See Amos 1859, p. 80, Lovegrove 2007, pp. 150–3, and *Statutes of the Realm* 3 (1817), p. 425.
5 The Act was 8 Eliz cap. 15, passed in 1566. As well as more birds and eggs, foxes, badgers, polecats, weasels, stoats, wild cats, otters, hedgehog, rats, mice and moles were also added (Anderson 2005, p. 210). The reward was still paid by landowners (Brushfield 1897, p. 295). See *Statutes of the Realm* 3 (1817), p. 426.
6 Bishop 1992, p. 187 (Chipping Camden); Nelson 1907a, p. 235 (Dent); *Chester Courant* 8 December 1880, p. 3. Thomas Wilkinson Norwood was born c.1829 and was vicar of Wrenbury from 1878 until his death in 1908 (see *Cheshire Observer* 30 May 1908). He was a member of the Cotteswold Naturalists Club (covering the Cotswold area).
7 Elliott 1936, p. 55.
8 Albin 1737, p. 16.
9 Anderson 2005, pp. 213–14, Brushfield 1897, p. 299, and Jones 1972, p. 110.
10 Anderson 2005, p. 223 and Jones 1972, p. 117.
11 Blunt 1877, p. 19 (Cam), Elliott 1936, p. 79 (Westoning). See also Anderson 2005, p. 224, and Lovegrove 2007, p. 82.
12 Howitt 1839, pp. 6–7. He was talking about Heanor in Derbyshire. It has been suggested (Anderson 2005, p. 213) that the term 'sparrow' was applied to any small bird, but Howitt's evidence disproves that notion. The term 'old birds' refers to ones that have raised young birds of their own.
13 Atkinson 1891, pp. 100–1.
14 Howitt 1839, pp. 113–14. By 'corn-field', he meant cereal crops such as wheat, barley and oats, rather than American Indian corn, which is known as maize in Britain.
15 See Copper 1971, p. 83 for an illustration of clappers and p. 84 for a description, and Borrow 1862, p. 443 for a sheepskin rattle.
16 Howitt 1839, p. 114.
17 Howitt 1839, pp. 114–15.
18 Arch 1898, pp. 27–8.
19 Smith 1988, p. 33, who says this one is probably from Somerset. See also Raymond 1910, p. 12.
20 Anon 1843, p. 236.

21 Glyde 1856, p. 362. Originally called Glide, he was also a hairdresser and
    antiquary. He lived 1823–1905.
22 Denton 1868, p. 59. He lived 1814–93. The meeting in London was on
    20 May 1868.
23 Haggard 1899, p. 105. This was February 1898. See also Sternberg 1851,
    pp. 33, 68.
24 Sterland 1869, p. 134.
25 Kalm 1892, pp. 376–7. Pehr Kalm had been a pupil of Carl Linnaeus at
    Uppsala. He lived 1716–79. He was commissioned by the Swedish Academy
    of Sciences to travel to North America and spent a few months in England
    beforehand.
26 Copper 1971, p. 86 (courtesy of Bob Copper, with permission of The
    Copper Family).
27 Haggard 1899, p. 105. This was 4 July 1898.
28 Copper 1971, p. 88 (courtesy of Bob Copper, with permission of The
    Copper Family).
29 Stephens 1844a, p. 676. He lived 1795–1874, a farmer who became an
    influential writer on agriculture. Robert Plot also mentioned gunpowder as
    a deterrent almost two centuries earlier (1677, p. 255).
30 Stephens 1844b, pp. 1302–3.
31 *The Garden: An Illustrated Weekly Journal* 2 (1873) p. 543 (for 28
    December 1872).
32 Haggard 1899, p. 90.
33 Haggard 1899, p. 91. This was 25 January 1898.
34 Haggard 1899, p. 278. This was 18 July 1898, and he went on to say that rooks
    were also taking everything as the ground was too hard for grubs and worms.
35 *The Times* 12 December 1862, p. 7.
36 Sterland 1869, p. 110.
37 Gibbs 1898, p. 111.
38 Jefferies 1889, pp. 179–80, 223. This publication was posthumously
    produced by his wife.
39 Haggard 1899, p. 91. This was 25 January 1898. See also Stevenson 1866,
    pp. 196, 209.
40 Pennant 1776, p. 130.
41 Shand 1905, pp. 293–4. Alexander Innes Shand was a writer and journalist,
    who lived 1832–1907.
42 Lower 1854, p. 168. Lower said the man was now dead, but had been a
    shepherd for many years and was talking about his life in the first half of the
    nineteenth century.
43 Atkinson 1891, pp. 329–30.
44 Andrews 1935, p. 251. This was on 12 July 1790. By swallows, Byng meant
    house martins, as they nest communally.
45 Andrews 1935, p. 251.
46 Blunt 1877, pp. 191, 193. He was a curate in many places, and this possibly
    happened between 1855 and 1861.
47 Gladstone 1910, p. 170. Capenoch House was near Penpont in the parish of
    Keir in Dumfriesshire.
48 Service 1903, p. 338. Robert Service lived 1854–1911.

THREE: HUNTING

1 Dixon 1890, p. 193.
2 Act 1, scene 3. Harting 1871, pp. 160–1.
3 Markham 1621, pp. 280–1. A second edition of *Hungers Prevention* was published in 1655, after his death.
4 Markham 1621, p. 281.
5 Markham 1621, pp. 281–2.
6 Markham 1621, pp. 284–5.
7 Nicholas Cox lived c.1650–1731. His book was first published in 1673, enlarged in 1677, with several subsequent editions (Roberts 1926).
8 Browne 1884, pp. 27–9.
9 Kear 1990, pp. 74–5.
10 Markham 1621, pp. 132–3.
11 Markham 1621, p. 133.
12 Shrubb 2013, pp. 33–7, 76–7, 102.
13 Early nets were often made from vegetable fibres that were twisted into ropes and cords.
14 Gurney 1876, p. 280. Frank Joseph Cresswell became an alderman after 1861, following the death of his father Cresswell Cresswell.
15 Gurney 1876, p. 282.
16 Dixon 1890, pp. 192–3. He wrote this book in 1890 when based at Torquay. John Stiff was married with six daughters by the time of the 1881 census, when he was at Friskney in Lincolnshire. In 1871 he was a coastguard at nearby Wrangle and by 1891 he was an innkeeper.
17 Dixon 1890, p. 193.
18 The Act was 25 Hen VIII cap. 11 (1533–4).
19 Payne-Gallwey 1886, pp. 54–5.
20 Whittington Parish Register of Baptisms (1753–1784) and Burials (1753–1790), Shropshire P305/A/1/5. Also in Davies c.1820, p. 39 (but not quite accurate). The estate of Aston Hall was owned by Colonel Richard Lloyd. This was a four-pipe decoy, located 2 miles from Whittington, near the River Perry.
21 Oldfield 1829, p. 179.
22 Oldfield 1829, p. 180.
23 Payne-Gallwey 1886, p. 158. He actually wrote J.G. Nicholson, which is incorrect. John Young Nicholson was rector. Some sources say the decoy was on Middle Moor, but it was actually Aller Moor. It is listed by Historic England, NGR ST 40136 28150. The decoy is shown on the Tithe Apportionment Map for 1838, portion 254, occupier Richard Evans and landowner Sir P.P.F.P. Acland. Hagglers were also known as higglers.
24 Payne-Gallwey 1886, p. 98.
25 Payne-Gallwey 1886, p. 98. George Chapman was still alive in 1886 and may well have been the person William Whitehead Gascoyne (probably after his father's death) took on.
26 Payne-Gallwey 1886, pp. 98–9.
27 Dutt 1903, pp. 117–18. William Alfred Dutt was born in 1870 at Ditchingham, Norfolk, and became a journalist in London, but returned to Norfolk to write due to ill-health.

28 Oggins 2004, p. 111.

29 Pennant 1768, p. 133.

30 Harting 1871, p. 49.

31 Oggins 2004, pp. 12–16.

32 Salvin and Brodrick 1855, p. 31. William Brodrick was an artist from Chudleigh in Devon, knowledgeable about falconry.

33 Lewis and Richardson 2017. Vervels were also known as varvels and hawking rings. Most were of silver, some of brass or gold.

34 Salvin and Brodrick 1855, p. 30.

35 Ward and Cunningham 1769, p. 372.

36 Coke 1797, pp. 97–8. The Acts were 34 Edw 3 cap. 22 and 37 Edw 3 cap. 19.

37 Sebright 1826, p. 30. He lived 1767–1846 and for many years was Member of Parliament for Hertfordshire.

38 The meaning of gentle changes, so first of all 'superior', but then 'pretty good'. Slight falcons are slender or slim. A male goshawk tended to be known as a goshawk tiercel, while male gyrfalcons were called jerkins.

39 Oggins 2004, p. 22.

40 Harting 1871, p. 51. *Richard II* act 1, scene 1, 118. This is in a scene where the king hears counter-accusations between Henry Bolingbroke, Duke of Hereford, and Thomas Mowbray, Duke of Norfolk. This is a comment on Bolingbroke accusing Mowbray of treason.

41 Cocker and Mabey 2020, pp. 145–6.

42 Sebright 1826, p. 27.

43 Sebright 1826, pp. 27–9.

44 Salvin and Brodrick 1855, p. 72.

45 Sebright 1826, pp. 39–40. It was the River Wissey. Colonel Wilson became the 9th Lord Berners. He lived 1761–1838. His original house has been demolished.

46 Salvin and Brodrick 1855, p. 75. He was Edward Clough Newcome of Feltwell Hall, near Brandon, Norfolk, 1810–71, who was hugely knowledgeable about falconry (Freeman and Salvin 1859, p. 329).

47 Harting 1871, p. 49.

48 Modern fox-hunting became particularly popular from the 1830s. See Mellor 2006, pp. xv–xvii, 55–62, 7. There was a revival in falconry in the mid-nineteenth century, such as with the Old Hawking Club that was set up in 1864.

49 Johann or Giovanni Faber was from Bamberg in Germany. This description was cited by the ornithologist Francis Willughby (published in Ray 1678, pp. 329–32), taken from Faber's 1628 book *Animalia Mexicana*.

50 Harting 1871, p. 262.

51 Freeman and Salvin 1859, pp. 327–52.

52 Anon 1846, p. 311. It was written about 1393.

53 Roth 2012, pp. 44–7, 180.

54 Howitt 1839, p. 123.

55 Beaumont 1824, p. 159.

56 Beaumont 1824, p. 161.

57 Beaumont 1824, p. 161.

58 Daniel 1802, pp. 501–2.

59 Gurney 1921, p. 128. Old Hunstanton Hall is a fifteenth-century moated house.

60 Gurney 1834, p. 530. Daniel Gurney's paper was read to the Society of Antiquaries of London in March 1833, by which time the house had been uninhabited for some years. Additional information was published in Gurney 1921, p. 130.

FOUR: SHOOTING

1   Prentis 1894, p. 32. Walter Prentis was a farmer of 350 acres at Congerton Farm in Rainham.

2   The Brown Bess military musket became the most famous flintlock gun. For the history of sporting guns, see Fremantle 1913.

3   Gladstone 1910, pp. 260–1. Haining was born in about 1823 and worked in farming all his life.

4   Southey 1813, p. 81. Nelson was on leave from December 1787 to December 1792, being at Burnham Thorpe from the summer of 1788. He obtained a licence costing 2 guineas in 1788, which is probably when these incidents occurred. At the end of 1792, relations with France deteriorated, triggering a mobilisation, and he was then senior enough to be given a ship.

5   *The Sporting Magazine* 8 (1796), p. 325.

6   *Bury and Norwich Post* 23 February 1803, p. 2.

7   *Kentish Gazette* 18 September 1801, p. 3.

8   Greener 1910, p. 374.

9   Hawker 1893a, p. 288. He was at the Rodney estate of Old Alresford House.

10 Hawker 1816, p. 67.

11 Daniel 1802, p. 507.

12 Hawker 1816, p. 67.

13 Greener 1910, p. 611. Shot could be hardened by mixing with other metals or by rapid cooling.

14 Arnold 1949, p. 190. This was 3 March 1764. Hayes merely grazed their feathers as they took off.

15 Pope 1969, pp. 158–65. Hawker 1893a, p. 31. This was 9 July 1811. Hawker was a colonel in the army.

16 For Forsyth's lock, see Greener 1910, pp. 112–16, in which he cites a piece by Hawker about Joe Manton, who in 1818 patented a tube lock, in which a tube of soft metal was filled with fulminate. One end was inserted into the touch-hole and the other was struck by the hammer.

17 Hawker 1893b, p. 207, 5 July 1841. Eley ammunition is still famous today.

18 13 Ric II stat. I cap.13. They had to own land worth 40 shillings a year or be a clergyman earning £10 per year (Kirby 1933, p. 240).

19 Kirby 1933, pp. 240–2. The 1671 Game Act was 22–3 Car II cap. 25. Game does not cover deer, though it once included rabbits.

20 Kirby 1933, pp. 251–2.

21 *The Sporting Magazine* 4 (1794), p. 251. Letter of 20 July 1794.

22 Hawker 1816, p. 294. He means £100 and £150 – *l.* was an abbreviation for *libra*, the Latin word for pound.

23 Kirby 1933, pp. 259–60.

24 Hawker 1893a, p. 89.
25 A/BTL/2/33, Somerset Archives and Local Studies, 1 September 1810.
26 A/BTL/2/9, Somerset Archives and Local Studies, Wednesday 20 August 1800. He is talking about the 1791 Act, 31 Geo cap. 21.
27 A/BTL/2/29, Somerset Archives and Local Studies, 24 December 1808.
28 Stevenson 1866, p. 265.
29 Palmer 1949, p. 132. See also Richardson 1997.
30 *The Sporting Magazine* 5, February 1795, p. 276.
31 Hawker 1893b, p. 120.
32 Harvie-Brown 1890, p. 250. A velvet scoter is a large black seaduck with an orange-yellow bill, but Graham said it had a red bill. He was living on Iona, but left there in 1854, married, went to Canada, then returned for a while to the Argyll coast, which is when this took place.
33 Kear 1990, pp. 16–18. What they feed on depends on the species.
34 Daniel 1802, p. 478.
35 Padley 1882, p. 47. James Padley was talking about the Fens of the eighteenth century. He died in August 1881, and his book was published in 1882.
36 Hawker 1893a, p. 87, 10 December 1813.
37 Hawker 1893b, p. 90.
38 Hawker 1816, pp. 240–1.
39 Hawker 1816, p. 242. The term 'fall out of bounds' refers to pigeon trap shooting, where a shot pigeon was disallowed if it dropped beyond the match boundary.
40 Hawker 1893b, p. 68. This was 22 November 1834.
41 Rowley 1875a, pp. 124–5.
42 Markham 1621, pp. 47–8.
43 Markham 1621, p. 50. He also suggested (p. 53) making a model of an ox, cow or bull.
44 Nelson 1907b, p. 552. He cites other recollections suggesting 1811 or 1816. The Wolds were then unenclosed and unploughed, and this shooting took place between Sledmere and Langtoft. About eleven great bustards were killed. Agars was gamekeeper to the St Quintin family at Lowthorpe Hall. The grandson was also called Robert, whose own father, John Agars, was also present.
45 Gray 1871, p. 317. The mouth of the River Tyne is close to Belhaven.
46 The dogs were usually pointers and setters, but also spaniels and Newfoundland dogs.
47 Markland 1727, p. 9. He was a former fellow of St John's College, Oxford (according to the title page). *Pteryplegia* is derived from the Greek, literally meaning 'feather strike'.
48 Yalden and Albarella 2009, pp. 101, 106–7.
49 Gent 1733, pp. 36–7. The poem, on the pleasures of country life, was 'by a Reverend Young Gentleman'. 'Fate' is the lead shot, and 'boots' means signifies.
50 MacGillivray 1837, p. 182.
51 MacGillivray 1837, p. 160. William Smellie Watson, who lived 1796–1874, had his own bird collection.
52 Greener 1910, pp. 589–96.

53 Kirby 1933, pp. 243–4.

54 Stevenson 1866, p. 370.

55 Stevenson 1866, p. 372.

56 Stevenson 1866, pp. 372–3.

57 Gibbs 1898, p. 218. Lord Eldon's estate was at Stowell Park in Gloucestershire. See also Carr 1981, p. 484.

58 Carr 1981, pp. 483–4.

59 Adkins and Adkins 2013, p. 240.

60 *Norfolk Chronicle* Saturday 16 November 1782, p. 3.

61 Cash 1933, p. 295.

62 Daniel 1802, p. 427.

63 Thompson 1981, p. 459.

64 Vesey-Fitzgerald 1969, pp. 108–9.

65 Hudson 1894, p. 28.

66 Lilford 1895, p. 63.

67 Atkinson 1891, p. 329. Robert Raw lived c.1808–79, a gamekeeper and farmer.

68 Sterland 1869, p. 142.

69 Gibbs 1898, p. 322

70 Gibbs 1898, p. 323. See also Williamson 2013, p. 109.

71 Gibbs 1898, p. 109. There was no such person as Tom Peregrine, but he was supposedly modelled on Tom Brown, a workman from the village.

72 Dixon 1899, p. 162. He called them ring-doves, which are wood pigeons.

73 Daniel 1802, p. 497. See also Greener 1910, p. 472.

74 Greener 1910, p. 482. For pigeon shooting, see also Hansell 1988b, pp. 32–5.

75 Waterton 1836, p. 344. Walton Hall, now Waterton Park Hotel, is south-east of Wakefield. Waterton had the grounds surrounded by a high wall, 3 miles in circumference, to protect the wildlife.

76 *Pall Mall Gazette* 22 June 1900, p. 10. There were two live pigeon events in June, as well as clay trap shooting in July. J. Banwell of Great Britain came seventh in the second. The prize money was in French francs (f.).

77 It has been stated that the IOC thought that Donald Mackintosh (who lived 1866–1951) won the archery and then game shooting, rather than live pigeon shooting, but he actually participated in two live pigeon events. Future Olympics used clay pigeons. See Mallon 1998, pp. 186–7, 188. It is usually said that 300 pigeons were killed, but it was probably a thousand or more. Mallon 1998, p. 188 lists 104 pigeons killed at the first event, but that was only the final.

78 Gibbs 1898, pp. 112–13.

79 Cash 1933, pp. 294–5, 297.

80 MacGillivray 1837, pp. 240–1. Ornithology derives from the ancient Greek word for bird, ὀρνισ.

81 MacGillivray 1837, pp. 242–3.

82 Moore 1837, p. 321; Rodd 1880, p. 126. Dr Edward Moore was a physician and eye surgeon at Plymouth.

83 Gray 1871, p. 124. He was probably William Craike Angus, a shoemaker of Aberdeen.

84 Gray 1871, p. 377.

85 Gray 1871, pp. 377–8.

FIVE: STUFFED

1   *Norwich Mercury* 2 August 1873, p. 3, on the opening of the Lombe
    collection of British and Foreign Birds and British mammals. The Reverend
    John Gunn, 1801–90, was President of the Norwich Geological Society,
    which he founded in 1864. Not to be confused with the ornithologist
    Thomas E. Gunn, also of Norwich.

2   Ikram and Dodson 1998, pp. 131–6, Wasef et al 2019 and Bailleul-Lesuer
    2016, pp. 435–43. Saqqara is a few miles south of Cairo. It was once thought
    that the ibis was farmed on a large scale to cater for the mummy industry,
    but recent genetic evidence suggests that the majority were migratory wild
    birds, from all over Africa. When these mummies began to be discovered,
    it was assumed they were bodies of children and that infant mortality was
    incredibly high. By about 1850, ibises were no longer seen in Egypt.

3   Harvie-Brown 1890, pp. 76, 89, letters of 5 April and 1 May 1852.

4   Harvie-Brown 1890, pp. 119–20, letter of 20 February 1853.

5   MacGillivray 1837, p. 619.

6   Dixon 1899, p. 171.

7   Hawker 1893a, p. 357. The weather was bitterly cold that day.

8   Hawker 1893b, pp. 9–10, 92. These incidents occurred on 1 January 1830
    and 16 December 1835.

9   Knox 1849, p. 49. Henry Swaysland was based in Queen Street, Brighton.

10 Hendrikson 2019, p. 1128. See an image and description in Rowley 1875b,
    opp. p. 164, 172–5.

11 Réaumur 1748, p. 306. The paper was read on 1 January 1748. He lived
    1638–1757.

12 Réaumur 1748, pp. 313, 317–18.

13 *Collection Forum* 2006, 21 (1–2), pp. 143–50. Bécoeur was the son of a
    pharmacist, studied pharmacy, but devoted himself to natural history,
    especially birds and insects. His recipe was lost, but later resurfaced. It was
    used in some museums up to the 1980s.

14 *Hardwicke's Science-Gossip* 1 March 1865, p. 63. The reader was J. Aspdin
    from Richmond in Yorkshire, who was probably James Aspdin. He gave
    suggestions based on James Gardner's book that had just been published.
    For early taxidermy, see Hendrikson 2019. For arsenical soap, see Ward
    1880, p. 16.

15 MacGillivray 1837, pp. 619–20. See also Morris 2018, p. 765.

16 Gardner 1865, p. 3.

17 Gardner 1865, p. 22.

18 *The Zoologist* 1 (2nd series) 1866, p. 523. Park Place is a massive country
    house built in the eighteenth century, supposedly Britain's most
    expensive house.

19 Gardner 1865, p. 17.

20 Gardner 1865, pp. 17–18.

21 Howse 1894, p. 346. It was shot in October 1892. The Natural History
    Society of Northumberland, Durham and Newcastle upon Tyne is now
    simply 'of Northumbria'.

22 Bull 1888, p. 239.

23 Harvie-Brown 1866, p. 67. His report was dated December 1865.

24 His late brother Francis was born on 14 March 1829, and after a career as a barrister, he became the recorder of Rangoon and died there on 18 July 1873. Woodchester House was built in 1740.

25 Housman 1881, p. 25.

26 Housman 1881, p. 32. Joseph Wise was born in 1826 and was fifteen years old in the 1841 census.

27 Housman 1881, pp. 32–3.

28 Housman 1881, p. 33.

29 Gardner 1865, p. 9.

30 Gardner 1865, pp. 12, 19, 22.

31 Housman 1881, pp. 34–5.

32 Mayhew 1851, p. 72.

33 Gardner 1865, pp. 25–6.

34 *Norwich Mercury* 2 August 1873, p. 3. Edward Lombe lived c.1800–47. His son Edward died in 1852 in Florence. Melton Hall was abandoned about the same time and is now in ruins. Julia Clarke died aged eighty-seven in 1894, and details about her and the bird collection are given in *Eastern Evening News* 16 February 1894, p. 4.

35 Clark Kennedy 1868 passim. Her husband was Matthew Theodosius Denis De Vitré, 1794–1870. She was Caroline Pratt De Vitré, 1825–98, previously married to Frederick Barlow. Odney Island is barely separated from the much larger Formosa Island.

36 D'Urban and Mathew 1892, p. lx.

37 D'Urban and Mathew 1892, p. lviii.

38 Henry Peek lived 1825–98. His father founded Peek Frean biscuits. His estate later became All Hallows School, but has since been converted to private housing.

39 Howse 1899, p. 12. John Hancock lived 1808–90. See Goddard 1929, pp. 171–6.

40 Howse 1899, p. 23.

41 Ward 1880, p. viii. Ward lived 1843–1912.

42 *The Zoologist* 1 (2nd series), 1866, p. 96. Dutton was at 51 Terminus Road, Eastbourne, and Caleb Luther Adams, a poulterer and fishmonger, was at number 91.

43 There were seventeen in all, and probably many more who were unlisted. Benjamin Leadbeater lived 1760–1837. White 1878–9, p. 878.

44 White 1861, p. 29. His trip to Portland was in July 1854.

45 Stevenson 1866, p. 285.

46 *The Zoologist* 21 (1863), p. 8766. William Whytehead Boulton was also a J.P. He died on 17 January 1897, aged sixty-three (*The British Medical Journal* 30 January 1897, p. 282).

47 *The Zoologist* 1 (2nd series) June 1866, p. 272. Cecil Smith lived at Lydeard House, Bishops Lydeard. He had previously been a barrister.

48 Jane Yandle lived 1844–1915. Mary Turle, born in 1821, was at 43 High Street, Taunton. Jane was unmarried and living with her in the 1861 census. See also Smith 1869, p. 35.

49 *The Zoologist* 1 (2nd series) June 1866, p. 269. Machin was an entomologist,

a compositor by trade, who lived 1821–94. Details and his photograph are in *The London Naturalist* 87 (2008).

50 D'Urban and Mathew 1892, p. 191.

51 D'Urban and Mathew 1892, p. 126.

52 Lilford 1895, pp. 25–6.

53 Service 1903, pp. 334–5.

54 George Bristow lived 1863–1947.

55 Nelder 1962; Nicholson and Ferguson-Lees 1962.

56 Old Bailey Proceedings Online (www.oldbaileyonline.org, version 8.0, 15 August 2020), September 1896, trial of WILLIAM ROLAND SEAR (36) (t18960908-680). Mrs Ellen Harper lived with her brother at 24 Chevalier Road, Hornsea Rise. George Benjamin Ashmead was from 35 Bishopgate Street Without. The case was held on 8 September 1896.

57 Gardner 1865, p. 23.

58 Fox 1827, pp. 136–7. George Townshend Fox was a landowner and member of the Zoological Society of London, who lived 1781–1848.

59 *The Leisure Hour* 15 (1866), p. 380.

60 *The Leisure Hour* 15 (1866), p. 381–2.

61 *The Leisure Hour* 15 (1866), p. 381–2. Ulisse Aldrovandi was one of the early naturalists who was deceived.

62 Lord 1865, p. 219. Lord was early on a veterinary surgeon in Tavistock, Devon, and lived 1818–72. His pseudonym was 'The Wanderer'.

63 Simeon 1860, pp. 207–9.

64 Johns 1873, p. 262. Simeon and Johns made this trip in August 1873.

65 Stevenson 1870, pp. 181–2.

66 *The Zoologist* June 1866, p. 271. This was dated 5 May 1866.

67 *The Western Daily Press* 8 January 1868, p. 2. George Harding was the secretary of the entomological branch of the society.

68 Harvie-Brown 1890, pp. 98–9. Letter of 26 October 1852.

69 Fowler 1886, p. 25.

70 Cash 1933, p. 258.

71 *Dover Telegraph and Cinque Ports General Advertiser* 13 May 1854, p. 8. The Plomley collection was previously displayed at Maidstone. Francis Plomley was attached to the West Kent Hospital and lived 1806–61.

72 D'Urban and Mathew 1892, p. 364.

73 Smith 1866, pp. 452–3. He spent the summer of 1866 in the Channel Islands.

74 The sale was at Steven's in London on 24 June 1890. Pidsley 1891, pp. 94, 148, 160, 188.

75 *The Western Morning News* 24 February 1949, p. 2. It would today have been one of the top tourist attractions in the West Country.

76 Hancock 1979.

77 Stevenson 1866, p. 311.

78 *Dover Express* 2 May 1941, p. 3, in a report on the previous year's bombing.

79 Smith 2012, p. 292. The diary entry was Saturday 3 October 1942.

80 Morris 2013; Creaney 2010.

SIX: CAGED

1 Valentine 1877, p. 208 (in a piece written by Miss Mary Anne Dyson).
2 Dixon 1851, pp. 303–5. Rayner died in 1872 at the age of seventy-two, having practised medicine since 1826 (*British Medical Journal* 30 March 1872, p. 356).
3 Anon 1784, pp. 70–2. The description was in a letter written by Robert Langham, a mercer, though its authenticity has been questioned.
4 Only royalty and the favoured few were allowed to drive through Birdcage Walk until 1828.
5 A/BTL/2/15, Somerset Archives and Local Studies, 14 and 15 July 1802. The common cuckoo spends the winter in central Africa and then flies northwards to Europe and Asia to breed.
6 Waugh 1867, p. 81. The cottage was in a court off Bell Street, in the Trinity district of Preston, near Holy Trinity church. Waugh lived 1817–90.
7 MacGillivray 1837, p. 395.
8 Mayhew 1851, pp. 63–4.
9 Mayhew 1851, pp. 58, 62, 66.
10 Mayhew 1851, p. 58.
11 Samuel 1975, p. 217 (courtesy of Alison Light and the Raphael Samuel Estate). This is the Buckland family who seem to have been showmen or gypsies but settled at Headington (see Samuel 1975, p. 210). Raphael Samuel interviewed many local people in 1969–70.
12 Belon 1555, p. 337. Belon was born in 1517 and murdered in 1564.
13 Russell 1756, p. 71. Alexander Russell was a Scottish naturalist and physician, who lived 1714–68.
14 *Chambers's Edinburgh Journal* (1847), p. 411.
15 *Chambers's Edinburgh Journal* (1847), p. 410.
16 D'Urban and Mathew 1892, p. 16.
17 D'Urban and Mathew 1892, p. 16.
18 Valentine 1877, p. 209. Mary Anne Dyson (1809–78) was the invalid sister of the Reverend Charles Dyson, rector of Dogmersfield. After he died, she and her mother went to live in nearby Crookham in Hampshire. Miss Dyson was an intellectual, earnest churchwoman who devoted herself and her means to the education of young girls.
19 Jesse 1847, pp. 33–4, 35, 40. He lived 1780–1868, though his daughter says 1782 (in *Sylvanus Redivivus*, p. 6).
20 Harting 1866a, pp. 48–9. He must have known the gamekeeper about 1860.
21 Jesse 1847, p. 34.
22 Albin 1737, unpaginated introduction.
23 Albin 1737, p. 25.
24 Valentine 1877, pp. 208, 212, 213.
25 Harting 1871, p. 274. *Henry IV, Part I*, act 1, scene 3.
26 MacGillivray 1837, pp. 605–6. Thomas Durham Weir was painted by William Smellie Watson, another keen ornithologist. Weir died in 1869. Boghead House was demolished in 1962.
27 Albin 1737, p. 11.
28 Albin 1737, pp. 13–14.
29 Mayhew 1851, p. 62.

30   Howitt 1839, p. 123. He lived 1792–1879, so this may just have been towards the end of the eighteenth century.

31   *Blackwood's Edinburgh Magazine* 60 (1846), pp. 287–8. Aird lived 1802–76. He was a Scottish poet of nature and at this time was editor of the *Dumfriesshire and Galloway Herald* (1823–63).

32   Greene 1897, p. 653. He was William Thomas Greene, FZS.

33   Mitford 1828, pp. 133–4. Mitford was a playwright and poet who lived 1787–1855. The sketches were based loosely on her Berkshire village of Three Mile Cross, south of Reading. 'Old Robin' was supposedly from the nearby village of Barkham.

34   Mayhew 1851, p. 59.

35   Jesse 1847, p. 258.

36   Albin 1737, p. 20.

37   Syme 1823, p. 183. He lived 1774–1845. His wife Elizabeth did the illustrations for the book. French prisoners-of-war were freed in 1814, which may have prompted this sale.

38   Syme 1823, p. 210.

39   Mayhew 1851, p. 62.

40   Mayhew 1851, p. 63.

41   Mayhew 1851, p. 59.

42   Mayhew 1851, p. 69.

43   Mayhew 1851, pp. 58–9.

44   Rossetti 1872, p. 21; Garlitz 1955, pp. 539–43.

45   Paul Laurence Dunbar lived 1872–1906. This is the last of three verses.

46   Mayhew 1851, p. 61.

47   Reiche 1871, p. 16. His book was first published in 1853, by which time he had been trading for about ten years.

48   Reiche 1871, p. 16.

49   Valentine 1877, pp. 205–6, 207. See also Pollock 2013.

50   Courtney 1890, p. 139. Miss Margaret Ann Courtney lived 1834–1920. See also Wright 1896, p. 122.

51   Berkeley 1854, pp. 245–6. Berkeley was MP for West Gloucestershire until 1852. He lived 1800–81.

52   *The Times* 21 November 1846; Hawker 1893b, p. 270, 20 November 1846.

53   *Daily Sketch* 3 July 1918, p. 2; Gladstone 1919, p. 134. Leslie Chapman lived in Linden Street, Romford.

54   Salvage money was a proportion of the value of the vessel and cargo relative to the work and danger involved in the recovery. See Kemp 1976, pp. 242, 747, 945.

55   Dixon 1851, p. 95.

56   Mayhew 1851, p. 68.

57   *The Zoologist* 2 (2nd series), September 1867, p. 913.

58   *Brisbane Courier* 24 December 1873, p. 2. They still thrive in New Zealand.

59   *Melbourne Punch* 17 February 1870, p. 7.

60   There are now about 200 million starlings in North America. A few starlings were supposedly introduced to Central Park in New York in 1890 as part of a programme to bring in all the birds mentioned by Shakespeare, but unfortunately this theory has no foundation. See Strycker 2014, pp. 40–3.

61   Mayhew 1851, p. 70.

62 *Morning Chronicle* 11 March 1850, p. 5. This was a piece by Mayhew on merchant shipping, and the seaman had been recommended to him as a trusty, well-educated, married man, a mariner for fifteen years and second mate on this last voyage, on board a brig of 200 tons, with a good captain.

63 Dundonald 1861, p. 58. Cochrane was on board the *Thetis*, having recently moved from the *Hind*.

64 Hay 1898, p. 48. The *Trinculo* reached Spithead on 17 May 1836. Hay lived 1821–1912. He joined the navy in 1834.

65 Mayhew 1851, p. 71.

66 Watson 1827, p. 28.

67 Adkins and Adkins 2008, p. 355. His captain was Sir Joseph Sydney Yorke.

68 Roud and Bishop 2012, p. 490; for 'May Colvin', see Herd 1776, pp. 93–5.

69 Herd 1776, p. 94 – 'the streen' means 'yesterday evening'

70 Mayhew 1851, p. 72.

71 Mayhew 1851, pp. 68, 69, 70.

72 Mayhew 1851, p. 62

73 *Manchester Evening News* 15 April 1905, p. 7.

74 *News of the World Household Guide and Almanac* 1952, p. 239.

75 Hansard vol. 86, 14 February 1933, col. 658. A century earlier, Lord Buckmaster's own father, John Charles Buckmaster (born in 1820), worked as a child agricultural labourer in Buckinghamshire, including a stint as a bird-scarer (Buckmaster 1897, p. 22).

76 Hansard vol. 86, 14 February 1933, col. 659.

77 Hansard vol. 86, 14 February 1933, col. 661.

78 She became an MP while Lady Grant (she was widowed in World War Two) and so is called Lady Grant of Monymusk in some sources.

79 Mayhew 1851, p. 60.

80 Dyson 1871, pp. 210–11.

SEVEN: MIGRATION

1 Hardy 1879, p. 48.

2 Hawker 1893b, pp. 130–1. This was on 12 September 1837.

3 Egerton-Warburton 1912, p. 82. Geoffrey Egerton-Warburton was known as a botanist. He lived 1847–1925, and most of the pieces relating to his *In A Cheshire Garden* were published April to June 1912 in *The Warrington Guardian*.

4 Nicholson 1890, p. 131. This is dialect for 'The cuckoo will soon be going, she hardly calls'.

5 Burne 1911, p. 238. He was Edward Chessall Scobell, 1850–1917. His original pamphlet was written some time between 1905 and 1911.

6 Ray 1678, p. 98. Willughby also gave a further eyewitness who claimed to have heard cuckoos and found feathers when cut willows were being put in a furnace.

7 Ray 1678, p. 98; Birkhead 2018, pp. 217–39. Francis Willughby lived 1635–72.

8 Harley 1840, p. 661. Harley lived 1801–60. He was a woolstapler and in 1839 was at Mill Stone Lane in Leicester.

9   MacGillivray 1840, pp. 571–2.

10 Fowler 1886, p. 5. He was writing about 1885.

11 A/BTL/2/30, Somerset Archives and Local Studies, 14 and 20 October 1809.

12 Carew 1749, p. 26. Richard Carew lived 1555–1620.

13 Hunt 1865, p. 226. Robert Hunt was an antiquarian, mineralogist and chemist who lived 1807–87.

14 A/BTL/2/32, Somerset Archives and Local Studies, 20 April 1810.

15 A/BTL/2/34, Somerset Archives and Local Studies, 15 October 1810.

16 The ability to migrate is highly complex and is summarised by Strycker 2014, pp. 16–20. See also Macdonald 2020, pp. 120–5.

17 Armstrong 2002, p. xxi.

18 Carew 1749, p. 26.

19 Hunt 1865, p. 227.

20 Letter from Collinson to Klein 6 March 1758, read to Royal Society 9 March 1758. Smith 1821, p. 46; Armstrong 2002, p. 211.

21 Letter from Collinson to Klein 6 March 1758, read to Royal Society 9 March 1758. Smith 1821, p. 47; Armstrong 2002, p. 211.

22 Adanson 1757, p. 67.

23 Letter from Collinson to Linnaeus 15 September 1763. Smith 1821, p. 61; Armstrong 2002, p. 249 .

24 Letter from Collinson to Klein 6 March 1758, read to Royal Society 9 March 1758. Smith 1821, p. 49; Armstrong 2002, p. 212.

25 *The Harleian Miscellany* 2 (1809), pp. 583–4. See also *Publications of the Colonial Society of Massachusetts* vol. 44 (1940). Charles Morton lived 1627–98.

26 *The Harleian Miscellany* 2 (1809), pp. 579, 582, 586; Harrison 1954.

27 *The Annals of Scottish Natural History* 43 (1903), p. 50.

28 Fowler 1886, p. 5 (written in 1885).

29 *The Zoologist* 12 (3rd series, 1888), p. 265. John Mansel-Pleydell was formerly known as John Clavell Mansel. He lived 1817–1902. The bird was named after the Prussian zoologist Peter Simon Pallas (1741–1811).

30 D'Urban and Mathew 1892, p. 260.

31 Hodgson 1910, p. 94. Thomlinson was the curate at Rothbury 1717–20.

32 *The Harleian Miscellany* 2 (1809), pp. 582–3. Travers heard the story from the captain or master.

33 Letter from Collinson to Klein 6 March 1758, read to Royal Society 9 March 1758. Smith 1821, p. 48; Armstrong 2002, p. 212. Sir Charles Wager was First Lord of the Admiralty and lived 1666–1743. Their friendship of Collinson and Wager is described in *Universal Magazine* 97 (1795), pp. 377–8.

34 Tyrer 1968, p. 55.

35 Arnold 1949, p. 122. Richard Hayes lived 1725–90 and owned Owletts, now a National Trust property.

36 A/BTL/2/32, Somerset Archives and Local Studies, 19 April 1810.

37 *Nicomachean Ethics* book 1, chapter 7 (Thomson 1955, p. 39). Aristotle lived 384–322 BC. It inspired Aesop's fable 'The Young Man and the Swallow'.

38 A/BTL/2/32, Somerset Archives and Local Studies, 19 April 1810.

39 Courtney 1890, p. 131. Miss Margaret Ann Courtney lived 1834–1920. See also Wright 1896, p. 122.

40 Arnold 1949, p. 123.

41 Fowler 1913, pp. 98–9.

42 Whittington Parish Register of Baptisms (1753–1784) and Burials (1753–1790), Shropshire P305/A/1/5, p. 64.

43 Egerton-Warburton 1912, p. 82.

44 *The English Mechanic and World of Science* vol. 13, 14 April 1871. The popular name was *English Mechanic*.

45 See 'Cuckoo Lore', *The Leisure Hour* (1877), pp. 285–7; Hardy 1879. They are a Red List species, with just 15,000 breeding pairs in the UK (RSPB website).

46 Marshall 1790, p. 148. He was an agricultural writer and land agent, who lived 1745–1818.

47 Rocke 1866, p. 78. This report was by by John Rocke of Clungunford House at Craven Arms, who was a landowner and magistrate and lived 1817–81.

48 *The Sussex Advertiser* 18 April 1825, p. 3.

49 *The Sussex Advertiser* 20 April 1858, p. 6.

50 Latham 1878, p. 17. Charlotte Latham was herself Sussex born, but lived in Wales during her first marriage. She moved to Torquay after Henry Latham's death.

51 Wymer 1950b, p. 243. The manuscript is in the British Library Harley MS 978.

52 Rocke 1866, p. 77. The wryneck is now only a scarce migrant.

53 *Hastings and St Leonards Observer* 13 April 1895, p. 7.

54 *The Times* 12 February 1913, p. 9. Lydekker lived at Harpenden Lodge in Harpenden. He lived 1849–1915 (Pycraft 1915).

55 Sternberg 1851, p. 159. His full name was Vincent Thomas Sternberg. He was only twenty years old when he published this book (he lived 1831–80). Soon after, Anne Baker published a vast book on the language of Northamptonshire, which she said was twenty years in the making. It was followed by Jabez Allies in 1852 (his earlier 1840 edition did not use the word folklore). See Roper 2014.

56 Nicholson 1890, p. 131.

57 Poole 1877, p. 40. This was a widespread superstition with peasants in England and Germany, according to *The Leisure Hour* 1877 p. 287.

58 Northall 1892, p. 118.

59 See Finlayson 1992.

60 Richter and Bick 2018, p. 585. The stork is a key attraction in the Zoological Collection of the university, and a perfect replica is now in Schloss Bothmer. There are twenty-five arrow storks in natural history collections, mostly in Germany and west-central Europe.

61 *The Sporting Magazine* 25 new series April 1830, p. 363.

62 Daniel 1802, p. 438. His first volume was published in 1801. The woodcock was caught and later shot in Clenston Wood. Edmund Morton Pleydell junior owned Whatcombe House (which is listed and still exists). He lived 1756–1835.

63 Hawker 1816, pp. 167–8.

64 Daniel 1812, p. 160.

65 Mansel-Pleydell 1888, p. 168. He inherited the Whatcombe estate in 1871.

66 MacGillivray 1840, pp. 592–3.
67 King 1836, pp. 96–7. He included his discovery in the publication of his Arctic expedition (under Captain Back) of 1833–6. King lived c.1810–76.
68 Nelson 1907a, p. 160. William Storey lived at 13 Fewston Lodge, and in 1893 was thirty-eight years old and the superintendent of the local waterworks.
69 Preuss 2001.
70 Gladstone 1919, p. 8.
71 Kear 1990, pp. 126–8.
72 Eagle Clarke 1912a, pp. 285–6; see also Love 2015, pp. 219–38. Eagle Clarke was at the lighthouse 18 September to 19 October 1901.
73 Eagle Clarke 1912b, p. 11. He was on board the lightship 17 September to 18 October 1903.
74 Eagle Clarke 1912b, pp. 14–15.

EIGHT: WEATHER

1 *The Zoologist* 14 (1856), p. 5159. There had been severe north-easterly gales.
2 Eagle Clarke 1912b, p. 12.
3 Coward 1923, p. 118. He was Thomas Alfred Coward, 1867–1933.
4 Underwood and Stowe 1984; 34,000 dead birds were recorded on the east coast of the British Isles, from the Orkney Isles to Kent, many emaciated but with no sign of infectious diseases. A deep depression had formed off southern Greenland, which crossed Iceland and then moved into the North Sea, followed by a week of gale-force winds. These bird wrecks can occur worldwide.
5 Simeon 1860, p. 235. He said it was no further westwards than Bournemouth, but meant eastwards. Simeon also said it was August and September 1859, but it appears to have been September only. The seabirds were emaciated, and he ruled out poison and weather conditions, because 'it seems scarcely possible that any storm could have at once so wide a range, and so extensively destructive an event'. In the end he concluded that their deaths were due to disease.
6 *Greenock Advertiser* 24 September 1859, p. 2.
7 Gray 1871, p. 440. The steam packets were the *Giraffe*, *Leopard*, *Stag* and *Lynx*.
8 *Belfast Daily Mercury* 19 September 1859, p. 2. The same story was repeated in other publications.
9 Johns 1862, p. 549.
10 Mathew 1894, p. 35.
11 A/BTL/2/14, Somerset Archives and Local Studies, 6 January 1802. He actually wrote 'but' not 'by'.
12 Hawker 1893b, p. 25. The word 'cottage' is a misnomer, as it was a substantial property.
13 Hawker 1893b, p. 259.
14 Smith 1856, p. 5249. Frederick Smith lived 1805–79. This event took place 8 July 1856.
15 Smith 1856, p. 5250. This part of Lower Street is known now as the High Street.

16 Jordan 1857, pp. 402–3.

17 Hawker 1893b, p. 5.

18 Hall 1910–11.

19 The diary is called 'Dr Taylor's Book' – see Chapter 13 for details. The church spire is 222 feet high.

20 Burton 1888, p. 43. The spire is 120 feet high, dominating the local landscape.

21 Oxberry 1818, p. 3. *The Jew of Malta* was written about 1589 or 1590. This is Act 1. Barabas then learns that his ships are safely moored.

22 March 1998, p. 40; March 2008, pp. 515–16.

23 Johnson 1802.

24 Wilkin 1835, p. 431. His *Vulgar Errors* was also called *Pseudodoxia Epidemica* and was first published in 1646.

25 Wilkin 1835, p. 432.

26 Smith 1807, pp. 88–9. She lived 1749–1806, and Jane Austen emulated parts of her novels and parodied others in *Northanger Abbey*.

27 Sternberg 1851, p. 34.

28 Palmer 1949, p. 63.

29 Torr 1918, p. 4.

30 Burne 1911, p. 238. Scobell was rector of Upton St Leonards near Gloucester and lived 1850–1917.

31 Dixon 1886, p. 253; White 1837, p. 195 – the letter was to Daines Barrington, 2 November 1769. For swallows, see Swainson 1873, p. 246.

32 Nicholson 1890, pp. 130, 131.

33 Gray 1871, p. 83.

34 *Notes and Queries* 1 (3rd series), 25 January 1875. The correspondent was 'Ina'.

35 Swainson 1885, p. 81. This was the parish church of St Peter Mancroft.

36 Simeon 1860, p. 160. He did not specify a location.

37 Johns 1862, p. 255.

38 'Eboracum' in *Notes and Queries* 2 (6th series), 1880, p. 165.

39 Nelson 1907a, p. 157.

40 Lubbock 1845, p. 62. He actually wrote *Charadius*, not *Charadrius*.

41 Simeon 1860, p. 223.

42 Savory 1920, p. 263.

43 Buckland 1860, pp. 286–7. This is not in his 1858 edition.

44 Harrison 1950.

45 Hunt 1881, p. 431.

46 Masefield 1913, p. 48.

47 Harvie-Brown 1890, pp. 96–7.

48 *The Bridlington Free Press* 9 November 1861, p. 5.

49 *The Times* 14 January 1869, p. 7. The Inchcape Rock, off the east coast of Scotland, is the site of the Bell Rock lighthouse.

50 Hansard HC Debate vol. 194, cols 404–6, February 1869.

51 The Act for the Preservation of Sea Birds 1869 was 32 & 33 Vict. cap. 17 (Rickards 1869, pp. 61–3).

52 Fowler 1909, p. 303.

53 *Notes and Queries* 4 (5th series), 31 July 1875, p. 84 (by a correspondent

called 'Ellcee' from Craven in Yorkshire); Rudkin 1939, p. 317. Ethel
Rudkin lived 1893–1985. The estate is probably Willoughton Manor.
Her records are in Lincolnshire Archives, and she lived at Rose Cottage,
Willoughton. Another long-range forecast was done by reading the furcula
(wishbone) of a goose, called the goose-bone oracle, known from the
fifteenth century (Easton 2014, p. 29).

54 Sykes 1833, p. 102.
55 *The Monthly Chronicle of North-Country Lore and Legend* 2 (1888), p. 181. It
    occurred 'a few years before', probably about 1880.
56 Ware 1994, p. 5.
57 Latham 1878, p. 18.

NINE: NESTS AND EGGS

1   de la Mare 1962, p. 547.
2   Stevenson 1866, p. 331. The process of nesting is referred to as nidification.
3   Kirkman 1910, pp. 366, 373. Frederick Bernulf Beever Kirkman, originally
    from South Africa, lived 1869–1945.
4   Gray 1871, p. 134.
5   Latham 1878, p. 44.
6   Courtney 1890, p. 156.
7   *Durham County Advertiser* 14 February 1851, p. 6.
8   Harting 1866b, p. 411. He was talking about *The Winter's Tale* act 4, scene 3
    (he wrongly says scene 2).
9   Lilford 1895, pp. 33–4. Lincoln was shot on 14 April 1865 and died the
    following morning.
10 Brabourne 1884, pp. 175–6, letter of 16–17 December 1816. Edward was
    later known as James Edward Austen-Leigh, who was Jane Austen's first
    biographer.
11 Gent 1690, column EF (no pagination).
12 Mayhew 1851, p. 72.
13 Mayhew 1851, pp. 73, 76.
14 Mayhew 1851, pp. 72–3.
15 Mayhew 1851, p. 72. He also mentioned nests of linnets, thrushes,
    moorhens, chaffinches, hedge-sparrows, bottletits, yellowhammers,
    water-wagtails, blackbirds, golden-crest wrens, bullfinches, crows,
    magpies, starlings, egg-chats, martins, swallows, butcher-birds, cuckoos,
    greenfinches, sparrowhawks, reed sparrows, woodpigeons, horned owls,
    woodpeckers, kingfishers, partridges [no nest], skylarks, robin-redbreasts,
    ring doves, tit larks and jays.
16 St John 1849, p. 15. Charles St John lived 1809–56.
17 McGowan 2009.
18 Harvie-Brown 1890, p. 103. The letter was dated 22 November 1852.
    Graham was sufficiently knowledgeable to blow the egg, but may have sent
    it unblown. He was probably not referring to a cormorant, but to a shag,
    which he called a green-crested cormorant.
19 Browne 1884, p. 227.
20 Gardner 1865, pp. 40–1.

21 Browne 1884, p. 226.
22 Gardner 1865, p. 41.
23 Browne 1884, pp. 226–7.
24 Knapp 1829a, p. 151. In the 1820s he was at Alveston, just north of Bristol.
25 Mayhew 1851, p. 73.
26 Howitt 1839, p. 49.
27 Southey 1813, p. 81.
28 Fowler 1886, pp. 138–9.
29 Wymer 1950a, p. 59; Browne 1884, p. 225.
30 Simeon 1860, p. 164.
31 By 1921 there were 400,000 members, but it then waned.
32 Adams 1887, p. 446.
33 Adams 1887, p. 446.
34 Adams 1887, p. 447.
35 *Newcastle Weekly Chronicle* 4 November 1876, p. 6.
36 Adams 1887, p. 446. William Stafford was born in Birmingham and Leila in Exeter, and they would soon depart for West Ham and subsequently Norwich.
37 Mayhew 1851, p. 73.
38 *The Book of Days* vol. 1 (1863), p. 678.
39 Nicholson 1890, pp. 131–2. The streets are today unrecognisable. Being a solitary nest, it probably belonged to a carrion crow, not a rook.
40 Wilkie 1916, p. 73.
41 *Notes and Queries* (4th series) 7 August 1869, p. 114, query by Edward J Wood, the author of a book on marriage customs published in 1869.
42 Knapp 1829b, pp. 228–9.
43 Sternberg 1851, p. 172; Latham 1878, p. 10.
44 Fowler 1909, p. 296; Gurdon 1893, p. 7.
45 Nicholson 1886, p. 218; Scot was a politician and landowner who lived c.1538–99.
46 *The Times* 14 May 1934, p. 10 (asked by William R. Power of 157 Stamford Hill, London).
47 *St James's Gazette* 3 December 1888, p. 12.
48 Sternberg 1851, p. 157; Richardson 1843, p. 280; Oliver 1832, p. 492.
49 Fowler 1909, p. 290; *Folklore* 5 (1894), p. 341, communicated by Mr. F. York Powell.
50 Sternberg 1851, p. 157; Courtney 1890, p. 139. Curiously, Sternberg's words were borrowed by Charles Poole for his 1877 book on Somerset.
51 Sternberg 1851, p. 157.
52 Wilkie 1916, pp. 60, 76, 105. He was writing in about the 1820s.
53 Norfolk Record Office parish register of Wells-next-the-Sea, PD 679/1. It is said that Mother Gabley was condemned to death, but nothing more can be found out about her, and most sources go round in circles, repeating the claim that there were thirteen men, when there were actually fourteen. The *Gentleman's Magazine* 62 (1792), p. 904, written by J.H., who was John Holt, gave a transcript of the parish register, but it is not entirely accurate and was repeated by Edward Peacock in *Folklore* 13, no. 4, December 25 (1902), p. 431. They both said that the men were 'Misled uppo' ye West Coaste', but

in fact it says 'Perished'. West Coast possibly means the West Sands. Most shipwreck victims were not buried in churchyards, but were left to rot on the coast or were buried in a communal grave on the coast. However, these were local men. The port was called Wells at that time, but is known now as Wells-next-the-Sea, though the coastline has changed owing to silting.

54 Mac Cárthaigh 1992–3.

55 *Bradford Daily Telegraph* 26 September 1904, p. 4 (where the father's name is wrongly stated as George); Peacock 1904. The house was number 5 Ewart Street.

56 They then moved to Hull and Myton-in-Swale. The 1911 census showed that in all they had ten children who were born alive – one more in Scarborough in 1903 and two in Hull. Five of the ten died very young.

57 Browne 1672, p. 310.

58 Loeb Classical Library, Pliny the Elder *Natural History*, LCL 418: 14–15, Book 28, paragraph 19.

59 Oliver 1832, p. 494. Oliver was vicar of Scopwick from 1831 to his death in 1867, though he left Scopwick in 1854 for Lincoln, as his health was failing. He wrote a detailed account (called a pamphlet) of village life called *Scopwickiana*.

60 Latham 1878, p. 10.

61 Coward 1919, p. 79.

62 Williams 1922, p. 289.

63 Couch 1871, p. 155. He lived 1789–1870. His book on the history of Polperro was published after his death by his son.

64 Sikes 1880, p. 304. A pullet is a hen less than one year old. Robert Ellis lived 1812–75. His *Manion hynafiaethol* was published in 1873, on antiquarian topics.

65 Sternberg 1851, pp. 187–8.

66 Bigge 1863, p. 95. He lived 1814–1901.

67 Lysons and Brand 1806, pp. 357–8. The king's family was at Langley in Buckinghamshire for seventeen weeks in the eighteenth year of his reign.

68 The Greek is το πασχα, which became pascha in Latin and then pascal. Hone 1825, pp. 425–6.

69 Harland and Wilkinson 1867, p. 230.

70 Nicholson 1890, p. 14. The bare legs and arms of boys were also beaten with nettles.

71 *The Times* 7 December 2019, p. 15. They were found at Berryfields near Aylesbury by Oxford Archaeology.

72 *Kentish Gazette* 13 October 1807, p. 3.

73 Harvie-Brown 1890, p. 268.

74 Romney 1984, pp. 41–2. Spike lived 1743–1806.

75 Romney 1984, pp. 47–8. This climb was on 12 June 1805.

76 Romney 1984, pp. 48–9. Eighteen months later, in December 1806, Bryan Spike died at the age of sixty-three.

77 Wilson 1842, p. 45. The party had set off from Edinburgh in June 1841. Lauder was an author, secretary of the Board of Fisheries and much else, who lived 1784–1848. James Wilson was a zoologist and ornithologist, who lived 1795–1856.

78 Wilson 1842, p. 50.

79 Wilson 1842, p. 52.

80 Emery 1996, pp. 1–2 and 101 – a report by J. Coulson in which information is also given on egg membrane found on other sites, including Roman ones.

81 *The Atlas* 18 June 1826, p. 6.

82 *Paisley Herald and Renfrewshire Advertiser* 3 June 1854, p. 2.

83 Ware 1994, p. 4.

84 Yarrell 1845, p. 483.

85 Yarrell 1882–4, p. 284. The editor was Howard Saunders.

86 Lubbock 1845, pp. 63–4.

87 *The Magazine of Domestic Economy* 3 (1838), p. 149.

88 Knapp 1829a, p. 255; Lubbock 1845, p. 64.

89 Hansard House of Lords 14 February 1928.

90 Lubbock 1845, p. 123. See also D'Urban and Mathew 1892, p. 384.

91 Gurney 1876, p. 292. His father was John Henry Gurney Sr, a banker, politician and ornithologist.

92 The salt killed some plants and was injurious to worms and various pests that the gulls fed on, but it was observed to improve productivity. Eventually, the use of salt waned, but not before affecting birds. See Smith and Secoy 1976 for the use of salt in agriculture. Pike were introduced in 1864 (Stevenson 1890, p. 327).

93 Romney 1984, p. 42.

94 Dodd 1843, p. 179. This was limestone, not citrus lime.

95 Gritt and Virgoe 2004, p. 6. Also known as Basil Thomas Scarisbrick and Thomas Eccleston, he lived about 1712–89.

96 Marshall 1796, pp. 340–1, 343. Isinglass is a type of gelatin. Another method of refining cider was to add bullock's blood, which was a term usually applied to sheep or cattle blood. He also mentioned that a few farmers thought that something should be added to cider after fermenting, including egg shells.

97 D'Urban and Mathew 1892, p. 428; Dodd 1843, pp. 89–110. A later development was to filter the sugar through bone-black or bone-charcoal (calcined bones), which made the sugar even whiter.

98 Baines 1823, p. 205.

99 Mathew 1894, p. 121.

100 Stevenson 1870, p. 429; Stevenson 1890, pp. 328–9. Robert Rising farmed 670 acres and employed 37 labourers. He lived 1802–85. He rebuilt Horsey Hall in 1845.

101 *The Field* 19 July 1862, pp. 53–4.

102 Coward 1923, pp. 23–4.

TEN: FOOD

1 Andrews 1934, p. 167. This was John Byng, on 14 July 1784.

2 Harvie-Brown 1890, p. 224.

3 Mathew 1894, p. 100. He was talking about a time before 1880.

4 Tufnell 1842, p. 38. The contents list calls him (incorrectly) Tuffnell, but he signed off the report as Tufnell, on 1 March 1840.

5   Samuel 1975, pp. 207–8 (courtesy of Alison Light and the Raphael Samuel Estate).

6   Williams 1922, pp. 138–9.

7   A/BTL/2/36, Somerset Archives and Local Studies, 15 January 1812.

8   Jefferies 1879, pp. 120–1. This is in his imagined village of Wick.

9   Opie 1997, no. 486 (pp. 470–2), first published 1951.

10   Tannahill 1988, p. 187. The pastry was a crust made with suet, so could not shatter by accident.

11   Jenks 1768, p. 87. The book was by James Jenks.

12   Fuller 1662, pp. 270–1. Fuller was a Church of England clergyman who lived 1607/8–61. The winter of 1644–5 was quiet, then the king was defeated at Naseby in June 1645. At Exeter many suburbs were destroyed as preparation for an attack or siege. Fairfax spent the winter at Ottery St Mary, as many in his army were sick, and the king's army fled to Cornwall, so Exeter was exposed, leading to its surrender. This skylark windfall was probably in February 1646.

13   Hawker 1893a, p. 91. This was 8 January 1814.

14   Mayhew 1851, p. 60.

15   Knox 1849, pp. 127, 129–30; Harting 1871, p. 136. Arthur Edward Knox died in 1880. He was an officer in the 2nd Life Guards. He lived in Petworth and later moved to Trotton House at Trotton, near Midhurst.

16   *The Leisure Hour* 9 (1860), p. 572.

17   Dixon 1890, pp. 129–30.

18   *The Sunday Herald Syracuse* 15 February 1891, p. 4. It was published on the same day in her own newspaper, the *Illustrated Buffalo Express (N.Y.)*. See also Reid 1896, pp. 119–22.

19   *The Sunday Herald Syracuse* 15 February 1891, p. 4.

20   *Dundee Courier* 3 March 1890, p. 3.

21   *The Sunday Herald Syracuse* 15 February 1891, p. 4.

22   Hawker 1893b, p. 49 (August 1833, no day given).

23   Andrews 1934, p. 359.

24   Andrews 1934, p. 360. The Swan Inn was demolished, rebuilt in 1889, then destroyed in World War Two.

25   Savory 1920, pp. 262–3.

26   Beeton 1861, p. 500. She was Mrs Isabella Beeton.

27   Dixon 1851, pp. 399–400.

28   Dixon 1851, pp. 400–1.

29   Gray 1871, pp. 133, 261.

30   MacGillivray 1837, p. 186.

31   *The Sporting Magazine* October 1794, p. 50.

32   *The Sporting Magazine* January 1795, p. 173.

33   Farley 1787, p. 8, first published in 1783. Supposedly written by John Farley, from the London Tavern, it is actually a compilation from earlier sources.

34   Raffald 1773, p. 66. She was a remarkable woman, who lived 1733–81. The birds were gutted after roasting and the meat served on the toast, with the butter poured over.

35   Daniel 1802, p. 462.

36   Hawker 1893b, p. 25. This was 2 February 1831.

37 Dixon 1890, p. 132.
38 Wymer 1950a, pp. 58–9; Raffald 1773, p. 168.
39 Meade-Waldo 1908; Ingram 1953; Meiklejohn 1954 (under 'Letters').
    Callahan (2014, pp. 24–5) says that fifty to sixty pots dating from 1500
    to 1850 are known to archaeologists, though these are probably mainly
    from Europe.
40 Parry 1907, p. 199. Thomas Randolph lived 1605–35. This poem was first
    published in 1640.
41 Olina 1684, p. 18 – his book was first published in 1622; Ray 1678, p. 196.
42 MacGillivray 1837, pp. 601–2.
43 Lilford 1895, p. 209.
44 Daniel 1802, p. 502; D'Urban and Mathew 1892, p. 96.
45 Raffald 1773, p. 141; Hartley 1954, p. 205.
46 Berkeley 1854, p. 255; D'Urban and Mathew 1892, pp. 180–1. Berkeley
    rented Beacon Lodge (now demolished) overlooking Christchurch Bay. He
    was accompanied on his expeditions by Stephen Hooper, who was once a
    smuggler, but by then a gamekeeper.
47 Harvie-Brown 1890, p. 262.
48 D'Urban and Mathew 1892, pp. 174–5. Shags are very similar to cormorants,
    but a little smaller and exclusively maritime.
49 Gray 1871, pp. 445–6.
50 Gray 1871, p. 381.
51 Harvie-Brown 1890, pp. 262–3.
52 Pennant 1776, p. 145.
53 D'Urban and Mathew 1892, p. 208.
54 Knox 1849, p. 236.
55 Johns 1862, p. 513.
56 Evans 1903, p. 205. John Caius was a scholar and physician who lived
    1510–73.
57 Bingley 1804, p. 354.
58 Wilson 1842, p. 5.
59 Wilson 1842, pp. 14–15.
60 Wilson 1842, pp. 45–6.
61 Bird 1866, p. 647. She later became Isabella Bishop and lived 1831–1904.
    See Stoddart 1906.
62 Bird 1866, pp. 647–8.
63 Blasco et al 2014; Hansell 1988b, pp. 39–40; McCann 2003, pp. 14–27.
64 Hansell 1988a, 1988b; McCann 2003, pp. 21–2.
65 Waterton 1836, p. 344.
66 Lysons 1810, p. 36. At Brentford in Middlesex, the chapelwardens' accounts
    have several entries recording payments for supplying games held on
    feast days, such as in 1643 when Thomas Powell was paid two shillings for
    pigeon-holes.
67 The Hansells (1988b, p. 25) say they produce two chicks eight to ten times
    a year for about seven years, while the McCanns (2003, p. 21) say five to six
    times a year. Dixon (1851, p. 52) reckoned that they produced chicks nine
    times a year. See also McCann 1991.
68 Torr 1931, p. 277; Maclachlan 1878, p. 278. Peter Vallavine was vicar of

Monkton from 1729 to his death in 1767. John Whaley lived 1710–45. By pigeons, they meant squabs.

69 Jenks 1768, pp. 13, 222. Pigeon pies became a traditional dish in the United States in the nineteenth century.

70 Hansell 1988b, p. 82; McCann 1991; McCann 2003, pp. 33–5.

71 Turkeys were supposedly introduced to England in 1550 by William Strickland, but probably came earlier than that. They may have been called turkeys because they were imported by Middle Eastern merchant ships or because of their resemblance to the guinea fowl that were imported to Europe through Turkey and so were called turkey fowl (Kear 1990, p. 40).

72 Jameson 2003, p. 39. This was 24 June 1791.

73 See Serjeantson 2002, pp. 40, 42.

74 Jameson 2006, p. 52. This was Saturday 29 September 1798.

75 Gurdon 1893, p. 132.

76 Brabourne 1884, pp. 175–6. Jane Austen was staying at Godmersham Park in Kent, the home of her wealthy brother Edward. Perhaps every author should eat goose on this day.

77 Serjeantson 2002, p. 42; Tyrer 1968, p. 188.

78 Riley 1868, p. 644 (4 Henry V A.D. 1416).

79 Pennant 1772, p. 9. Also spelled cagg maggs.

80 *The Leisure Hour* 15 (1863), p. 807.

81 Hartley 1979, p. 232; Serjeantson 2002, p. 51; Jones 1989, p. 204.

82 *Country Life* 1 March 1946, p. 404. She signed the letter as Madge S. Smith, Soberton House, Soberton by Southampton. She and her younger sister Cicely Fox Smith both wrote fiction and non-fiction.

83 *The Leisure Hour* 15 (1863), p. 806. Convivial goose clubs also met to eat and drink (see *North Devon Magazine* 1824, pp. 250–1).

84 *The Leisure Hour* 15 (1863), p. 807.

85 *The Leisure Hour* 15 (1863), p. 807.

86 *The Leisure Hour* 15 (1863), p. 807. Albatrosses rarely venture into the North Atlantic.

87 Mundy 1885, p. 366. James Frampton became notorious for his part in the Tolpuddle Martyrs in 1834.

88 Rodd 1880, pp. 147–8; *The Zoologist* 14 (1856), p. 5160.

89 Harting 1866a, pp. 149–53 for a discussion of these plovers. He was Robert Henry Mitford, son of John Mitford, editor of the *Gentleman's Magazine*. R.H. Mitford lived 1816–1905. See also Baines 1890, pp. 125–6.

90 *The Zoologist* 3 (3rd series, 1879), p. 173. Robert Charles Latham Browne lived in the vicarage at North Curry. He was thirty-three years old in 1878 and married Florence Chard the following year.

91 Sneyd 1847, pp. 10–11. This was probably written 1496–7, though it might have been the work of the secretary of Francesco Capello, the earliest Venetian ambassador on record to visit England in 1500.

92 Madden 1831, p. 355.

93 Kear 1990, pp. 166–70; Gurney 1921, pp. 109–15; Stevenson 1890, pp. 88–111.

94 Winstanley 1984, p. 12. This was 28 January 1780.

95 Clark Kennedy 1868, p. 119.

ELEVEN: SICKNESS AND DEATH

1   *The Newcastle Weekly Chronicle* 11 February 1899, p. 7.
2   Shaw 1776, p. 311.
3   Markham 1615, p. 23.
4   Ray 1678, p. 282; *Oswestry Advertizer* 18 September 1872, cited in *Bye-Gones* 1872, p. 88.
5   Ray 1678, pp. 211–12.
6   Harting 1867.
7   *The Zoologist* 1 (2nd series), 1866, p. 523. George Alexander Louis Lebour was educated at the Royal School of Mines and lived 1847–1918. See also Duffin 2013.
8   Burton 1676, p. 230. He lived 1577–1640, in Oxford, as a writer, scholar, librarian and clergyman.
9   Pliny the Elder, *Naturalis Historia*, book 30, chapter 27; Ray 1678, p. 211.
10 Shaw 1738, p. 306. Shaw lived 1694–1751.
11 Mitchell 1863, pp. 273–4. He lived 1826–1909 and was a commissioner in lunacy from at least 1863.
12 Mitchell 1863, p. 276; 1881, pp. 163–4.
13 Mitchell 1863, p. 274. The man most likely lived in a single-storey dwelling with an earthen floor. Grigor was born 1814 in Elgin and retired to Nairn.
14 Pennant 1810, p. 15. Socrates possibly had temporal lobe epilepsy. Pennant's tours of Wales were in the early 1770s, published in three volumes, 1778–83. He died in 1798, and a new edition appeared in 1810.
15 Pennant 1810, p. 16.
16 Sikes 1880, p. 350.
17 Farrer 1876, p. 54.
18 Black 1883, p. 34–5. He lived 1857–1932.
19 Markham 1631, p. 10; 1615, p. 9 (where he refers to a little pigeon rather than a live pigeon).
20 Wheatley 1893, p. 307. Catherine of Braganza actually lived to 1705.
21 Wheatley 1896, p. 288. This was 21 January 1667/8.
22 Heber 1828, p. 290. Taylor mentioned this while discussing the consumption of 'living blood' as opposed to slaughtered meat.
23 Pratt 1801, pp. 167–8.
24 Nicholson 1890, pp. 139–40.
25 *Gentleman's Magazine* 57 (1787), p. 719 fn. The dog was referred to as Tray. Polwhele lived 1760–1838. The poem was written in 1786 and published in 1787.
26 Polwhele 1826, p. 607 fn.
27 Blagrave 1671, p. 159. He lived 1610–c.1682.
28 *New Statistical Account* 15 (1845), p. 125.
29 Black 1883, p. 158. William George Black was an antiquary and lawyer from Glasgow, who lived 1857–1932.
30 Ray 1678, p. 211. Squinancy, also known as quinsy, is a throat infection, as is inflammation of the uvula. Quartan ague was malaria.
31 Markham 1615, pp. 23–4, and 1631, p. 57.
32 Harris 1917, pp. 83–4. James Rendel Harris (1852–1941) was a biblical scholar

and owned this manuscript. It is now in the Cadbury Research Library at the University of Birmingham.

33 Ray 1678, p. 211.
34 Markham 1615, p. 26. 'Plaister' was a layer of healing material applied to the wound or pain and often made according to a family recipe. Some of them were small adhesive plasters. See Adkins and Adkins 2013, p. 304.
35 Anon 1677, p. 51.
36 Ray 1678, p. 187.
37 A/BTL/2/37, Somerset Archives and Local Studies, 8 October 1812.
38 Markham 1631, p. 17.
39 Harland and Wilkinson 1867, p. 121.
40 Martin 1753, pp. 31–2.
41 Swainson 1885, p. 145.
42 Fox 1827, p. 300. The letter was dated 10 December 1826, written at his property of Coupland Castle in Northumberland.
43 Swainson 1885, p. 145; *Oswestry Advertizer* 18 September 1872, cited in *Bye-Gones* September 1872, p. 86.
44 Nelson 1907a, p. 204; Dyer 1878, p. 154.
45 *The Leisure Hour* 1877 p. 286. See Nicholson 1890, p. 131, and Armstrong 1970, p. 201.
46 Sikes 1880, pp. 212–13.
47 Bourne 1725, pp. 70–1.
48 Roberts 1831, p. 62. She was born in 1788 and moved with her parents to Painswick in 1790, but went to London when her father Daniel Roberts, a merchant of London, died in 1811.
49 Nelson 1907a, p. 294.
50 Latham 1878, pp. 54–5.
51 Hawker 1893b, p. 55; *Gentleman's Magazine* 50 (1834), p. 452. Leech's burial was on 20 March 1834. He was seventy-nine or eighty years old.
52 Hawker 1893b, p. 55.
53 Hawker 1893b, p. 55 fn.
54 Winstanley and Rose 1928, pp. 173, 174. Mathry is 6 miles south-west of Fishguard.
55 *Notes and Queries* 12 (3rd series), 1867, p. 477 (by G.A.C.).
56 Swainson 1885, p. 55.
57 Mudie 1835a, pp. 268–9.
58 Alston 1867, p. 1007. Edward R. Alston was born in 1845 in Stockbriggs, Lesmahagow, in Lanarkshire, and died very young in 1881.
59 Wilkie 1916, p. 60. He died in 1838, but William Henderson thought he had died much earlier in India.
60 Wilkie 1916, p. 81.
61 Lee 1887, p. 375. He also published *Historical Notes of Haydon Bridge* (1876) and was living at New Benwell in 1887.
62 Henderson 1866, p. 94 fn.
63 Lee 1887, p. 375. East Brokenheugh was 1½ miles north-east of Haydon Bridge.
64 MacGillivray 1837, p. 509.
65 Latham 1878, p. 9.

66 Hunt 1865, pp. 238–9.

67 McKerrow 1910, p. 172. This piece was in *Christ Teares over Ierusalem*. Nashe lived 1567–1601. The church was destroyed in the Great Fire of London in 1666.

68 *Lincoln Herald* cited in *The Field* 22 September 1860, p. 16. Mr Hackwood lived in Church Close and had the cormorant preserved.

69 Swainson 1885, p. 143; *The Ulverston Mirror* 29 September 1860, p. 3; *Illustrated London News* 6 October 1860, p. 307; *Lincolnshire Standard and Boston Guardian* 21 April 1928, p. 9.

70 Nicholson 1890, p. 130; Fowler 1909, pp. 292–3. Mary Wilhelmina Emily Fowler was born in 1867 at Wakefield and grew up there, but later moved to Hampshire, where she died, at Odiham, in 1944.

71 *Notes and Queries* 3, 28 June 1851, p. 517. Eastwood lived 1823–64.

72 Wilkie 1916, p. 81; Sternberg 1851, p. 156.

73 *Newcastle Weekly Chronicle* 11 February 1899, p. 7; Ross et al 1877, p. 43.

74 Wymer 1950a, p. 250.

75 Henderson 1866, p. 92.

76 Henderson 1866, pp. 92–3 fn. Baring-Gould (1834–1924) was at the college c.1856–64, and then went to Horbury near Wakefield (Graebe 2011).

77 Capern 1865, p. 123. He lived 1819–94.

78 Capern 1870, p. 279.

TWELVE: MYTHS AND MAGIC

1 Nelson 1907a, pp. 46–7.

2 Merrifield 1969, p. 104.

3 Bourne 1725, pp. 37–8. The parochial chapel of All Saints in Newcastle upon Tyne was also known as All Hallows church. Bourne lived 1694–1733. His book was included and expanded by Brand in his *Observations on Popular Antiquities* (1810).

4 Alexander 1951, p. 1030, spoken by Horatio.

5 Hunt 1881, p. 187. The white rock at Looe was at the harbour's edge in Saunder's Lane at Looe and was submerged at each high tide. This was in the 1881 updated edition of Robert Hunt's book, but it was not in his earlier 1865 version. See also Grinsell 1976, pp. 56–9.

6 Addy 1895, pp. 55–6. Hollow Meadows is a hamlet on the Manchester Road, 7 miles from Sheffield. Another cock-crowing stone in Derbyshire was at Ashover, 8 miles from Curbar. Sidney Oldall Addy studied classics at Oxford and became a writer and solicitor. He lived 1848–1933.

7 Buckler 1751, pp. 15–16. The first printing of this book was in 1750. The original account came from *Historia Brevia* of c.1440, by Thomas Walsingham, a monk at St Alban's. The church is known today as the University Church of St Mary the Virgin.

8 Buckler 1751, p. 18. The date was February 1437–8, but the centenary celebrations take place in January.

9 Buckler lived 1716/17–1780.

10 Hone 1832, p. 46. The terms 'swapping' and 'schwoppinge' mean very large. The procession takes place on 14 January.

11 The priests were known as augurs. By the first century BC and throughout the Roman Empire, divination was done by interpreting the entrails of sacrificed animals rather than by observing birds. Adkins and Adkins 1996, pp. 23–4.

12 Mundy 1830, p. 375. It was the Battle of the Saintes or Saints. The bantam-cock may have been on board Rodney's flagship, HMS *Formidable*, but many ships fired into the *Ville de Paris*, which surrendered to the *Barfleur*.

13 Knight 1861, pp. 116–17.

14 Mackenzie 1913, p. 167.

15 Romney 1984, pp. 82–3. Captain William Thornton was of Thornville Royal, and he raised the force at his own expense after Prestonpans, with Blind Jack Metcalf from Knaresborough as the fiddler. They returned home after the Battle of Falkirk in January 1746, due to billeting disputes, so would not have been at Culloden (Reid 2012, p. 45). John Spooner, famous for his hunting skills, died, aged ninety-five, on 22 November 1819 and was buried at Hartwith. The sparrow battle was on 16 April 1745.

16 Cam 1916.

17 Wright 1863, p. 150. Africa probably meant merely a distant place – Norway was more likely. Gerald of Wales, who lived c.1146–c.1223, was archdeacon of Brecon in Wales.

18 Wright 1863, p. 150. Thomas Wright lived 1810–77. See also Breeze 2011, pp. 178–9.

19 Wright 1872, p. 80 fn.

20 Drinkwater 1785, p. 265; Adkins and Adkins 2017, p. 316.

21 Nicolay 1907, pp. 241, 364.

22 Nigg 2016, p. 138.

23 Alexander 1951, p. 785. Cranmer's speech is in act 5, scene 5, at the end of the play. For the phoenix, see Nigg 2016.

24 Cox 1915; Oman 1930.

25 Smalley Law 1922, p. 22.

26 Spraggon 2003, p. 112.

27 Evans 1816, p.314. Reverend John Evans kept a school in Bristol and later Euston Square in London, where he died in 1832.

28 *The Sun* 30 August 1802, p. 4.

29 *The Sun* 30 August 1802, p. 4.

30 Sampson 1904, p. 162. The poem is from John Webster's play *The White Devil*, act 5, scene 4.

31 *The Book of Days* vol. 1 (1863), p. 678, by C.W.J. of Suffolk.

32 *Notes and Queries* 29 February 1868, pp. 193–4. The farrier was summoned for veterinary advice. Charles Clough Robinson was then living as a boarder at 6 St James Street, Leeds. He was born in July 1836 in Leeds and was a self-educated warehouseman. By 1901 he was living in London, described as 'editor author'.

33 *Notes and Queries* 4 April 1868, p. 329. Federer was a schoolmaster and prolific writer who came to England from Switzerland in 1857.

34 *The Book of Days* vol. 1 (1863), p. 678, by C.W.J. of Suffolk; *Newcastle Weekly Chronicle* 11 February 1899, p. 7.

35 Tweddell 1875, p. 2 'And if you've nothing new to put on, that is a fine to

do, For the crows are sure to find it out and will soil your old clothes more.'
She wrote under the nom-de-plume of Florence Cleveland. Her poem is
called 'T'Awd Cleveland Custums'. See also Nelson 1907a, p. 253.

36 Nelson 1907a, pp. 242, 253–4.

37 Gurdon 1893, p. 133; Sternberg 1851, pp. 160–1.

38 Clare 1821, p.125. He lived 1793–1864 and was from Helpston in
Northamptonshire.

39 Winstanley and Rose 1928, pp.173–4. The male student was talking about
the neighbourhood of Mathry.

40 *Oswestry Advertizer* 18 September 1872, cited in *Bye-Gones* (1872), p. 86.

41 Waterton 1838, p. 233.

42 Latham 1878, p. 9. This was in the 1860s or earlier at Fittleworth.

43 Armstrong 1970, p. 114.

44 Henderson 1866, p. 95. He was the Reverend Canon Humble of St Ninian's
Cathedral in Perth.

45 Savory 1921, p. 257.

46 Nelson 1907a, p. 231.

47 Nelson 1907a, p. 231. This is also recorded in *Newcastle Weekly Chronicle* 11
February 1899.

48 Hawker 1893b, p. 264.

49 Henderson 1866, p. 96.

50 *The Times* 13 December 1825, p. 2.

51 Savory 1921, p. 257.

52 *Notes and Queries* 9 (3rd series) 20 January 1866, p. 59. The writer was 'A.A'.
They were a few miles beyond Reading, on the Berkshire side of the river.

53 *Notes and Queries* 9 (3rd series) 3 February 1866, p. 109.

54 *Notes and Queries* 9 (3rd series) 3 March 1866, p. 187.

55 D'Urban and Mathew 1894, p. 231.

56 Wood 1995, pp. 201–3.

57 Merrifield 1987, p. 129, fig. 41. The shoes were dated by the shoe expert
June Swann (2015, p. 128).

58 Smith 1874, p. 12. Porch House was built about 1480.

59 Howard 1951, p. 149; Merrifield 1987, pp. 129–30. Hay Hall, Tyseley, is now
in a huge industrial area. It is owned by Reynolds Tube Co Ltd and used as
offices. This deposit was found in refurbishment work in 1948.

60 Howard 1951, p. 150.

61 Easton 2014, p. 29; Easton 2015, pp. 157–8; Davies and Easton 2015, p. 228.
Timothy Easton has studied and published several examples of 'spiritual
middens' in houses in Suffolk.

62 Wood 2011; *The Times* 29 December 2014, p. 23; Hutton 2015, pp. 4–5. The
excavator is Jacqui Wood.

63 Black 1883, p. 117.

64 Nicholson 1890, pp. 91–2. This area is now Humberside. Dr Wood was the
medical officer from about 1885 to 1900. He later moved to Robin Hood's
Bay and died in 1908 aged eighty-eight.

65 *Western Chronicle* 28 January 1887, p. 8.

66 *Newcastle Courant* 22 January 1869, p. 2. See also Davies and Easton 2015, p.
215. Many newspapers carried the story.

67 *Folklore* 54 (1943), p. 408. It was reprinted in *The Cornishman* on 12 November 1931. The grandson was L. Pengelly from Bowenlands (probably Bowerlands) of Newton Abbot, and the ritual must have been practised in the first half of the nineteenth century.

THIRTEEN: A NATIONAL SPORT

1 *Kentish Gazette* 23 April 1799, p. 4.
2 The diary is called 'Dr Taylor's Book' after the donor of the manuscript to Shrewsbury School which still owns it (MS Mus X.31). The chronicler was possibly a draper. See a discussion in Walsham 2018 and excerpts in Leighton 1880.
3 See Thomas 1983.
4 White 1837, p. 47. His letter was written 9 September 1767.
5 D'Urban and Mathew 1892, pp. lxxvi, 368.
6 Carnival means literally 'farewell to meat', as no meat could be eaten during Lent.
7 *Folk-Lore Record* 2 (1879), p. 213.
8 Anon 1772, pp. 45–6.
9 Wilkie 1916, p. 122. Domenie is a slightly contemptuous name for a minister, or alternatively a schoolmaster.
10 Miller 1869, p. 49. He lived 1802–56, was probably at this school around 1812–15 and was a geologist, folklorist and man of faith.
11 Miller 1869, p. 50.
12 Hingeston 1703, p. 23; Fox 1874, pp. 44–5.
13 Leighton 1880, p. 338.
14 van Muyden 1902, p. 281.
15 van Muyden 1902, p. 282; Hodgson 1910, p. 62, from the diary of Jacob Bee of Durham (1636–1711).
16 Bruce 1904, p. 82. Lord Redesdale lived 1748–1830 and was probably talking of London.
17 Jobey 1992, pp. 3–6.
18 Gilbey 1912, pp. 40–1; Jobey 1992, p. 6.
19 Burton 1888, p. 131.
20 Leighton 1880, p. 338.
21 van Muyden 1902, p. 281. The cockpit is unnamed. See also Kielmansegge 1902, p. 241.
22 van Muyden 1902, pp. 281–2.
23 van Muyden 1902, pp. 280–1.
24 Waugh 1855, p. 45 fn. He lived 1817–90. The alehouse is the Crown & Shuttle on the Rochdale Road. Waugh was living in Manchester when this took place, but was revisiting places around Rochdale that he knew as a child in the early nineteenth century.
25 Burne 1886, p. 590. Charlotte Burne published Jackson's notes on proverbs, along with much of her own work. See Ashman and Bennett 2000, p. 200.
26 Brand 1813, p. 482. This was written some years earlier.
27 Hingeston 1703, pp. 17–18; Fox 1874, pp. 39–40.
28 Blencoe and Lower 1859, p. 207.

29 Hackwood 1907, pp. 245–8.
30 *The Monthly Chronicle of North-Country Lore and Legend* June 1887, p. 188.
31 *Newcastle Chronicle* 20 April 1765, p. 3. The Flesh Market in Newcastle was on the site of the medieval Cloth Market, and in 1808 a purpose-built Butcher Market on Grey Street replaced it.
32 Hodgson 1910, p. 202. This was in October 1757 and is from the diary of Thomas Gyll.
33 Leighton 1880, p. 338.
34 Tyrer 1968, pp. 221–2.
35 Nicholson 1890, pp. 98, 149.
36 Lackington 1794, pp. 50–1.
37 Kielmansegge 1902, pp. 241–2 This was 1 February 1762. He came to London with his brother Carl to see the coronation of George III. Friedrich lived 1728–1800.
38 *The Monthly Chronicle of North-Country Lore and Legend* (March 1887), p. 43. The piece was written by an anonymous writer known only by the initials M.H.
39 Gilbey 1912, pp. 101–2.
40 Owen 1896, p. 312.
41 Owen 1896, pp. 267, 312; *Bye-Gones* September 1872, p. 88.
42 *The Literary Panorama* 1 December 1806 (1807), p. 642. Roberts died in November 1806. The suspicion was that he had never travelled beyond his own parish.
43 Randall 1862, pp. 203–4. He lived 1810–1910. Joe Crow was probably a nickname.
44 Hackwood 1907, p. 277. George Thomas Lawley lived at Priestfield House in Priestfield, where he died. He was very interested in local history and its cockfighting history.
45 Palmer 1976, p. 117 (for Warwickshire); Hackwood 1907, p. 282 (for the Black Country).
46 Hackwood 1907, p. 282.
47 Waugh 1855, pp. 110–11.
48 Strutt 1810, p. 251, which was first published in 1801. Strutt died in 1802.
49 Strutt 1810, p. 251. The incident must have been about 1760.
50 Lysons 1810, p. 386.
51 Tyrer 1968, p. 163 (17 February 1708); Tyrer 1972, p. 6 (1 March 1720).
52 Hackwood 1907, p. 292. It was dated 27 February 1754, which was a Wednesday, so he wrote this the day after Shrove Tuesday.
53 Brewster 1829, pp. 260, 316. Skelly became vicar in 1742 and died in 1772, Brewster was curate there from 1776.
54 *Pope's Bath Chronicle and Weekly Gazette* 10 February 1763, p. 4.
55 Grose 1785, p. 176.
56 *Newcastle Courant* 13 May 1758, p. 3.
57 Grose 1785, p. 112.
58 *Derby Mercury* 7 October 1840, p. 4.
59 Brand 1813, p. 379 (which was published after his death). These would have been red squirrels.
60 Armstrong 1970, pp. 141–66; see Gardner-Medwin 1970 for American versions.

61 Borrer 1891, p. 80. He was the son of the famous botanist William Borrer. He lived 1814–98.
62 Moore 1891, p. 139.
63 Aubrey 1784, p. 65. John Aubrey lived 1626–97.
64 Waldron 1744, p. 97. He lived c.1690–1730. He was a commissioner for trade on the Isle of Man for the British government.
65 Bullock 1816, pp. 370–1. Hannah Ann Brooks married Stanley Bullock in 1790 at Bloomsbury, and they moved to the Isle of Man and farmed, leaving the Isle of Man in 1815.
66 Thomas 1983, p. 160; Gilbey 1912, p. 11. The 1833 Act put an end to houses for cockfighting within 5 miles of Temple Bar, London (3 Wm. 4 cap. 19). The Cruelty of Animals Act of 1835 made it illegal to keep a cockpit in England and Wales (5 & 6 Wm. 5 cap. 59), while the Prevention of Cruelty to Animals Act 1849 (12 & 13 Vict. cap. 92) made cockfighting illegal in England and Wales.
67 Miller 1869, p. 50.
68 Bullock 1816, p. 371.
69 Harrison 1869, pp. 151–2. William Harrison lived 1802–84.

FOURTEEN: BY-PRODUCTS

1   Walter 1922, p. 118.
2   Serjeantson 2002, p. 43.
3   Pennant 1772, pp. 8–9.
4   Hutton 1804, pp. 315–16. Hutton was a dissenter, historian and bookseller, 1723–1815.
5   Serjeantson 2002, p. 43. The Latin word *penna* (or *pinna*) means a feather and the plural *pennae* a wing. The word 'pen' has long been used for a female swan (and 'cob' for a male), and James Edmund Harting (1895) thought it was from her habit, when sailing with her brood, of arching her wings above her back to display her feathers. The Latin *pluma* was a small, soft feather, such as that used in the plumage trade (millinery). From the medieval period, plumers traded in feathers (Serjeantson 2002, pp. 43–4).
6   Serjeantson 2002, p. 43. *The Sporting Magazine* 4 (1794), p. 214 mentioned bustard quills.
7   MacGillivray 1837, p. 551.
8   Hutton 1804, p. 300. The poem is dated 22 June 1796.
9   Anon 1677, pp. 16–17.
10 Torr 1923, p. 110.
11 Dixon 1890, pp. 27–8.
12 *The English Mechanic and World of Science* 13 (1871), p. 96.
13 *The Leisure Hour* 1 September 1869, p. 571.
14 *The Leisure Hour* 1 September 1869, p. 571.
15 *The Leisure Hour* 1 September 1869, p. 572.
16 Blencowe 1849, p. 255. John Dudeney lived 1782–1852.
17 Hutton 1804, p. 302.
18 Dixon 1890, p. 28.
19 *Warragul Guardian and Buln Buln and Narracan Shire Advocate* 23

November 1888, p. 1. Warragul is 60 miles south-east of Melbourne in Australia. Joinville-le-Pont is now a suburb of Paris.

20 Hawker 1816, p. 167.

21 Clutterbuck 1898, p. 29.

22 *The Sporting Magazine* 4 (1794), p. 214; Palmer 1949, p. 131.

23 A/BTL/2/14, Somerset Archives and Local Studies, 6 January 1802.

24 *Encyclopedia Londiniensis* 21 (1826), p. 567.

25 *The Liverpool Standard, and General Commercial Advertiser* 31 May 1836, p. 1; *Glasgow Evening News* 5 December 1889, p. 2; *Dundee Evening Telegraph* 22 May 1899, p. 4 (part of a 'Ladies Column'); *The Cardiff Times and South Wales Weekly News* 26 November 1898, p. 1, report by 'Coquette' (the word 'panne' was a type of crushed velvet).

26 Gladstone 1919, pp. viii–ix.

27 Serjeantson 2002, p. 43; Dixon 1890, p. 27; Fox 1827, pp. 252–3, where the surname is given as Brickford. Mewburn salvaged much of the bird three weeks later, including one wing, and it ended up in the museum at Newcastle.

28 Daniel 1802, p. 512.

29 Markland 1727, p. 4.

30 Kear 1990, p. 51; Serjeantson 2002, p. 43 (who says that one goose can yield 150 grams of down feathers).

31 Children's Employment Commission 1842, f. 257, nos 707 and 708. The manager was James Halkinston of Messrs Blyth, Hamilton and Hughes in Elizabeth Street, off Hackney Road. In March 1840, their premises in Upper Thames Street had a devastating fire.

32 Anon 1808, p. viii.

33 Wilkie 1916, p. 84.

34 *Notes and Queries* 10 April 1852, p. 341, by a correspondent called 'Nedlam'.

35 Brontë 1848, p. 107. The novel was published in 1847.

36 *Notes and Queries* 8 (4th series, 1871), p. 470. Grey lived 1819–72.

37 Newman 1942, p. 109 (the paper was read on 26 July 1926). Leslie F. Newman was an agricultural chemist with a keen interest in folklore. For Sussex, see Latham 1878, p. 11. This was from Charlotte Latham in Fittleworth, West Sussex, before 1868.

38 Worsley 1781, p. 194. The book was written by Sir James Worsley, expanded on his death by his son Thomas and published by Richard.

39 Wilson 1842, p. 26.

40 Kearton 1897, p. 119.

41 Shrubb 2013, pp. 88–9; Kear 1990, pp. 170–2.

42 *The Morning Post* 30 December 1850, p. 1.

43 Lord 1866a, p. 183.

44 Lord 1866a, p. 183.

45 *The Hendon and Finchley Times* 20 November 1908, p. 4; *Sheffield Evening Telegraph* 20 January 1909, p. 8.

46 Creighton et al 2020, p. 183. The burials were associated with the thirteenth-century Dominican friary.

47 From the ballad 'The King's disguise, and friendship with Robin Hood', in which the king is present in disguise, so the verses are ironic. See *Robin*

Hood's Garland, c.1775, pp. 74–6, no. 25, published at Leeds. Garlands were first published from the seventeenth century and repeatedly reissued, with additions and removals of individual ballads. This one was a late addition around 1740–60. See Basdeo 2017, pp. 19, 34–42.

48 Williams 1850, pp. 52–4.

49 *The Boy's Hand Book of Sports, Pastimes and Amusements* 1863, p. 66. These were wooden cotton reels. See also Kear 1990, p. 49.

50 Dixon 1890, p. 27; Simeon 1860, p. 56; Hawker 1893a, p. 94. This was 12 April 1814.

51 Walton 1653, pp. 110–12. The book was reissued and revised many times. He actually wrote 'ingenuous' for 'ingenious'.

52 Walton 1653, p. 117.

53 Nelson 1907b, pp. 568–9; Anon 1804, p. 60. The traveller, probably a geologist, was about to go to the Continent, but changed his plans with the outbreak of war and was at Skiddaw in July 1803.

54 Yarrell 1843, pp. 397–8. Thomas Coulthard Heysham was a keen naturalist, who lived 1791–1857.

55 Nelson 1907b, p. 568.

56 *Manchester Times* 26 May 1888, p. 7. Gilpin lived at Over Staveley, Cumberland.

57 Knapp 1829a, p.192; Gray 1871, p. 188.

58 It was also traditional for the first sportsman each season to shoot a woodcock to put a feather in his hat. For feathers in military headgear, see *Notes and Queries* 22 April 1854, p. 378, and Dillon 1896.

59 Dillon 1896, p. 138. The master of the armoury was Sir George Howard (or Howarde).

60 Sternberg 1851, p. 70.

61 *The Field* 26 March 1864, p. 216.

62 Lord 1866b, p. 464. See also Shrubb 2013, pp. 192–202.

63 Lord 1866b, p. 464.

64 Lord 1866b, p. 464.

65 Cordeaux 1867, pp. 1009–10.

66 Yarrell 1882–4, p. 653 (Yarrell died in 1856, and this edition was revised and expanded by Saunders).

67 Yarrell 1882–4, p. 653.

68 The SPB moved to London in 1891.

69 RSPB 1911, pp. 16, 39.

70 Shrubb 2013, p. 202.

71 Mundy 1885, p. 2.

72 *Gentleman's Magazine* 58 (1788), p. 542.

73 *Notes and Queries* 9 (1866), p. 109 (for 3 February).

74 *Notes and Queries* 9 (1866), p. 187 (for 3 March). Jewitt was an antiquary, illustrator and prolific writer, who lived 1816–86.

75 Nicholson 1890, pp. 132–3.

76 *Newcastle Weekly Chronicle* 11 February 1899, p. 7.

77 Nicholson 1890, p. 3.

78 Isham 1875, p. 84. The original diary was written in Latin.

79 Colles 1887, pp. 2–3. Colles was in the High Street in 1884 and at Clifford House, Wellington, in 1894.

80 Colles 1887; Frazer 1887.
81 *The Folk-Lore Journal* 5 (1887), pp. 351–6. See Wingfield 2010 and 2011. The Pitt Rivers Museum accession number is 1911.32. A sewel was a length of rope to which feathers were secured or some other light material such as bark, then usually suspended above the ground to frighten deer away. Although Sir John Popham (who may have owned the farmhouse) died of sickness aged seventy-two, one legend claims that he was killed in 1607 while hunting near Wellington and fell into a bottomless pit. His wife's prayers enabled him to crawl towards his tomb in the church, 3 miles away, at a 'cockstride' each year, but one farmhouse was so disturbed that a white witch was consulted, sending him back to the pit (Westwood and Simpson 2005, p. 654).
82 Kear 1990, p. 250. Williams 1922, p. 104 says that the Romans made musical pipes out of the leg bones of herons.
83 Ward 1880, pp. 45, 46.
84 Ward 1880, pp. 45–6.
85 Gladstone 1910, p. 472.
86 Martin 1753, p. 57.
87 Nicholson 1890, p. 140; Hansell 1988b, p. 31.
88 Paton 1940, p. 51, taken from *Manx Notes and Queries* 1904, p. 44, note 92, derived from information given to the German Charles Roeder at Manchester provided by the Manx farmer Edward Faragher (1831–1908). Chemerly was urine collected from chamber pots.
89 Saltpetre was imported from Germany and the Low Countries. A deal was done for a short time with Morocco in about 1577, to exchange munitions for saltpetre (Cressy 2011, pp. 84–5). In Elizabeth I's reign, saltpetre was imported from India and Persia into Europe, and England obtained most of it through Germany and the Low Countries. Nowadays, it is known as potassium nitrate, a crystalline salt, KNO_3. The brimstone was largely obtained from Italian volcanic areas (Cressy 2011, p. 74 fn).
90 The manufacture of saltpetre had been known from the early sixteenth century, but it was ad hoc. Bovill 1947, p. 183; Giuseppi 1905. Cressy 2011, p. 78 says it had reached England by 1540. A deal was done with a German captain (Gerrard Honrick) for £300 to give them the recipe and train people.
91 Giuseppi 1905. Saltpetremen started about 1590 and were employed by the commissioners of saltpetre and gunpowder on behalf of the Ordnance Office. The role petered out about 1640 when the industry was opened up, no longer a monopoly.
92 15 November 1634, *Calendar of State Papers, Charles I, Domestic*, vol. 277, no. 52; Giuseppi 1905, p. 308.
93 Giuseppi 1905, p. 308. 3 June 1637, *Calendar of State Papers, Domestic, Charles I, 1637*, vol. 361, June 1–18, p. 187. The collapse occurred in March 1637. East Knoyle was known then as Knoyle Magna.
94 Cressy 2011, pp. 78, 107–8.
95 Dixon 1851, p. 51; Campbell 1774, p. 318 fn. He was a historian, 1708–75, and trained as a lawyer.
96 Davy 1813, p. 258. His book was published in many languages and became a bestseller. The word guano came from the Andean indigenous language Quechua, which refers to any form of dung used as an agricultural fertiliser.

97   Beatson 1816, p. 22 fn.
98   The French scientist Jean-Baptiste Boussingault did research on guano and other manures. For the history of guano in Peru, see the extensive research presented in Cushman 2013.
99   Johnson 1843. He lived 1799–1878. Guano was used mainly in Britain for turnips and other root crops for animal feed, leading to a decrease in crops such as oats for animal feed.
100 *The Morning Chronicle* 11 March 1850, p. 5. The islands are generally spelled Chincha, but the seaman spelled it as Chinqua. Other islands, such as those off the coast of South Africa, also mined guano, and it became known as a white gold rush.
101 *The Morning Chronicle* 11 March 1850, p. 5.
102 *The Morning Chronicle* 11 March 1850, p. 5.
103 *The Leisure Hour* 15 (1866), pp. 735–6. See also Cushman 2013, p. 55 for coolie labourers from 1847 to 1874. The correspondent, probably a merchant, vowed never again to sail in a vessel associated with the guano trade.
104 *The Morning Chronicle* 11 March 1850, p. 5.
105 Mathew 1884; 1894, p. 112. The friend was Mr Mortimer Probert of St David's, an enthusiastic oologist. In 1894, Mathew described 'cartloads' of cockles ploughed in as manure.
106 Wilson 1842, pp. 27–8.
107 Gray 1871, p. 502, report to Gray by Edward Charles Coulis Stanford, who was born in Worthing as Edward Coulis, the son of a chemist and pharmacist, lived 1837–99.
108 D'Urban and Mathew 1892, p. 401. Robert Mudie (1835b, p. 391) refers to storm petrels being converted by the Faroese into lamps by drawing a wick through them.
109 Buckland 1879, p. 137.
110 Worsley 1781, p. 272.
111 Hawker 1893a, pp. 293–4. This was 17 February 1826.
112 *The Graphic* 4 August 1883, p. 131.

FIFTEEN: AT WORK

1   *Nottingham Evening Post* 13 May 1936, p. 7.
2   Because human passengers are carried on transport, some confusion arises between carrier and passenger pigeons. The French word 'passager' means migratory, and millions of passenger pigeons, also known as 'Wild Pigeons', once existed in North America, but were hunted to extinction more than a century ago.
3   Dixon 1851, p. 129.
4   Dixon 1851, p. 127.
5   Loeb Classical Library, Pliny the Elder *Natural History*, LCL 353: 362–363, Book 10, paragraph 53.
6   Russell 1756, pp. 66–7. The earliest carrier pigeons were used for sending messages, but carrier pigeons became so selectively bred that this function ceased.

7   In 1870, the book *Voyages aériens* was published, and an expanded English version of the book, *Travels in the Air*, appeared in 1871.

8   France declared war in July 1870 and the defeat at Sedan was in September. See Horne 1965, p. 258 for the move to Bordeaux.

9   Holmes 2013, pp. 291–3.

10 Allatt 1886. He was Henry Thomas William Allatt, born in 1847 and joined the Duke of Cornwall's Light Infantry (46th Foot) in 1866.

11 *Daily Telegraph* 27 February 1871, p. 5. Microphotography was a process devised over three decades earlier, in 1839, and collodion microfilm was first used in England in 1852. See Luther 1950 and Holmes 2013, pp. 284–8. The initial microphotography was printed on small sheets of thin paper, but they were too heavy for the pigeons, so collodion was used instead.

12 Allatt 1886.

13 *Daily Telegraph* 27 February 1871, p. 5. See Davidson 1990, pp. 379–83 for photomicrography and electricity.

14 Allatt 1886 says that there were nine in September, twenty-one in October, twenty-four in November, thirteen in December, three in January and three in February, a total of seventy-three, plus several others that had no dispatches.

15 Whitehurst 1875, p. 269.

16 Whitehurst 1875, pp. 80–1.

17 *The Beverley Recorder and Independent* 15 January 1916, p. 5. The piece is by a 'special correspondent'. See also Johnes 2007.

18 The magpies were black-and-white pigeons, named after wild magpies. See Lyell 1881.

19 *Country Life Illustrated* 10 April 1897, p. 369.

20 *The Beverley Recorder and Independent* 15 January 1916, p. 5. See also Johnes 2007.

21 Allatt 1886. This was at the United Service Institution talk in London.

22 Dixon 1890, p. 36.

23 Wilson 1899, p. 222.

24 Wilson 1899, p. 223.

25 Wilson 1899, p. 223.

26 Wilson 1899, p. 223.

27 Osman 1907, p. 149. Alfred Henry Osman lived 1864–1930. He started his working life as a solicitor's clerk. *The Racing Pigeon* is still published.

28 *Nottingham Evening Post* 12 August 1907, p. 7; Phillips 2018, p. 84.

29 Phillips 2018, p. 88.

30 *The Yarmouth Independent* 20 July 1918, p. 7; Anon 1918, p. 26.

31 The Royal Air Force Museum (London) accession number is 72/C/465. For full details of this mission, see Hobbs 2017, pp. 246–55.

32 Connolly 2018, pp. 12, 15; Phillips 2018, pp. 87–9, 92.

33 Gosse 1934, pp. 31–2.

34 Connolly 2018, pp. 267–8; Hansell 1988b, p. 21; Anon 1918, p. 25.

35 *Lancashire Evening Post* 5 August 1940, p. 5.

36 Gunston 1944. He was David J. Gunston, who lived at 89 Madeira Road, North End, Portsmouth, and was editor of *The Country Journal*, as well as an author of books.

37 Gunston 1944.

38 Corera 2018.

39 Gunston 1944. See also Corera 2018, pp. 91–2.

40 Phayre 1940, p. 273. Ignatius Phayre was a pseudonym of William George Fitzgerald.

41 Levaillant 1796, pp. 474–6. Levaillant is also known as Le Vaillant, a French explorer, naturalist and collector. He started off by working with Becoeur who discovered arsenical soap. He made three expeditions to South Africa from 1780, which was then relatively unknown. This was the final trip. He lived 1753–1824.

42 *Staffordshire Advertiser* 19 September 1795, p. 2. The hawk was found by a Mr Malcolm from Kingshorn, on his way to Gothenburg, probably on board a merchant ship.

43 Horsfall 1981.

44 *Evening Telegraph and Post (Dundee)* 23 December 1948, p. 2.

45 *Burnley News* 8 May 1915, p. 11.

46 *The Observer* 6 January 1918, p. 8.

47 Kennedy 1917, p. 34. They were sailing 15 miles west of Norway at 59° 12′N and 4° 33′E. John Noble Kennedy lived 1893–1970. He was in the Royal Navy in 1914 and transferred to the Royal Garrison Artillery in 1915. From October 1916 to the end of March 1917 he was at the Somme. He was elected a member of the British Ornithologists' Union in 1914.

48 *Dublin Daily Express* 13 February 1915, p. 7.

49 *Portadown News* 29 May 1915, p. 2.

50 Savory 1920, p. 255. This was on 1 February 1901.

51 *The Mail* 4 October 1916, p. 2. Zeppelins were used over Britain from 1915 to 1917.

52 *Bird Notes and News* 7, summer 1916, p. 15. Taxi whistles were a type of metal whistle originally used by doormen to summon cabs.

53 Harting 1871, p. 133.

54 Stevenson 1866, pp. 175–6.

55 Egerton-Warburton 1912, p. 16.

56 Jefferies 1879, p. 256.

57 Gosse 1934, p. 88. Sassoon was then a transport officer.

58 Pouncett 1998, p. 47.

59 Dixon 1890, pp. 40–1.

60 Neale 1828, pp. 193–4. See also Adkins and Adkins 2006, pp. 258–66. For bloated horse carcasses, see Anon 1827, pp. 63–4 (written by a Scottish soldier of the 71st Highlanders).

61 Rocca 1828, p. 85. The Battle of Medellín was on 28 March 1809. Albert Jean Michel de Rocca lived 1788–1818 and was a lieutenant in the 2nd regiment of Hussars.

62 Malcolm 1828, p. 252. John Malcolm, who lived 1795–c.1835, was from Orkney and also published poetry.

63 Also known as El-Obeid. William Hicks lived 1831–83.

64 *Galloway News and Kirkcudbrightshire Advertiser* 27 March 1885, p. 6. He passed the battlefield in mid-February 1885. Charles George Gordon lived 1833–85. About 50,000 Mahdists stormed Khartoum.

65 *Sheldrake's Aldershot and Sandhurst Military Gazette* 9 May 1885, p. 3. See also Galloway 1887.

66 *Army and Navy Gazette* 9 May 1885, p. 1.

67 Dixon 1890, p. 41.

68 Salvin 1859, p. 184. Salvin was travelling with the Reverend Henry Baker Tristram.

69 Harting 1871, pp. 43–5.

70 Gurney 1921, p. 82. The secretary was called Schaschek, and he wrote a journal in Latin.

71 Sneyd 1847, pp. v–vi, 11. This was most likely 1496–7, though may have been the work of Francesco Capello's secretary in about 1500.

72 Evans 1903, p. 117. William Turner lived c.1509/10–68.

73 Ticehurst 1920, p. 36.

74 Lovegrove 2007, pp. 118–19.

75 Smith 1772, pp. 208–10.

76 MacGillivray 1837, p. 506.

77 MacGillivray 1837, pp. 539, 557.

78 Dixon 1899, p. 147.

79 Dixon 1899, pp. 203–4.

80 Blake-Knox 1867, pp. 625–6.

81 Gray 1871, p. 285.

82 Haldane 1896, p. 35. The red and pink colouring was due to the fact that carbon monoxide binds with the haemoglobin in the blood. John Scott Haldane, 1860–1936, was a Scottish medical researcher whose work included the investigation of gases in coal mines, starting with the Albion Colliery explosion in South Wales on 23 June 1894, which killed 290 men. For a life of Haldane, see Goodman 2007.

83 Haldane 1896, pp. 7–8, 22–3.

84 Haldane 1894–5, p. 252; 1895, p. 449; Goodman 2007, p. 134. Mice and canaries are good 'sentinel species' – indicating changes in the environment. Haldane went down the Staffordshire pits in October 1894.

85 *The Times* 8 August 1895, p. 7. Four of the men drowned, and the other was pulled out alive but died.

86 Haldane 1896, pp. 39–40. The Brancepeth explosion was 20 April 1896.

87 Haldane 1896, pp. 20, 31. The miners were supposedly not keen on the idea of mice, as pits tended to be overrun with rodents.

88 Burton (2014, p. 146) states that miners had taken small birds down the mines since the early nineteenth century.

89 Brand 1813, p. 482. Brand's manuscript was believed to have been ready for publication by 1795.

90 *The Fife Free Press* 23 February 1901, p. 2. The disaster took place on 15 February 1901. Henry Rowan was the manager of a nearby district. It was an old fire in part of the pit that had been cut off, but continued to smoulder (generating carbon monoxide). See also Pollock 2016.

91 *Nottingham Evening Post* 21 December 1906, p. 5. William Cuthbert Blackett was a distinguished mining engineer who served in World War One. He lived 1859–1935. Blackett helped in the evening when others were exhausted. The Urpeth explosion was 17 December 1906. See also

Haldane 1905, p. 175 fn. for mice. The Courrières disaster was on 10 March 1906.

92 *Nottingham Evening Post* 21 December 1906, p. 5.

93 *Nottingham Evening Post* 21 December 1906, p. 4.

94 *Sheffield Daily Telegraph* 13 June 1917, p. 6.

95 The Bureau of Mines was formed in 1910. See 'Canary Birds in Mine-Rescue Work' *Popular Mechanics Magazine* 17 (1912), p. 355 and Page 1912, pp. 549–53, 561.

96 Burrell 1912, p. 12. This appeared in April 1912, and George Arthur Burrell cited as his source Haldane 1895, p. 448, which made no mention of birds, as his research centred on mice. In 1915, Haldane gave lectures to Yale University, which were published in an expanded form after World War One, in which he mentioned birds: 'In order to give miners a practical means of detecting dangerous percentages of CO, I introduced the plan of making use of a small warm-blooded animal such as a mouse or small bird' (Haldane 1922, p. 317). Burrell became the director of research into war gases for the Bureau of Mines in World War One.

97 Burrell 1912, p. 13.

98 *Daily Mirror* 15 October 1913, p. 4. There were 379 dead men and one rescuer. Sixty of the dead were under twenty.

99 Barton et al 2004, pp. 31–3; Jones 2010, p. 182. A memorial to mice and canaries was erected in 1927 at Edinburgh, to The Tunnellers' Friends.

100 *Nottingham Evening Post* 13 May 1936, p. 7.

101 *Nottingham Evening Post* 13 May 1936, p. 7.

102 Haldane 1922, pp. 317–18.

POSTSCRIPT

1 Holmes 1974, p. 599.

2 Shelley 1868, p. 213.

3 Hardy 1923, p. 92. He was in the neighbourhood of Livorno in March 1887.

4 Kearton 1898, p. 284.

5 Meredith 1881, p. 588. The poem is 224 lines of rhyming couplets, first published in May 1881. George Meredith was a prolific novelist and poet who lived 1828–1909 and rose to fame at the end of the nineteenth century.

6 Vaughan Williams composed this music in 1914 and kept revising and changing it until the first performance in 1920.

7 *Burnley News* 8 May 1915, p. 11.

8 Macdonald 2020, pp. 52–8, where he says that the number of individual birds of all species lost in the last six decades is more than 44 million, but then gives very positive suggestions on how the landscape can be reclaimed.

9 Farmland and farm buildings have, for example, been turned into shooting establishments, solar power stations, landfills and vast farm shops, in spite of local objections.

10 The CPRE is known now as 'The countryside charity'.

BIBLIOGRAPHY

—◆—

Adams, W. 1887 'The Dicky Bird Society' *The Monthly Chronicle of North-Country Lore and Legend*, pp. 443–51

Adanson, M. 1757 *Histoire naturelle du Sénégal. Coquillages. Avec la Relation abrégée d'un Voyage fait en ce pays, pendant les années 1749, 50, 51, 52 & 53* (Paris)

Addy, S.O. 1895 *Household Tales with other Traditional Remains collected in the counties of York, Lincoln, Derby, and Nottingham* (London, Sheffield)

Adkins, L. and Adkins, R.A. 1996 *Dictionary of Roman Religion* (New York, Oxford)

Adkins, R. and Adkins, L. 2006 *The War for All the Oceans: From Nelson at the Nile to Napoleon at Waterloo* (London)

Adkins, R. and Adkins, L. 2008 *Jack Tar: Life in Nelson's Navy* (London)

Adkins, R. and Adkins, L. 2013 *Eavesdropping on Jane Austen's England* (London)

Adkins, R. and Adkins, L. 2017 *Gibraltar: The Greatest Siege in British History* (London)

Albin, E. 1737 *A Natural History of English Song-Birds, and such of the Foreign as are usually brought over and esteemed for their Singing* (London)

Alexander, P. (ed.) 1951 *William Shakespeare: The Complete Works* (London, Glasgow)

Allatt, H.P. 1886 'Pigeons as Messengers in War' *Pall Mall Gazette* 30 January, p. 6

Alston, E.R. 1867 'Notes on the Folk-lore of Zoology' *The Zoologist* 2 (2nd series), pp. 1005–8

Amos, A. 1859 *Observations on the Statutes of the Reformation Parliament in the reign of King Henry the Eighth* (London)

Anderson, D. 2005 'Noyfull Fowles and Vermyn: Parish Payments for Killing Wildlife in Hampshire 1533–1863' *Proceedings of Hampshire Field Club and Archaeological Society* 60, pp. 209–28

Andrews, C.B. (ed.) 1934 *The Torrington Diaries containing the tours through England and Wales of the Hon. John Byng (later fifth Viscount Torrington) between the years 1781 and 1794 volume 1* (London)

Andrews, C.B. (ed.) 1935 *The Torrington Diaries containing the tours through*

England and Wales of the Hon. John Byng (later fifth Viscount Torrington) between the years 1781 and 1794 volume 2 (London)

Anon 1677 *The Compleat Servant-Maid; or, the Young Maidens Tutor* (London)

Anon 1772 *Fitz-Stephen's Description of the City of London, newly translated (from the Latin Original; with a necessary commentary)* (London)

Anon 1784 *A Letter Whearin, part of the Entertainment untoo the Queenz Majesty, at Killingsworth Castl in Warwick Sheer, in this Soomerz Progrest 1575, is signified* (Warwick)

Anon 1804 *Observations, Chiefly Lithological, made in a five weeks' tour to the principal lakes in Westmoreland and Cumberland* (London)

Anon 1808 *A New System of Domestic Cookery, formed upon principles of economy, and adapted to the use of private families* (Exeter)

Anon 1827 *Vicissitudes in the Life of a Scottish Soldier. Written by Himself* (London)

Anon 1843 *Reports of Special Assistant Poor Law Commissioners on the Employment of Women and Children in Agriculture* (London)

Anon 1846 *Le Ménagier de Paris. Traité de morale et d'économie domestique composé vers 1393 par un bourgeois Parisien, tome second* (Paris)

Anon 1918 'The First Flying Corps' *Bird Notes & News* 8, pp. 25–7

Arch, J. 1898 (ed. Countess of Warwick) *Joseph Arch: The Story of his Life, Told by Himself* (London)

Armstrong, A.W. (ed.) 2002 *"Forget not Mee & My Garden ---" Selected Letters, 1725–1768 of Peter Collinson, F.R.S.* (Philadelphia)

Armstrong, E.A. 1970 (2nd edn) *The Folklore of Birds: An Enquiry into the Origin & Distribution of some Magico-Religious Traditions* (New York)

Arnold, R. 1949 *A Yeoman of Kent. An Account of Richard Hayes (1725–1790) and of the village of Cobham in which he lived and farmed* (London)

Ashman, G. and Bennett, G. 2000 'Charlotte Sophia Burne: Shropshire Folklorist, First Woman President of the Folklore Society, and First Woman Editor of Folklore. Part 1: A Life and Appreciation' *Folklore* III, pp. 1–21

Atkinson, J.C. 1891 *Forty Years in a Moorland Parish: Reminiscences and researches in Danby in Cleveland* (London, New York)

Aubrey, J. 1784 (new edn) *Miscellanies upon Various Subjects* (London)

Bailleul-Lesuer, R.F. 2016 *The Exploitation of Live Avian Resources in Pharaonic Egypt: A Socio-Economic Study* (Chicago PhD dissertation)

Baines, E. 1923 *History, Directory & Gazetteer of the County of York vol. II East and North Riding* (Leeds)

Baines, F.E. (ed.) 1890 *Records of the Manor, Parish, and Borough of Hampstead, in the county of London, to December 31st, 1889* (London)

Baker, A.E. 1854 *Glossary of Northamptonshire Words and Phrases with examples of their colloquial use, and illustrations from various authors: to which are added, the customs of the county* (London, Northampton)

Barton, P., Doyle, P. and Vandevalle, J. 2004 *Beneath Flanders Fields: The Tunnellers' War 1914–1918* (Stroud)

Basdeo, S. 2017 *The Changing Faces of Robin Hood, c.1700–c.1900: Rethinking Gentrification in the Post-Medieval Tradition* (University of Leeds PhD thesis)

Beatson, A. 1816 *Tracts relative to the Island of St. Helena; written during a residence of five years* (London)

Beaumont, F. 1824 'Personal Recollections of Rook-shooting, preparatory to the month of May' *The North Devon Magazine, containing The Cave and Lundy Review* 1, pp. 158–63

Beeton, I. 1861 *The Book of Household Management* (London)

Belon, P. 1555 *L'Histoire de la Nature des oyseaux, avec leurs descriptions, & naïfs portraicts retirez du naturel* (Paris)

Berkeley, G.F. 1854 *Reminiscences of a Huntsman* (London)

Bigge, J.F. 1863 'Local Superstitions at Stamfordham' *Transactions of the Tyneside Naturalists' Field Club* 5, pp. 88–98

Bingley, W. 1804 *North Wales; including its scenery, antiquities, customs, and some sketches of its natural history vol. 1* (London)

Bird, I. 1866 'Pen and Pencil Sketches among the Outer Hebrides. Chapter IV. – The Cruise of the "Shamrock" – South Bernera' *The Leisure Hour* 15, pp. 646–50

Birkhead, T. 2018 *The Wonderful Mr Willughby: The First True Ornithologist* (London)

Bishop, L. 1992 *The General Accounts of the Churchwardens of Chipping Camden 1626 to 1907* (Campden Record Series)

Black, W.G. 1883 *Folk-Medicine; a chapter in the history of culture* (London)

Blagrave, J. 1671 *Blagraves Astrological Practice of Physick* (London)

Blake-Knox, H. 1867 'A natural history of the Common Gull, being an account of its habit, food, nidification, cry, flight, &c' *The Zoologist* 2 (2nd series), pp. 625–31

Blasco, R., Finlayson, C., Rosell, J. et al 2014 'The earliest pigeon fanciers' *SciRed* 4, 5971, https://doi.org/10.1038/srep05971

Blencowe, R.W. 1849 'South-Down Shepherds, and their songs at the sheepshearings' *Sussex Archaeological Collections* 2, pp. 247–56

Blencowe, R.W. and Lower, M.A. 1859 'Extracts from the Diary of a Sussex tradesman, a hundred years ago' *Sussex Archaeological Collections* 11, pp. 179–220

Blunt, J.H. 1877 *Dursley and its neighbourhood; being historical memorials of Dursley, Beverston, Cam, and Uley* (London)

Borrer, W. 1891 *The Birds of Sussex* (London)

Borrow, G. 1862 *Wild Wales: Its People, Language, and Scenery vol. 3* (London)

Bourne, H. 1725 *Antiquitates Vulgares; or, the antiquities of the common people* (Newcastle)

Bovill, E.W. 1947 'Queen Elizabeth's Gunpowder' *The Mariner's Mirror* 33, pp. 179–86

Brabourne, E. (ed.) 1884 *Letters of Jane Austen vol. II* (London)

Brand, J. 1810 *Observations on Popular Antiquities: including the whole of Mr. Bourne's Antiquitates Vulgares* (London)

Brand, J. 1813 (rev. edn by H. Ellis) *Observations on Popular Antiquities: chiefly illustrating the origin of our vulgar customs, ceremonies and superstitions vol. 1* (London)

Breeze, A. 2011 'Worcester's Roman Name, Buildwas, and Uricon' *The Housman Society Journal* 37, pp. 169–81

Brewster, J. 1829 (2nd edn) *The Parochial History and Antiquities of Stockton-upon-Tees* (Stockton-upon-Tees)

Brontë, E. 1848 *Wuthering Heights. A Novel* (New York)

Browne, M. 1884 (2nd edn) *Practical Taxidermy: A Manual of Instruction to the Amateur in collecting, preserving, and setting up natural history specimens of all kinds* (London)

Browne, T. 1672 (6th edn) *Pseudodoxia Epidemica: or, enquiries into very many received tenents and commonly presumed truths, together with the Religio Medici* (London)

Bruce, J.C. 1904 *Old Newcastle: Lectures by the late Dr. J. Collingwood Bruce* (Newcastle-upon-Tyne)

Brushfield, T.N. 1897 'On the destruction of "Vermin" in rural parishes' *Report and Transactions of the Devonshire Association* 29, pp. 291–349

Buckland, F. 1860 (2nd series) *Curiosities of Natural History* (London)

Buckland, F.T. 1879 *Curiosities of Natural History* (London)

Buckler, B. 1751 (2nd edn) *A complete vindication of the mallard of All-Souls College, against the injurious suggestions of the Rev. Mr. Pointer, Rector of Slapton in the County of Northampton and Diocese of Peterborough* (London, Oxford)

Buckmaster, J.C. (ed.) 1897 *A Village Politician: The Life-Story of John Buckley* (London)

Bull, H.G. (ed.) 1888 *Notes on the Birds of Herefordshire, contributed by members of the Woolhope Club, collected and arranged by the late Henry Graves Bull, M.D.* (London, Hereford)

Bullock, H.A. 1816 *History of the Isle of Man, with a comparative view of the past and present state of society and manners* (London)

Burne, C.S. (ed.) 1886 *Shropshire Folk-Lore: A Sheaf of Gleanings Part III* (London, Shrewsbury, Chester)

Burne, M. 1911 'Parish Gleanings from Upton St. Leonard's, Gloucestershire' *Folklore* 22, pp. 236–9

Burrell, G.A. 1912 *The use of birds and mice for detecting carbon monoxide after mine fires and explosions* (Washington)

Burritt, E. 1864 *A Walk from London to John O'Groats, with notes by the way* (London)

Burton, C. 2014 'Life and Wing: Victorian and Edwardian Conceptions of Coal-Mine Canaries' *Victorian Review* 40, pp. 143–59

Burton, R. 1676 (8th edn) *The Anatomy of Melancholy* (London)

Burton, T. 1888 (ed. J. Raines) *The History and Antiquities of the parish of Hemingbrough in the county of York* (York)

Callahan, D. 2014 *A History of Birdwatching in 100 Objects* (London)

Cam, H.M. 1916 'The Legend of the Incendiary Birds' *The English Historical Review* 31, pp. 98–101

Campbell, J. 1774 *A Political Survey of Britain: being a series of reflections on the situation, lands, inhabitants, revenues, colonies, and commerce of this island vol. 1* (London)

Capern, E. 1865 *Wayside Warbles* (London)

Capern, E. 1870 (2nd edn) *Wayside Warbles* (London, Birmingham)

Carew, R. 1749 (new edn) *The Survey of Cornwall and an epistle concerning the excellencies of the English Tongue with the Life of the author* (London)

Carr, R. 1981 'Country Sports' pp. 475–87 in G.E. Mingay 1981 *The Victorian Countryside vol. 2* (London, Boston, Henley)

Cash, J.J. 1933 *In the Season of the Year: A Record of Country Life throughout the months* (London)

Champlin, J.D. 1886 *Chronicle of the Coach: Charing Cross to Ilfracombe* (London)

Children's Employment Commission 1842 *Appendix to the Second Report of the Commissioners: Trades and Manufactures Part I* (London)

Clare, J. 1821 *The Village Minstrel, and other Poems* (London)

Clark Kennedy, A.W.M. 1868 *The Birds of Berkshire and Buckinghamshire: A Contribution to the natural history of the two counties* (Eton, London)

Clutterbuck, R.H. 1898 (rev. and ed. E.D. Webb) *Notes on the parishes of Fyfield, Kimpton, Penton Mewsey, Weyhill and Wherwell, in the county of Hampshire* (Salisbury)

Cobbett, W. 1830 *Rural Rides in the Counties of Surrey, Kent, Sussex* (London)

Cocker, M. and Mabey, R. 2020 (2nd edn) *Birds Britannica* (London)

Coke, E. 1797 *The Third Part of the Institutes of the Laws of England: concerning High Treason, and other pleas of the Crown, and criminal causes* (London)

Colles, A. 1887 'A Witches' Ladder' *The Folk-Lore Journal* 5, pp. 1–5

Connolly, J.E. 2018 *The experience of occupation in the Nord, 1914–18: Living with the enemy in First-World-War France* (Manchester)

Copper, B. 1971 *A Song for Every Season: A Hundred Years of a Sussex Farming Family* (London)

Cordeaux, J. 1867 'Notes from Flamborough' *The Zoologist* 2 (2nd series), pp. 1008–11

Corera, G. 2018 *Secret Pigeon Service: Operation Columba, Resistance and the Struggle to Liberate Europe* (London)

Cornish, J.G. 1939 (ed. V. Cornish) *Reminiscences of Country Life* (London)

Couch, J. 1871 *The History of Polperro, a fishing town on the south coast of Cornwall* (Truro, London)

Coues, E. 1876 'The destruction of birds by telegraph wire' *The American Naturalist* 10, pp. 734–6

Courtney, M.A. 1890 *Cornish Feasts and Folk-Lore* (Penzance)

Coward, T.A. 1919 *The Birds of the British Isles and their Eggs: First Series comprising families Corvidae to Sulidae* (London, New York)

Coward, T.A. 1923 *Birds and Their Young* (London)

Cox, J.C. 1915 *Pulpits, lecterns and organs in English churches* (Oxford)

Creaney, C. 2010 'Paralytic Animation: The Anthropomorphic Taxidermy of Walter Potter' *Victorian Studies* 53, pp. 7–35

Creighton, O.H., Evis, L., Kingdom, M., McKenzie, C.J., Watt, I. and Outram, A. K. 2020 'The Face of Battle? Debating Arrow Trauma on medieval human remains from Princesshay, Exeter' *The Antiquaries Journal* 100, pp. 165–89

Cressy, D. 2011 'Saltpetre, state security and vexation in early modern England' *Past and Present* 212, pp. 73–111

Cushman, G.T. 2013 *Guano and the opening of the Pacific World: A Global Ecological History* (Cambridge)

Daniel, W.B. 1802 *Rural Sports vol. 2* (London)

Daniel, W.B. 1812 *Rural Sports vol. 3* (London)

Davidson, B.M. 1990 'Sources of Illumination for the Microscope 1650–1950' *Microscopy* 36, pp. 369–86

Davies, O. and Easton, T. 2015 'Cunning-Folk and the Production of Magical Artefacts' pp. 209–31 in R. Hutton (ed.) *Physical Evidence for Ritual Acts, Sorcery and Witchcraft in Christian Britain: A Feeling for Magic* (Basingstoke)

Davies, W. 1820 *The History of Whittington* (Oswestry)

Davy, H. 1813 *Elements of Agricultural Chemistry, in a course of lectures for the Board of Agriculture* (London)

de la Mare, W. 1962 *Come Hither: A Collection of Rhymes & Poems for the Young of All Ages* (London)

Denton, J.B. 1868 *The Agricultural Labourer* (London)

Dillon, H.A. 1896 'Feathers and Plumes' *Archaeological Journal* 53, pp. 126–39

Dixon, C. 1890 *Stray Feathers From Many Birds: being leaves from a naturalist's note-book* (London)

Dixon, C. 1899 *Bird-Life in a southern county being eight years' gleanings among the birds of Devonshire* (London)

Dixon, E.S. 1851 *The Dovecote and the Aviary* (London)

Dixon, J.H. 1886 *Gairloch in north-east Ross-shire, its records, traditions, inhabitants, and natural history* (Edinburgh)

Dodd, G. 1843 *Days at the Factories; or, the manufacturing industry of Great Britain described, series I. London* (London)

Drinkwater, J. 1785 *A History of the Late Siege of Gibraltar* (London)

Duffin, C.J. 2013 'Chelidonius: The Swallow Stone' *Folklore* 124, pp. 81–103

Dundonald, T. 1861 (2nd edn) *The Autobiography of a Seaman vol. 1* (London)

D'Urban, W.S.M. and Mathew, M.A. 1892 *The Birds of Devon* (London)

Dutt, W.A. 1903 *The Norfolk Broads* (London)

Dyer, T.F.T. 1878 *English Folk-Lore* (London)

Eagle Clarke, W. 1912a *Studies in Bird Migration vol. 1* (London, Edinburgh)

Eagle Clarke, W. 1912b *Studies in Bird Migration vol. 2* (London, Edinburgh)

Easton, T. 2014 'Spiritual Middens in Mid Suffolk, England, ca. 1650 to 1850' *Historical Archaeology* 48, pp. 10–34

Easton, T. 2015 'Spiritual Middens' pp. 147–63 in R. Hutton (ed.) *Physical Evidence for Ritual Acts, Sorcery and Witchcraft in Christian Britain: A Feeling for Magic* (Basingstoke)

Egerton-Warburton, G. 1912 *In A Cheshire Garden: Natural History Notes* (London)

Elkins, N. 2010 (3rd edn) *Weather and Bird Behaviour* (London)

Elliott, J.S. 1936 *Bedfordshire 'Vermin' Payments: concerning the destruction of 'vermin' by parish officials during the XVI–XIX centuries, with extracts from their accounts* (Luton)

Emery, N. 1996 *Archaeological Excavations on Hirta 1986–1990* (Edinburgh)

Evans, A.H. (ed.) 1903 *Turner on Birds: A Short and Succinct History of the Principal Birds noted by Pliny and Aristotle, first published by Doctor William Turner, 1544* (Cambridge)

Evans, J. 1816 *The History of Bristol, Civil and Ecclesiastical; including biographical notices of eminent and distinguished natives, volume 2* (Bristol)

Evelyn, J. 1697 *Numismata: A Discourse of Medals Antient and Modern* (London)

Farley, J. 1787 (4th edn) *The London Art of Cookery, and Housekeeper's Complete Assistant* (London)

Farrer, J.A. 1876 'Comparative Folk-Lore' *Cornhill Magazine* 33, pp. 41–60

Finlayson, C. 1992 *Birds of the Strait of Gibraltar* (London)

Fitter, R.S.R. 1942 (1943) 'The Starling Roosts of the London Area' *The London Naturalist* pp. 3–23

Fowler, M.W.E. 1909 'Yorkshire folk-lore' pp. 286–305 in T.M. Fallow (ed.) *Memorials of Old Yorkshire* (London)

Fowler, W.W. 1886 *A Year with the Birds* (Oxford, London)

Fowler, W.W. 1913 *Kingham Old and New: Studies in a rural parish* (Oxford, London)

Fox, G.T. 1827 *Synopsis of the Newcastle Museum, late the Allan, formerly the Tunstall, or Wycliffe Museum* (Newcastle)

Fox, S.P. 1874 *Kingsbridge and its surroundings* (Plymouth)

Frazer, J.G. 1887 'A Witches' Ladder' *The Folk-Lore Journal* 5, pp. 81–3

Freeman, G.E. and Salvin, F.H. 1859 *Falconry: its claims, history and practice* (London)

Fremantle, T.F. 1913 'The Evolution of Guns and Rifles' pp. 281–321 in J.G. Millais, W. Baxendale, J.E. Harting, W.C. Oates, M. Portal and T.F. Fremantle *The Gun at Home and Abroad: British Deer & Ground Game, Dogs, Guns & Rifles* (London)

Fuller, T. 1662 *The History of the Worthies of England* (London)

Galloway, W. 1887 *The Battle of Tofrek fought near Suâkin, March 22nd, 1885* (London)

Gardner, J. 1865 *Bird, Quadruped, and Fish Preserving: A Manual of Taxidermy for Amateurs* (London)

Gardner-Medwin, A. 1970 'The Wren Hunt Song' *Folklore* 81, pp. 215–18

Garlitz, B. 1955 'Christina Rossetti's Sing-Song and Nineteenth-Century Children's Poetry' *PMLA* 70, pp. 539–43

Gent, B.E. 1690 *A New Dictionary of the Terms Ancient and Modern of the Canting Crew, in its several tribes, of Gypsies, Beggers, Thieves, Cheats, &c.* (London)

Gent, T. 1733 *The Antient and Modern History of the Loyal Town of Rippon* (York)

Gibbs, J.A. 1898 *A Cotswold Village: or country life and pursuits in Gloucestershire* (London)

Gilbey, W. 1912 *Sport in the Olden Times* (London)

Giuseppi, M. 1905 'Gunpowder' pp. 306–29 in H.E. Malden (ed.) *The Victoria History of the counties of England: Surrey* (London)

Gladstone, H.S. 1910 *The Birds of Dumfriesshire: A Contribution to the Fauna of the Solway Area* (London)

Gladstone, H.S. 1919 *Birds and the War* (London)

Glyde, J. 1856 *Suffolk in the Nineteenth Century: Physical, Social, Moral, Religious, and Industrial* (London)

Goddard, T. R. 1929 *History of the Natural History Society of Northumberland, Durham and Newcastle upon Tyne 1829–1929* (Newcastle)

Goodman, M. 2007 *Suffer & Survive: Gas Attacks, Miners' Canaries, Spacesuits and the Bends: The Extreme Life of Dr J.S. Haldane* (London, New York, Sydney, Toronto)

Gosse, P. 1934 *Memoirs of a Camp-Follower* (London, New York, Toronto)

Graebe, M. 2011 'A Devonshire Man goes North' *The Transactions of the Sabine Baring-Gould Appreciation Society* 11, pp. 34–48

Gray, R. 1871 *The Birds of the West of Scotland including the Outer Hebrides* (Glasgow)

Gray, R. and Anderson, T. 1869 *The Birds of Ayrshire and Wigtownshire* (Glasgow)

Greene, W.T. 1897 'Some uncommon bird pets' *The Boy's Own Paper* 19, pp. 653–4

Greener, W.W. 1910 (9th edn) *The Gun and its Development* (London)

Grinsell, L.V. 1976 *Folklore of Prehistoric Sites in Britain* (Newton Abbot, North Pomfret)

Gritt, A.J. and Virgoe, J.M. (eds.) 2004 *The Memoranda Books of Basil Thomas Eccleston, 1757–1789* (Record Society of Lancashire and Cheshire)

Grose, F. 1785 *A Classical Dictionary of the Vulgar Tongue* (London)

Gunston, D.J. 1944 'Pigeons of the R.A.F.' *Country Life* 17 March, p. 462

Gurdon, E.C. 1893 *County Folk-Lore Printed Extracts No 2 Suffolk* (London)

Gurney, D. 1834 'Extracts from the Household and Privy Purse Accounts of Lestranges of Hunstanton, from *A.D.* 1519 to *A.D.* 1578' *Archaeologia* 25, pp. 411–569

Gurney, J.H. 1876 *Rambles of a Naturalist in Egypt & Other Countries* (London)

Gurney, J.H. 1921 *Early Annals of Ornithology* (London)

Hackwood, F.W. 1907 *Old English Sports* (London)

Haggard, H.R. 1899 *A Farmer's Year being his commonplace book for 1898* (London, New York, Bombay)

Haldane, J. 1894–5 'Notes of an Enquiry into the Nature and Physiological Action of Black-Damp, as Met with in Podmore Colliery, Staffordshire, and Lilleshall Colliery, Shropshire' *Proceedings of the Royal Society of London* 57, pp. 249–57

Haldane, J. 1895 (November) 'The Action of Carbonic Oxides on Man' *Journal of Physiology* 18, pp. 430–62

Haldane, J. 1896 *Report to the Secretary of State for the Home Department on the causes of death in colliery explosions and underground fires, with special reference to the explosions at Tylorstown, Brancepeth, and Micklefield* (London)

Haldane, J. 1905 'The Examination of Mine Air' pp. 95–156 in C. Le Neve Foster and J.S. Haldane (eds) 1905 *The Investigation of Mine Air: An Account by several authors of the nature, significance, and practical methods of measurement of metalliferous mines* (London)

Haldane, J.S. 1922 *Respiration* (New Haven)

Hall, J.J. 1910–11 'The Trumpeting Cock of St. Mary of Ottery' *Devon & Cornwall Notes & Queries* 6, pp. 1–2

Hancock, E.G. 1979 'Edward Hearle Rodd (1810–1880) – Problem Solved' *Biology Curators Group Newsletter* 2, pp. 91–2

Hansell, P. and J. 1988a *Dovecotes* (Princes Risborough)

Hansell, P. and J. 1988b *Doves and Dovecotes* (Bath)

Hardy, J. 1879 'Popular History of the Cuckoo' *The Folk-Lore Record* 2, pp. 47–91

Hardy, T. 1923 *Collected Poems of Thomas Hardy* (London)

Harland, J. and Wilkinson, T.T. 1867 *Lancashire Folk-Lore: illustrative of the superstitious beliefs and practices, local customs and wages of the people of the county Palatine* (London)

Harley, J. 1840 'Catalogue of the Land Birds of Leicestershire' pp. 646–64 in MacGillivray 1840

Harris, R. 1917 *The Ascent of Olympus* (Manchester)

Harrison, T.P. 1950 'The Whistler, Bird of Omen' *Modern Language Notes* 65, pp. 539–41

Harrison, T.P. 1954 'Birds in the Moon' *Isis* 45, pp. 323–30

Harrison, W. 1869 'Customs and Superstitions' *Mona Miscellany* 16, pp. 135–75

Harting, J.E. 1866a *The Birds of Middlesex: A Contribution to the Natural History of the County* (London)

Harting, J.E. 1866b 'The Birds of Shakespeare' *The Zoologist* 1 (2nd series), pp. 393–424

Harting, J.E. 1867 'An Inquiry into the Nature and Properties of the Swallow-stone and Swallow's herb' *The Zoologist* 2 (2nd series), pp. 744–7

Harting, J.E. 1871 *The Ornithology of Shakespeare, critically examined, explained and illustrated* (London)

Harting, J.E. 1895 'On the origin of the terms "cob" and "pen"' *The Zoologist* 19 (3rd series), pp. 372–4

Hartley, D. 1954 *Food in England* (London, Sydney)

Hartley, D. 1979 *The Land of England* (London)

Harvie-Brown, J.A. 1866 'A List of Birds observed in the two adjoining Parishes of Dunipace and Larbert, in Stirlingshire' *The Zoologist* 1 (2nd series), pp. 67–72

Harvie-Brown, J.A. (ed.) 1890 *The Birds of Iona and Mull 1852–70 by the late Henry Davenport Graham author of 'The Antiquities of Iona' with a memoir of the author* (London)

Hawker, P. 1816 (2nd edn) *Instructions to Young Sportsmen, or the choice, care, and mangement of guns; hints for the preservation of game; directions for shooting wildfowl, &c &c &c* (London)

Hawker, P. 1893a *The Diary of Colonel Peter Hawker author of 'Instructions to Young Sportsmen' 1802–1853 vol 1* (London)

Hawker, P. 1893b *The Diary of Colonel Peter Hawker author of 'Instructions to Young Sportsmen' 1802–1853 vol 2* (London)

Hay, J.C.D. 1898 *Lines from my Log-Books* (Edinburgh)

Heber, R. (ed.) 1828 *The Whole Works of the Right Rev. Jeremy Taylor, D.D. vol. XII* (London)

Henderson, W. 1866 *Notes on the Folk Lore of the northern counties of England and the Borders* (London)

Hendriksen, M.M.A. 2019 'Animal Bodies between Wonder and Natural History: Taxidermy in the Cabinet and Menagerie of Stadholder Willem V (1748–1806)' *Journal of Social History* 52, pp. 1110–31

Herd, D. 1776 *Ancient and Modern Scottish Songs, Heroic Ballads, Etc. vol. 1* (Edinburgh)

Hingeston, H. 1703 *An Address to England, &c. containing sundry warnings and admonitions to the inhabitants thereof, of all degrees and perswasions, but more especially to those of the Church of England, and more particularly to those of the town of Kings-Bridge in Devon, (the Place of my Nativity and Abode) and parts adjacent* (Exeter)

Hobbs, D. 2017 *The Royal Navy's Air Service in the Great War* (Barnsley)

Hodgson, J.C. (ed.) 1910 *Six North Country Diaries* (Durham)

Holland, H. 1808 *General View of the Agriculture of Cheshire; with observations drawn up for the consideration of the Board of Agriculture and internal improvement* (London)

Hollowell, S. 2000 *Enclosure Records for Historians* (Chichester)

Holmes, R. 1974 *Shelley: The Pursuit* (London)

Holmes, R. 2013 *Falling Upwards: How We Took to the Air* (London)

Hone, W. 1825 *The Every-Day Book; or, the Guide to the Year: relating the popular amusements, sports, ceremonies, manners, customs, and events, incident to the 365 days in past and present times* (London)

Hone, W. 1832 *The Year Book of Daily Recreation and Information* (London)

Horne, A. 1965 *The Fall of Paris: The Siege and the Commune 1870–71* (London)

Horsfall, N. 1981 'From History to Legend: M. Manlius and the Geese' *The Classical Journal* 76, pp. 298–311

Housman, H. 1881 *The Story of Our Museum: showing how we formed it, and what it taught us* (London, Brighton, New York)

Howard, M.M. 1951 'Dried cats' *Man* 51, pp. 149–51

Howitt, W. (ed.) 1839 *The Boy's Country-Book: being the real life of a country boy, written by himself; exhibiting all the amusements, pleasures, and pursuits of children in the country* (London)

Howse, R. 1894 'On the occurrence on the North-East Coast of Yorkshire of a Ruffed-Bustard, Otis houbara, *Gmelin*, commonly known as Macqueen's Bustard, Otis Macqueenii, *J.E. Gray*' *Natural History Transactions of Northumberland, Durham, and Newcastle-on-Tyne* 11, pp. 345–50

Howse, R. 1899 *Index-Catalogue of the Birds in the Hancock Collection* (Newcastle)

Hudson, W.H. 1894 *Lost British Birds* (London)

Hudson, W.H. 1909 *Afoot in England* (London)

Hunt, R. 1865 (2nd series) *Popular Romances of the West of England; or, The Drolls, Traditions, and Superstitions of Old Cornwall* (London)

Hunt, R. 1881 (3rd edn) *Popular Romances of the West of England; or, The Drolls, Traditions, and Superstitions of Old Cornwall* (London)

Hutton, R. 2015 'Introduction' pp. 1–14 in R. Hutton (ed.) *Physical Evidence for Ritual Acts, Sorcery and Witchcraft in Christian Britain: A Feeling for Magic* (Basingstoke)

Hutton, W. 1804 *Poems; chiefly Tales* (London)

Ikram, S. and Dodson, A. 1998 *The Mummy in Ancient Egypt: Equipping the Dead for Eternity* (London)

Ingram, C. 1953 'The Use of Earthenware Pots as Nesting-Holes for House Sparrows' *British Birds* 46, p. 352

Isham, R. 1875 *The Journal of Thomas Isham, of Lamport, in the County of Northampton, from 1st Nov., 1671, to 30th Sept. 1673, written by him when a boy, in Latin, by command of his father, Sir Justinian Isham* (Norwich)

Jameson, P. 2003 *The Diary of James Woodforde, Volume 13 1791–1793* (Parson Woodforde Society)

Jameson, P. 2006 *The Diary of James Woodforde, Volume 16 1798–1800* (Parson Woodforde Society)

Jefferies, R. 1879 *Wild Life in a Southern County* (Boston)

Jefferies, R. 1889 *Field and Hedgerow, being the last essays* (London, New York)

Jenks, J. 1768 *The Complete Cook: Teaching the Art of Cookery in all its branches* (London)

Jesse, E. 1847 *Favorite Haunts and Rural Studies; including visits to spots of interest in the vicinity of Windsor and Eton* (London)

Jobey, G. 1992 'Cock-Fighting in Northumberland and Durham' *Archaeologica Aeliana* 20 (5th series), pp. 1–25

Johnes, M. 2007 'Pigeon Racing and Working Class culture in Britain, c.1870–1950' *Cultural & Social History* 4, issue 3, pp. 361–83

Johns, C.A. 1862 *British Birds in their Haunts* (London)

Johns, C.A. 1873 (2nd edn) *A Week at the Lizard* (London)

Johnson, C.W. 1843 *On Guano as a fertilizer* (London)

Johnson, S. 1802 (12th edn) *A Dictionary of the English Language* (Montrose)

Jones, E.L. 1972 'The Bird Pests of British Agriculture in Recent Centuries' *The Agricultural Review* 20, pp. 107–25

Jones, M. 1989 'Folklore Motifs in Late Medieval Art I: Proverbial Follies and Impossibilities' *Folklore* 100, pp. 200–17

Jones, S. 2010 *Underground Warfare 1914–1918* (Barnsley)

Jordan, J. 1857 *A Parochial History of Enstone, in the county of Oxford* (London, Oxford)

Kalm, P. 1892 (trans. J. Lucas) *Kalm's account of his visit to England on his way to America in 1748* (London, New York)

Kear, J. 1990 *Man and Wildfowl* (London)

Kearton, R. and C. 1897 *With nature and a camera. Being the adventures and observation of a field naturalist and an animal photographer* (London, Paris, Melbourne)

Kemp, P. (ed.) 1976 *The Oxford Companion to Ships and the Sea* (London, New York, Melbourne)

Kennedy, J.N. 1917 'Notes on Birds Observed in the North Sea and North Atlantic Ocean during the Autumn and Winter of 1914' *Ibis* 5, pp. 30–41

Kielmansegge, F. von 1902 *Diary of a Journey to England in the years 1761–1762* (London)

Kirby, C. 1933 'The English Game Law System' *The American Historical Review* 38, pp. 240–62

Kirkman, F.B. (ed.) 1910 *The British-Bird Book Section XI vol 4 part 2* (London, Edinburgh)

Knapp, J.L. 1829a *The Journal of a Naturalist* (London)

Knapp, J.L. 1829b (2nd edn) *The Journal of a Naturalist* (London)

Knight, C. 1861 *Autobiography of Miss Cornelia Knight, Lady companion to the Princess Charlotte of Wales, with extracts from her journals and anecdote books volume 1* (London)

Knox, A.E. 1849 *Ornithological Rambles in Sussex with a systematic catalogue of the birds of that county, and remarks on their local distribution* (London)

Lackington, J. 1794 (new edn) *Memoirs of the forty-five first years of the Life of James Lackington* (London)

Latham, C. 1878 'Some West Sussex Superstitions lingering in 1868' *The Folk-Lore Record* 1, pp. 1–67

Lee, W. 1887 'Northern Superstitions' *The Monthly Chronicle of North-Country Lore and Legend* 1, pp. 374–5

Leighton, W.A. 1880 'Early Chronicles of Shrewsbury' *Transactions of the Shropshire Archaeological and Natural History Society* 3, pp. 239–352

Levaillant, F. 1796 *New Travels into the interior parts of Africa, by the way of the Cape of Good Hope, in the years 1783, 84 and 85 vol III* (London)

Lewis, M. and Richardson, I. 2017 'Inscribed Vervels' *Post-Medieval Archaeology* 31, pp. 194–200

Lilford, Lord 1895 *Notes on the Birds of Northamptonshire and neighbourhood vol 1* (London)

Lord, J.K. 1865 (October) 'Humming Birds' *Hardwicke's Science Gossip* 1, pp. 218–20

Lord, J.K. 1866a 'First Arrivals of Migrating Birds' *The Leisure Hour* 15, pp. 181–3

Lord, J.K. 1866b 'A Ramble through the London Fur Stores' *The Leisure Hour* 15, pp. 460–4

Love, J.A. 2015 *A Natural History of Lighthouses* (Dunbeath)

Lovegrove, R. 2007 *Silent Fields: The long decline of a nation's wildlife* (Oxford, New York)

Lower, M.A. 1854 *Contributions to Literature, Historical, Antiquarian, and Metrical* (London)

Lubbock, R. 1845 *Observations on the Fauna of Norfolk, and more especially on the district of the Broads* (Norwich)

Luther, F. 1950 'The Earliest Experiments in Microphotography' *Isis* 41, pp. 277–81

Lyell, J.C. 1881 *Fancy Pigeons: containing full directions for their breeding and management* (London)

Lysons, D. 1810 (2nd edn) *The Environs of London: Being an Historical Account of the Towns, Villages, and Hamlets, within twelve miles of that capital Volume II: County of Middlesex* (London)

Lysons, S. and Brand, J. 1806 'Extracts from the *Rotulus Familiae* in the eighteenth year of the Reign of K. Edw. I. preserved among his Majesty's Records in the *Tower' Archaeologia* 15, pp. 350–62

McCann, J. 1991 'An Historical Enquiry into the Design and Use of Dovecotes' *Transactions of the Ancient Monuments Society* 35, pp. 89–160

McCann, J. and P. 2003 *The Dovecotes of Historical Somerset* (Somerset Vernacular Building Research Group)

Mac Cárthaigh, C. 1992–3 'The Ship-Sinking Witch: A Maritime Folk Legend from North-West Europe' *Béoloideas*, pp. 267–86

M'Culloch, J.R. and Reid, H.G. 1881 'Corn Laws' pp. 651–64 in J. Labor (ed.) *Cyclopedia of Political Science, Political Economy, and of the Political History of the United States vol. 1* (Chicago)

Macdonald, B. 2020 (rev. edn) *Rebirding: Restoring Britain's Wildlife* (Exeter)

MacGillivray, W. 1837 *A History of British Birds, indigenous and migratory vol. 1* (London)

MacGillivray, W. 1840 *A History of British Birds, indigenous and migratory vol. 3* (London)

McGowan, B. 2009 'The decline of the Scottish Ospreys' *Scottish Birds* 29, pp. 55–8

Mackenzie, R.H. 1913 *The Trafalgar Roll* (London)

McKerrow, R.B. 1910 *The Works of Thomas Nashe edited from the original texts vol. II* (London)

Maclachlan, E.H. 1878 'Monkton Manor and Church' *Archaeologia Cantiana* 12, pp. 269–82

Madden, F. 1831 'Narrative of the Visit of the Duke de Nájera to England in the year 1543–4; written by his Secretary, Pedro de Gante' *Archaeologia* 23, pp. 344–57

Mag Fhloinn, B. 2007 'Martinmas Tradition in South-West County Clare: A Case Study' *Béaloideas* 75, pp. 79–108

Malcolm, J. 1828 'Reminiscences of a campaign in the Pyrenees and south of France in 1814' pp. 235–307 in *Memorials of the Late War vol. I* (Edinburgh)

Mallon, B. 1998 *The 1900 Olympic Games: Results for All Competitors in All Events, with Commentary* (Jefferson, London)

Mansel-Pleydell, J.C. 1888 *The Birds of Dorsetshire: A contribution to the natural history of the county* (London, Dorchester)

March, J. 1998 *Cassell Dictionary of Classical Mythology* (London)

March, J. 2008 *The Penguin Book of Classical Myths* (London, New York)

Markham, G. 1615 *The English Hus-wife, contayning the inward and outward vertues which ought to be in a compleat woman* (London)

Markham, G. 1621 *Hungers Prevention: or, the whole arte of fowling by water and land* (London)

Markham, G. 1631 *The English House-Wife, containing the inward and outward vertues which ought to be in a compleate woman* (London)

Markland, G. 1727 *Pteryplegia: Or, the Art of Shooting-Flying. A Poem* (London)

Marshall, W. 1790 *The Rural Economy of the Midland Counties vol. 1* (London)

Marshall, W. 1796 (2nd edn) *The Rural Economy of Gloucestershire vol. II* (London)

Martin, M. 1753 (4th edn) *A Voyage to St Kilda. The remotest of all the* Hebrides, *or Western Isles of* Scotland (London)

Masefield, J. 1913 *Salt-Water Ballads* (London)

Massingham, H.J. 1936 *English Downland* (London)

Mathew, G.F. 1872 'Ornithological Notes from Devonshire, &c.' *The Zoologist* 7 (2nd series), pp. 2917–21

Mathew, M. 1884 'A Visit to Skomer Island' *The Zoologist* 8 (3rd series), pp. 433–8

Mathew, M.A. 1894 *The Birds of Pembrokeshire and its Islands* (London)

Mayhew, H. 1851 *London Labour and the London Poor vol 2 The Street-Folk* (London)

Meade-Waldo, E.G.B. 1908 'Old English Nesting Bottles' *British Birds* 2, pp. 164–5

Meiklejohn, M.F.M. 1954 'The use of Earthenware Pots as Nesting-Holes for Starlings and House-Sparrows' *British Birds* 47, pp. 95–6

Mellor, G.T. 2006 *Falconry in Britain between 1750 and 1927: the survival, organisation and development of the sport* (De Montfort University PhD thesis)

Meredith, G. 1881 (May) 'The Lark Ascending' *The Fortnightly Review* 29 (new series), pp. 588–91

Merrifield, R. 1969 'Folk-Lore in London Archaeology: Part 2: The Post-Roman Period' *London Archaeologist* 1 (5), pp. 99–104

Merrifield, R. 1987 *The Archaeology of Ritual and Magic* (London)

Mill, J.S. 1848 *Principles of Political Economy with some of their applications to social philosophy vol 2* (London)

Miller, H. 1869 (14th edn) *My Schools and Schoolmasters or the story of my education* (Edinburgh)

Mitchell, A. 1863 'On various superstitions in the north-west Highlands
    and islands of Scotland, especially in relation to lunacy' *Proceedings of
    the Society of Antiquaries of Scotland* 4, pp. 251–88
Mitchell, A. 1881 *The Past in the Present: What is Civilization?* (New York)
Mitford, M.R. 1828 'Village Sketches. No IX The Bird-Catcher' *The
    Monthly Magazine* 5, pp. 130–4
Moore, A.W. 1891 *The Folk-Lore of the Isle of Man, being an account
    of its myths, legends and superstitions, customs, & proverbs* (Isle of
    Man, London)
Moore, E. 1837 'On the Wading Birds of Devonshire' *The Magazine of
    Natural History, and Journal of Zoology, Botany, Mineralogy, Geology
    and Meteorology* 1 (new series), pp. 319–23
Morris, P. 2013 *Walter Potter's Curious World of Taxidermy* (London)
Morris, P. 2018 'Reflections on Some Practical Aspects of Collecting
    During the Nineteenth and Twentieth Centuries' pp. 759–74 in
    A. Macgregor (ed.) *Naturalists in the Field: Collecting, Recording and
    Preserving the Natural World from the Fifteenth to the Twenty-First
    Century* (London, Boston)
Moss, S. 2018 *Mrs Moreau's Warbler: How Birds Got Their Names* (London)
Mudie, R. 1835a (2nd edn) *The Feathered Tribes of the British Islands volume
    the first* (London, Edinburgh, Dublin)
Mudie, R. 1835b (2nd edn) *The Feathered Tribes of the British Islands volume
    the second* (London, Edinburgh, Dublin)
Mundy, G.B. 1830 *The Life and Correspondence of the late Admiral Lord
    Rodney* (London)
Mundy, H.G. (ed.) 1885 *The Journal of Mary Frampton, from the year 1779,
    until the year 1846* (London)
Neale, A. 1828 'The Spanish Campaign of 1808' pp. 143–209 in *Memorials
    of the Late War vol. I* (Edinburgh)
Nelder, J.A. 1962 'A statistical examination of the Hastings Rarities'
    British Birds 55, pp. 283–98
Nelson, T.H. 1907a *The Birds of Yorkshire, being a historical account of the
    avi-fauna of the county vol 1* (London, Hull, York)
Nelson, T.H. 1907b *The Birds of Yorkshire, being a historical account of the
    avi-fauna of the county vol 2* (London, Hull, York)
Newman, L.F. 1942 'Some Notes on the Folklore of Poultry' *Folklore* 53,
    pp. 104–11
Nicholson, B. (ed.) 1886 *The Discoverie of Witchcraft by Reginald Scot,
    Esquire, being a reprint of the first edition published in 1584* (London)
Nicholson, E.M. and Ferguson-Lees, I.J. 1962 'The Hastings Rarities'
    British Birds 55, pp. 299–384
Nicholson, J. 1890 *Folk Lore of East Yorkshire* (London, Hull, Driffield)
Nicolay, F. 1907 (trans. G.L. Davis) *Napoleon at the Boulogne Camp (based*

on numerous hitherto unpublished documents) (London, Paris, New York, Melbourne)

Nigg, J. 2016 *The Phoenix: An Unnatural Biography of a Mythical Beast* (Chicago, London)

Northall, G.F. 1892 *English Folk-Rhymes: A Collection of Traditional Verses relating to places, persons, customs, superstitions* (London)

Oggins, R.S. 2004 *The Kings and their Hawks: Falconry in Medieval England* (New Haven, London)

Oldfield, E. 1829 *A topographical and historical account of Wainfleet and the Wapentake of Candleshoe, in the county of Lincoln* (London)

Olina, G.P. 1684 *Uccelliera overo discorso della natura* (Rome)

Oliver, G. 1832 'Old Christmas customs and popular superstitions of Lincolnshire' *Gentleman's Magazine* 102, pp. 491–4

Oman, C.C. 1930 'Medieval Brass Lecterns in England' *The Archaeological Journal* 87, pp. 117–49

Opie, I. and P. (eds) 1997 (new edn) *The Oxford Dictionary of Nursery Rhymes* (Oxford)

Osman, A.H. 1907 *The Pigeon Book* (London)

Owen, E. 1896 (rev. edn) *Welsh Folk-Lore: A Collection of the Folk-Tales and Legends of North Wales, being the prize essay of the National Eisteddfod 1887* (Oswestry, Wrexham)

Oxbery, W. 1818 *The Rich Jew of Malta, A Tragedy by Christopher Marlow* (London)

Padley, J.S. 1882 *The Fens and Floods of mid-Lincolnshire; with a description of the River Witham* (Lincoln)

Page, A.W. 1912 '"Safety First" Underground' *The World's Work* 23, pp. 549–63

Palmer, R. 1976 *The Folklore of Warwickshire* (London)

Palmer, W.T. 1949 *Wanderings in Surrey* (London)

Parry, J.J. (ed.) 1907 *The Poems and Amyntas of Thomas Randolph* (New Haven, London, Oxford)

Paton, C.I. 1940 'Manx Calendar Customs' *Folklore* 51, pp. 43–63

Payne-Gallwey, R. 1886 *The Book of Duck Decoys: Their Construction, Management, and History* (London)

Peacock, E. 1904 'Eggs in Witchcraft' *Folklore* 15, pp. 463–4

Pennant, T. 1768 *British Zoology: Class I. Quadrupeds. II. Birds. Vol I* (London)

Pennant, T. 1772 (2nd edn) *A Tour in Scotland 1769* (London)

Pennant, T. 1776 (4th edn) *A Tour in Scotland: MDCCLXIX* (London)

Pennant, T. 1810 (new edn) *Tours in Wales vol 2* (London)

Phayre, I. 1940 (September) '"Doves of War" The Training and Use of Carrier-Pigeons by the Royal Air Force' *Country Life*, pp. 272–3

Phillips, G. 2018 'Technology, "Machine Age" Warfare, and the

Military Use of Dogs, 1880–1918' *The Journal of Military History* 82, pp. 67–94

Pidsley, W.E.H. 1891 *The Birds of Devonshire* (London, Exeter)

Plot, R. 1677 *The Natural History of Oxford-shire, Being an Essay towards the Natural History of England* (Oxford, London)

Pollock, C.G. 2013 'Companion Birds in Early America' *Journal of Avian Medicine and Surgery* 27, pp. 148–51

Pollock, C. 2016 'The Canary in the Coal Mine' *Journal of Avian Medicine and Surgery* 30, pp. 386–91

Polwhele, R. 1826 *Traditions and Recollections; domestic, clerical, and literary, vol 2* (London)

Poole, C.H. 1877 *The Customs, Superstitions, and Legends of the County of Somerset, collected from various sources* (London)

Pope, D. 1969 *Guns* (London)

Potts, G.R. 1967 'Urban Starling Roosts in the British Isles' *Bird Study* 14, pp. 25–42

Pouncett, J. 1998 'The Neolithic Period' pp. 36–62 in R. Adkins, L. Adkins and V. Leitch *The Handbook of British Archaeology* (London)

Pratt, S. 1801 *Gleanings in England; Descriptive of the Countenance, Mind and Character of the country vol 2* (London)

Prentis, W. 1894 *Notes on the Birds of Rainham including the district between Chatham and Sittingbourne* (London)

Preuss, N.O. 2001 'Hans Christian Cornelius Mortensen: Aspects of his Life and the history of bird ringing' *Ardea* 89, pp. 1–6

Pycraft, W.P. 1915 'Richard Lydekker, F.R.S., J.P.' *Science Progress in the Twentieth Century (1906–1916)* 10, pp. 142–5

Rackham, O. 1986 *The History of the Countryside* (London)

Raffald, E. 1773 (3rd edn) *The Experienced English Housekeeper, for the use and ease of Ladies, Housekeepers, Cooks, &c Wrote purely from practice* (London)

Raithby, J. (ed.) 1811 *The Statutes at Large, of England and of Great Britain: from Magna Carta to the Union of the Kingdoms of Great Britain and Ireland vol. III. From: Hen. VIII. A.D. 1509–10 – Edw. VI. A.D. 1533* (London)

Randall, J. 1862 *The Severn Valley: A Series of Sketches, Descriptive and Pictorial of the course of the Severn* (London)

Ray, J. 1678 *The Ornithology of Francis Willughby of Middleton in the County of Warwick, Esq* (London)

Raymond, W. 1910 *English Country Life* (London, Edinburgh)

Réaumur, R.-A.F. de 1748 'Divers means for preserving from corruption dead birds, intended to be sent to remote countries, so that they may arrive there in good condition' *Philosophical Transactions* 45, pp. 304–20

Reiche, C. 1871 (10th edn) *The Bird Fancier's Companion; or, natural history of cage birds* (New York, Boston)

Reid, S. 2012 *Cumberland's Culloden Army 1745–46* (Oxford)

Reid, T.W. 1896 (3rd edn, rev. W.H. Graham) *The Book of the Cheese being traits and stories of 'Ye Olde Cheshire Cheese' Wine Office Court, Fleet Street London E.C.* (London)

Richardson, M.A. 1843 *The Local Historian's Table Book* (London)

Richardson, S. 1997 *The Forgotten Man of Lakeland* (Workington)

Richter, S. and Bick, A. 2018 'Rostock: The Zoological Collection of the University of Rostock' pp. 583–9 in L.A. Beck (ed.) *Zoological Collections of Germany: The Animal Kingdom in its Amazing Plenty at Museums and Universities* (Cham)

Rickards, G.K. 1869 *The Statutes of the United Kingdom of Great Britain and Ireland, 32 & 33 Victoria, 1868–9* (London)

Riley, H.T. 1868 *Memorials of London and London Life, in the XIIITH, XIVTH, AND XVTH CENTURIES, being a series of extracts, local, social, and political, from the early archives of the City of London. A.D. 1276–1419* (London)

Roberts, F.G. 1926 (December) 'Nicholas Cox: A Literary Manciple' *St Edmund Hall Magazine* 11, pp. 26–31

Roberts, M. 1831 *Annals of My Village: being a Calendar of Nature, for every month in the year* (London)

Rocca, A.J.M. de 1828 'Memoirs of the War of the French in Spain' pp. 13–206 in *Memorials of the Late War vol. II* (Edinburgh)

Rocke, J. 1866 'Ornithological Notes from Shropshire' *The Zoologist* 1 (2nd series), pp. 76–84

Rodd, E.H. (ed. J.E. Harting) 1880 *The Birds of Cornwall and the Scilly Islands* (London)

Romney, P. (ed.) 1984 *The Diary of Charles Fothergill 1805: An Itinerary to York, Flamborough and the North-Western Dales of Yorkshire* (Leeds)

Roper, J. 2014 'Sternberg, the Second Folklorist' *Folklore* 125, pp. 202–17

Ross, F., Stead, R. and Holderness, T. 1877 *A Glossary of Words Used in Holderness in the East-Riding of Yorkshire* (London)

Rossetti, C.G. 1872 *Sing-Song. A Nursery Rhyme Book* (London)

Roth, E. 2012 *With a Bended Bow: Archery in Medieval and Renaissance Europe* (Stroud)

Roud, S. and Bishop, J. (eds) 2012 *The New Penguin Book of English Folk Songs* (London, New York)

Rowley, G.D. 1875a 'British Birds' *Ornithological Miscellany* 1, pp. 51–139

Rowley, G.D. 1875b 'Psittacus Erithacus, *Linn*. The Common Grey Parrot' *Ornithological Miscellany* 1, pp. 164–75

RSPB 1911 *Feathers and Facts: A Reply to the Feather-Trade, and review of facts with reference to the persecution of birds for their plumage* (London)

Rudkin, E.H. 1939 'Willoughton, Lincolnshire: The Breast-Bone of a Christmas Goose' *Folklore* 50, p. 317

Russell, A. 1756 *The Natural History of Aleppo, and parts adjacent* (London)

St John, C. 1849 *A Tour in Sutherlandshire, with extracts from the field-books of a sportsman and naturalist vol. 1* (London)

Salvin, F.H. and Brodrick, W. 1855 *Falconry in the British Isles* (London)

Salvin, O. 1859 'Five Months' Birds'-nesting in the Eastern Atlas' *Ibis* 1, pp. 174–91

Sampson, M.W. (ed.) 1904 *The White Devil and the Duchess of Malfy by John Webster* (Boston, London)

Samuel, R. 1975 '"Quarry roughs": life and labour in Headington Quarry, 1860–1920. An essay in oral history', pp.139–263 in R. Samuel (ed.) *Village Life and Labour* (London, Boston)

Savory, A.H. 1920 *Grain and Chaff from an English Manor* (Oxford)

Sebright, J.S. 1826 *Observations upon hawking describing the mode of breaking and managing the several kinds of hawks used in falconry* (London)

Serjeantson, D. 2002 'Goose husbandry in Medieval England, and the problem of ageing goose bones' *Acta zoologica cracoviensia* 45 (special issue), pp. 39–54

Service, R. 1903 (1906) 'The Diurnal and Nocturnal Raptorial Birds of the Solway Area' *The Transactions and Journals of Proceedings of the Dumfriesshire and Galloway Natural History & Antiquarian Society* 17, pp. 327–39

Shand, A.I. 1905 *Days of the Past: A Medley of Memories* (New York)

Shaw, L. 1776 'Of Elgin and the Shire of Murray' pp. 287–312 in T. Pennant 1776 (4th edn) *A Tour in Scotland: MDCCLXIX* (London)

Shaw, T. 1738 *Travels, or Observations relating to several parts of Barbary and the Levant* (Oxford)

Shelley, P.B. 1868 *The Minor Poems of Percy Bysshe Shelley. A New Edition* (London)

Shrubb, M. 2003, *Birds, Scythes and Combines: A history of birds and agricultural change* (Cambridge)

Shrubb, M. 2013 *Feasting, Fowling and Feathers: A history of the exploitation of wild birds* (London)

Sikes, W. 1880 *British Goblins: Welsh Folk-Lore, Fairy Mythology, Legends and Traditions* (London)

Simeon, C. 1860 *Stray notes on fishing and natural history* (Cambridge, London)

Smalley Law, W. 1922 *Oundle's Story: A History of Town and School* (London, New York, Toronto)

Smith, A.C. (rev. edn) 1874 *On the Old Porch-House at Potterne* (Devizes)

Smith, A.E. and Secoy, D.M. 1976 'Salt as a Pesticide, Manure, and Seed Steep' *Agricultural History* 50, pp. 506–16

Smith, C. 1807 *A Natural History of Birds, intended chiefly for young persons* *vol. 1* (London)

Smith, C. 1866 'List of Birds observed during a Six Weeks' Summer Visit to the Channel Islands, exclusive of Jersey' *The Zoologist* 1 (2nd series), pp. 447–53

Smith, C. 1869 *The Birds of Somersetshire* (London)

Smith, F. 1856 'Extraordinary Effect of Sudden Cold on Swifts' *The Zoologist: A Popular Miscellany of Natural History* 14, pp. 5249–50

Smith, J.B. 1988 'Chew-hallaw and Buckalee: A comparative study of some bird-scaring and herding rhyme' *Folk Life* 27, pp. 26–43

Smith, J.E. 1821 *A Selection of the Correspondence of Linnaeus, and other naturalists, from the original manuscripts, vol. 1* (London)

Smith, M. 2012 *These Wonderful Rumours! A Young Schoolteacher's wartime diaries* (London)

Smith, R. 1772 *The Universal Directory for Taking Alive and Destroying Rats, and all other kinds of four footed and winged vermin* (Dublin)

Sneyd, C.A. 1847 *A Relation, or rather a true account, of the island of England; with sundry particulars of the customs of these people, and of the royal revenues under King Henry the Seventh, about the year 1500* (London)

Southey, R. 1813 *The Life of Nelson vol. 1* (London)

Spraggon, J. 2003 *Puritan Iconoclasm during the English Civil War* (Woodbridge)

Stephens, H. 1844a *The Book of the Farm, detailing the labours of the farmer, farm-steward, ploughman, shepherd, hedger, cattle-man, field-worker, and dairy-maid vol. 2* (Edinburgh, London)

Stephens, H. 1844b *The Book of the Farm, detailing the labours of the farmer, farm-steward, ploughman, shepherd, hedger, cattle-man, field-worker, and dairy-maid vol. 3* (Edinburgh, London)

Sterland, W.J. 1869 *The Birds of Sherwood Forest, with notes on their habits, nesting, migrations* (London)

Sternberg, T. 1851 *The Dialect and Folk-Lore of Northamptonshire* (London, Northampton)

Stevenson, H. 1866 *The Birds of Norfolk, with remarks on their habits, migration, and local distribution vol. I* (London)

Stevenson, H. 1870 *The Birds of Norfolk, with remarks on their habits, migration, and local distribution vol. II* (London)

Stevenson, H. 1890 *The Birds of Norfolk, with remarks on their habits, migration, and local distribution vol. III* (London, Norwich)

Stoddart, A.M. 1906 *The Life of Isabella Bird (Mrs Bishop)* (London)

Strutt, J. 1810 (2nd edn) *Sports and Pastimes of the People of England* (London)

Strycker, N. 2014 *The Thing with Feathers: The Surprising Lives of Birds and What They Reveal About Being Human* (New York, London)

Swainson, C. 1873 *A Handbook of Weather Folk-Lore* (Edinburgh, London)

Swainson, C. 1885 *Provincial Names and Folk Lore of British Birds* (London)

Swann, J. 2015 'Shoes Concealed in Buildings' pp. 118–30 in R. Hutton *Physical Evidence for Ritual Acts, Sorcery and Witchcraft in Christian Britain: A Feeling for Magic* (Basingstoke)

Sykes, J. 1833 *Local Records; or Historical Register of Remarkable Events, which have occurred in Northumberland and Durham, Newcastle upon Tyne, and Berwick upon Tweed vol. 2* (Newcastle)

Syme, P. 1823 *A Treatise on British Song-Birds, including observations on their natural habits, manner of incubation, &c. with remarks on the treatment of the young and management of the old birds in a domestic state* (Edinburgh, London)

Tannahill, R. 1988 (rev. edn) *Food in History* (London)

Thomas, K. 1983 *Man and the Natural World: Changing Attitudes in England 1500–1800* (London)

Thompson, F.M.L. 1981 'Landowners and the Rural Community' pp. 457–74 in G.E. Mingay 1981 *The Victorian Countryside vol. 2* (London, Boston, Henley)

Thomson, J.A.K. 1955 *The Ethics of Aristotle: The Nicomachean Ethics translated* (Harmondsworth)

Ticehurst, N.F. 1920 'On the former abundance of the kite, buzzard, and raven in Kent' *British Birds* 14, pp. 34–7

Torr, C. 1918 *Small Talk at Wreyland vol. 1* (Cambridge)

Torr, C. 1923 *Small Talk at Wreyland vol. 3* (Cambridge)

Torr, V.J.B. 1931 'A Tour through Kent in 1735' *Archaeologia Cantiana* 43, pp. 267–80

Tufnell, E.C. 1842 'Report on the State of the Dwellings and General Economy of the Labouring Classes in Kent and Sussex' pp. 36–56 in *Local Reports on the Sanitary Condition of the Labouring Population of England* (London)

Tweddell, E. (Mrs G.M.) 1875 *Rhymes and Sketches, to illustrate the Cleveland Dialect* (Stokesley)

Tyrer, F. 1968 *The Great Diurnal of Nicholas Blundell of Little Crosby, Lancashire volume one 1702–1711* (Chester)

Tyrer, F. 1972 *The Great Diurnal of Nicholas Blundell of Little Crosby, Lancashire volume three 1720–1728* (Chester)

Underwood, L.A. and Stowe, T.J. 1984 'Massive wreck of seabirds in eastern England, 1983' *Bird Study* 31, pp. 79–88

Valentine, L. 1877 *The Young Woman's Book: A Useful Manual for Everyday Life* (London)

van Muyden, Madame 1902 *A foreign view of England in the reigns of George I. & George II. The letters of Monsieur de Saussure to his family* (London)

Vesey-Fitzgerald, B. 1969 *The Vanishing Wild Life of Britain* (London)

Waldron, G. 1744 (2nd edn) *The History and Description of the Isle of Man* (London)

Walsham, A. 2018 'Chronicles, memory and autobiography in Reformation England' *Memory Studies* 11, pp. 36–50

Walter, L.E. (ed.) 1922 *Mother Goose's Nursery Rhymes* (London)

Walton, I. 1653 *The Compleat Angler or the Contemplative Man's Recreation* (London)

Ward, J. and Cunningham, T. 1769 *The Law of a Justice of Peace and Parish Officer: containing all the Acts of Parliament at Large concerning them, and the cases determined on those Acts in the Court of King's Bench vol. II* (London)

Ward, R. 1880 *The Sportsman's Handbook to Practical Collecting, preserving, and artistic setting-up of trophies and specimens* (London)

Ware, E. 1994 'Seventy Years of Woodbury Life' pp. 1–7 in S. and R. Elliott (eds) *Woodbury Parish 1894–1994: Living a Century of Change* (Exeter)

Wasef, S., Subramanian, S., O'Rorke, R., Huynen, L., El-Marghani, S., Curtis, C., et al 2019 'Mitogenomic diversity in Sacred Ibis Mummies sheds light on early Egyptian practices' *PLoSONE* 14 (11): e0223964. https:/doi.org/10.1371/journal.pone.0223964

Waterton, C. 1836 'Notes on the Habits of the Dovecot Pigeon' *The Magazine of Natural History, and Journal of Zoology, Botany, Mineralogy, Geology, and Meteorology* 9, pp. 343–6

Waterton, C. 1838 (2nd edn) *Essays on Natural History, chiefly ornithology* (London)

Watson, G. 1827 *A Narrative of the Adventures of a Greenwich Pensioner written by himself* (Newcastle)

Waugh, E. 1855 *Sketches of Lancashire Life and Localities* (London, Manchester)

Waugh, E. 1867 *Home-Life of the Lancashire Factory Folk during the cotton famine* (London, Manchester)

Westwood, J. and Simpson, J. 2005 *The Lore of the Land: A Guide to England's Legends, from Spring-Heeled Jack to the Witches of Warboys* (London)

Wheatley, H. 1893 *The Diary of Samuel Pepys M.A. F.R.S.* vol. 3 (London)

Wheatley, H. 1896 *The Diary of Samuel Pepys M.A. F.R.S.* vol. 7 (London)

White, G. 1837 *The Natural History and Antiquities of Selborne by the Rev. Gilbert White, M.A. with the Naturalist's Calendar and Miscellaneous Observations extracted from his papers* (London)

White, W. 1861 (2nd edn) *A Londoner's Walk to the Land's End and a trip to the Scilly Isles* (London)

White, W. 1865 *Eastern England from the Thames to the Humber vol. 2* (London)

White, W. 1878–9 (2nd edn) *History, Gazetteer and Directory of the County of Devon including the City of Exeter* (Sheffield, London)

Whitehurst, F.M. 1875 *My Private Diary During the Siege of Paris vol. 2* (London)

Wilkie, T. 1916 (1919) (ed. J. Hardy) 'Old Rites, Ceremonies, and Customs of the Inhabitants of the Southern Counties of Scotland' *History of the Berwickshire Naturalists' Club* 23, pp. 50–146

Wilkin, S. (ed.) 1835 *Sir Thomas Browne's works including his life and correspondence volume IV* (London)

Williams, A. 1923 *Folk-Songs of the Upper Thames with an essay on folk-song activity in the Upper Thames neighbourhood* (London)

Williams, B. 1850 *Henrici Quinti, Angliae Regis, Gesta, cum chronica neustriae, Gallicè, ab anno M.CCCC.XIV.* AD *M.CCCC.XXII* (London)

Williamson, T. 2013 *An Environmental History of Wildlife in England 1650–1950* (London, New York)

Wilson, J. 1842 *A voyage round the coasts of Scotland and the Isles vol. 2* (Edinburgh)

Wilson, J. 1899 'The Naval Pigeon-Post' *The Navy and Army Illustrated* 7, pp. 222–3

Wingfield, C. 2010 'A case re-opened: the science and folklore of a Witch's Ladder' *Journal of Material Culture* 15, pp. 302–22

Wingfield, C. 2011 'Is the Heart at Home? E.B. Tylor's Collections from Somerset' *Journal of Museum Ethnography* 22, pp. 22–38

Winstanley, L. and Rose, H.J. 1928 'Welsh Folklore Items III' *Folklore* 39, pp. 171–8

Winstanley, R.L. 1984 *The Diary of James Woodforde (The first six Norfolk years 1776–1781). Volume 3 1780–1781* (Parson Woodforde Society)

Wood, E.S. 1995 *Historical Britain: A comprehensive account of the development of rural and urban life and landscape from prehistory to the present day* (London)

Wood, J. 2011 'The lost witches of Saveock', pp. 117–42 in D. Gheorgiu (ed.) *Archaeology Experiences Spirituality?* (Newcastle)

Worsley, J. and T. 1781 *The History of the Isle of Wight* (London)

Wright, T. (ed.) 1863 (rev. edn) *The Historical Works of Giraldus Cambrensis* (London)

Wright, T. 1872 *A Historical Account of the Ancient City, and of the Excavations made upon its site at Wroxeter in Shropshire* (London, Shrewsbury)

Wright, W.H.K. 1896 *West-Country Poets: Their Lives and Works* (London)

Wymer, N. 1950a *Green Hills and Grey Spires: The West Midlands* (London)

Wymer, N. 1950b *Companion into Sussex* (London)

Yalden, D.W. and Albarella, U. 2009 *The History of British Birds* (Oxford, New York)

Yarrell, W. 1843 *A History of British Birds vol. II* (London)

Yarrell, W. 1845 (2nd edn) *A History of British Birds vol. 2* (London)

Yarrell, W. 1882–4 (4th edn, rev. and enlarged by H. Saunders) *A History of British Birds vol. 3* (London)

LIST OF ILLUSTRATIONS

————•◆•————

So-called vermin, mainly birds, suspended as a deterrent by a gamekeeper (Harting 1871, p. 82, drawn by John Gerrard Keulemans)

Andrew Carnegie's coach tour to the West Country in the summer of 1884 (*Harper's Monthly Magazine* 1884–5, p. 23)

An idyllic Victorian view of swallows migrating over Gibraltar to their old home at a Surrey cottage (drawn by Charles Whymper, from Dixon 1890, opp. p. 75)

Catching birds with birdlime on a bush (right) and a call bird in a cage (left) (Gent 1733, p. 13)

Migrating birds at night round the Eddystone lighthouse, 12 October 1901 (Eagle Clarke 1912a frontispiece, from a painting by Marian Eagle Clarke)

Retrieving birds caught in nets on the mudflats of the Wash, Lincolnshire (by Charles Whymper, from Dixon 1890, opp. p. 192)

Sparrow catchers in winter, hunting with nets for sparrows roosting in hay-ricks (Kearton 1897, pp. 237, 241–2; photographer Cherry Kearton)

A duck decoy pipe with a detachable net at the end (Kearton 1897, p. 297; photographer Cherry Kearton)

A decoyman killing wild ducks trapped in netting at an East Anglian duck decoy (Kearton 1897, pp. 296, 307; photographer Cherry Kearton)

Hoodwinked falcons being carried by a cadger on a wooden frame (cadge) (Harting 1871, p. 63, drawn by John Gerrard Keulemans)

A jesse with a bell ((Harting 1871, p. 60, drawn by John Gerrard Keulemans)

A pigeon badly stuffed by the Housman boys (Housman 1881, p. 53)

Starting to remove the skin of a starling, with cotton wool placed in its mouth (Browne 1884, p. 95)

The starling skin with internal organs and bones removed, prepared for a collection (Browne 1884, p. 103)

A Yorkshire 'climmer' preparing to descend the cliffs to collect eggs (Nelson 1907b, opp. p. 726)

'Climmer' William Wilkinson descending the cliffs at Jubilee Corner, Bempton, Yorkshire, to collect eggs (Nelson 1907b, opp. p. 732, photographer E.W. Wade)

Wren, nightingale, thrush and blackbird (left to right) (Harting 1871, p. 123, drawn by John Gerrard Keulemans)

Dicky Bird Society medal of 1886 with 'Uncle Toby' (left) and the members' pledge (right) (author collection)

Dovecote of circular, beehive shape, sixteenth century, at Aberdour, Scotland (author collection)

Nesting boxes ('pigeon-holes') at a dovecote, possibly fourteenth century, with roof missing, Stoke-sub-Hamdon, Somerset (author collection)

Caged canaries and white mice from World War One carved on the Scottish National War Memorial at Edinburgh Castle (sculptor: Miss Phyllis Bone)

Advertisement for brooches made from the claws of black grouse shot in the Highlands (*The Graphic* 4 August 1883, p. 131)

The guano workings on the Chincha Islands off Peru (*Illustrated London News* 21 February 1863, p. 201)

The vast accumulation of guano on the Chincha Islands off Peru (*Illustrated London News* 21 February 1863, p. 197)

Unlucky peacock feathers (author collection)

Pigeon, peacock and turkey (left to right) (Harting 1871, p. 167, drawn by John Gerrard Keulemans)

Pheasant (left) and two partridges (Harting 1871, p. 209, drawn by John Gerrard Keulemans)

A gamekeeper shooting magpies (by Charles Whymper, from Dixon 1890, opp. p. 110)

A boy robbing a nest (*The Boy's Own Paper* 1897, p. 472)

Eggs from a hedge sparrow's nest were sought after by boys (*The Boy's Own Paper* 1897, p. 489)

Young cormorants (Nelson 1907b, p. 98)

Trying to shoot birds with an early flintlock gun (Gent 1733, p. 36)

ACKNOWLEDGEMENTS

We are indebted to various people and organisations for their support and assistance during the research and writing of *When There Were Birds*, including the Society of Authors, the London Library, Jim Bond, Peter Lockyer, Martin Graebe and Guy de la Bédoyère. It is a pleasure to thank The Copper Family for permission to use the words of Bob Copper from his book *A Song for Every Season*. For permission to quote from the 1805 diary of Charles Fothergill, we are grateful to the Yorkshire Archaeological Society, and for quotations about Headington Quarry, we would like to thank Alison Light and the Raphael Samuel Estate. For quoting from the diaries of the Reverend James Woodforde, we would like to thank the Parson Woodforde Society, and for quoting from the William Holland diaries, we are indebted to Somerset Archives.

At our publisher Little, Brown, we are grateful, as ever, to Richard Beswick for commissioning this book and for his insightful editing, and for everyone else involved in the various publishing processes, including Zoë Gullen (excellent, as always). Thank you as well to James Gurbutt. We were especially fortunate that Nigel Redman undertook the copyediting, with his unparalleled ornithological expertise. Many thanks as well to Richard Collins for his invaluable proofreading. A final vote of thanks should go to Nico Taylor and James Weston Lewis for the superb cover design and artwork.

INDEX